Service Industries

A geographical appraisal

P.W. DANIELS

METHUEN
London and New York

First published in 1985 by
Methuen & Co. Ltd
11 New Fetter Lane, London EC4P 4EE

Published in the USA by
Methuen & Co.
in association with Methuen, Inc.
29 West 35th Street, New York, NY 10001

Printed in Great Britain at the
University Press, Cambridge

British Library Cataloguing in Publication Data

Daniels, P. W.
Service industries: a geographical appraisal.
1. Service industries 2. Geography, Economic
I. Title
338.4 HD9980.5

ISBN 0 416 34530 1

Library of Congress Cataloging in Publication Data

Daniels, P. W.
Service industries.
Bibliography: p.
Includes index.
1. Service industries—Location. I. Title.
HD9980.5.D35 1985 338.6′042 85-13880

ISBN 0 416 34530 1

For my parents

Contents

List of figures

List of tables

Acknowledgements

Although some of the material in this book is new, it is inevitable that its contents depend a great deal on the work already undertaken by others; hopefully the interpretation of their findings is representative and accurate. Inspiration has also come from my good fortune in being able to share ideas with colleagues and the opportunity of conducting seminars and giving lectures at several North American and Australian universities during the past five years. Those who have been especially helpful include Ian Alexander (Western Australia Institute of Technology), John Clapp (University of Connecticut, Storrs), William Code (University of Western Ontario), Kenneth Corey (University of Maryland), Brian Holly (Kent State University), Mario Polese (University of Quebec), Michael Taylor (Australian National University), Peter Wilde (University of Tasmania), Ann Witte (University of North Carolina, Chapel Hill); and all those, including a number of graduate students, who kindly agreed to present papers and to participate in a sequence of special sessions convened by Brian Holly and myself on a variety of issues connected with office and service industry location at the Association of American Geographers' Annual Meetings between 1981 and 1984.

Since all of the manuscript has been prepared and edited on a word processor, I am responsible alone for the time it has taken to complete in final form. But thanks are due to Sandra Mather, Julie Isaac, Alan Hodgkiss, of the Drawing Office, Department of Geography, Liverpool University, for very patiently redrawing or converting sketches and statistics of such diverse detail (and legibility) into the diagrams used throughout the book. The assistance is gratefully acknowledged of Ian Qualtrough and Susan Yee, of the Faculty Photographic Unit, in expediting the preparation of photographic prints of the diagrams. As always, the Department of Geography of Liverpool University has provided unstinting material support, together with some assistance with travel costs, and generally helped to make it possible to complete the manuscript. Particular thanks are due to all my colleagues who have 'covered' for me while gathering material overseas on leave of absence, and for discussing some of the topics covered in the book. These visits abroad would not have been possible without the generous support, sometimes at very short notice, of the Liverpool University Staff Travel Fund. Last but by no means least, I should like to thank my wife, Carole, and our two children, Paul and Charlotte, for their seemingly infinite patience and understanding. Frequent and sometimes lengthy absences from Liverpool, interspersed with unrelenting toil at a word processor in my study, represent considerable disruption to family life; the value of their love, tolerance, interest, constant encouragement and willingness to hold the fort while I have been away is beyond estimation.

P.W.D.
Liverpool
February 1985

The author and the publishers would like to thank the following copyright holders for permission to reproduce material:

Editors

Urban Studies for Table 7.1: *Ekistics* for Table 8.7; *Economic Geography* for Figure 6.5; *Transactions of the Institute of British Geographers* for Figures 6.9, 6.10.

Publishers

Liverpool University Press for Table 2.1; Basic Books Inc. for Table 3.7; Australian Government Publishing Service for Table 3.2; Pion Ltd for Figures 5.3, 6.1, 6.2, 6.3, 6.7 and Table 7.3; Her Majesty's Stationery Office for Figures 2.6, 3.2, 4.5; Hutchinson Publishing Group Ltd for Figure 4.2; Methuen Ltd for Figure 4.1; Edward Arnold (Publishers) Ltd for Figures 4.3, 4.4, 6.8; Cambridge University Press for Figures 6.4, 7.4, 9.2; Harper and Row Ltd for Figure 7.1; Macmillan Ltd and Holmes and Meier Publishing Inc. for Figure 8.3; Pergamon Press Ltd for Figure 8.4: Longman Group Ltd for Figure 8.5; Times Newspapers Ltd for Figures 9.1, 10.3, 10.5; Frances Pinter (Publishers) Ltd for Figure 10.1.

Organizations

Kent County Council for Figure 4.6.

Individuals

J. Whitelegg and Straw Barnes Press for Table 6.2 and Figure 6.6; B. Moore for Table 8.1; R.L. Davies for Table 8.5; W.R. Code for Figure 7.2; R. Barras for Figure 9.3; J.M. Nilles for Figure 10.6; M. Polese for Figure 7.5.

Preface

Service industries are quite as heterogeneous as the topics now encompassed in the research undertaken by contemporary geographers. Consequently the task of devising a consistent framework around which to write a geographical text about service industries is far from straightforward. As a geographer one is, of course, interested in the spatial pattern of phenomena and the processes responsible for their growth, location, accessibility and changing distribution. But such is the diversity of the service sector that it defies application of a principal theory, a particular analytical method, or a dominant mode of interpretation. This problem exists in most lines of geographical enquiry but it seems particularly acute when service industries are being considered. Indeed this may explain their long-standing neglect both within geography and by other disciplines.

Courses which explore themes in economic or urban geography must, in general, rely heavily on texts in which manufacturing industries receive prominent coverage, while services are only selectively discussed, most notably retailing and transport which have long been a target for intensive research by urban/economic geographers. *Service Industries: A geographical appraisal* has, therefore, been put together in the belief that while the significance of service industries for many facets of socio-economic development and spatial organization is now much more readily acknowledged than at any time in the past, there is room for an undergraduate and graduate text which attempts to demonstrate in a structured and systematic way the diverse and pervasive role of service industries in modern economic systems. The present book follows from my preliminary, and highly distilled, text produced for use in sixth forms and at first-year undergraduate level (*Service Industries: Growth and Location*, Cambridge University Press, 1982), but it is longer, more comprehensive and substantive.

A number of broad objectives – each of which is not mutually exclusive – have guided the preparation of this book. It is important to state at the outset that an attempt has not been made to put forward a new geographical interpretation of the recent expansion of service industries based on original research; the intent is much more modest, and it is essentially to synthesize recent work by geographers and others in a way that may be helpful to students encountering the geography of services for the first time. Initially, then, it seems important to acquaint students with the basic characteristics, purpose and diversity of service sector activities and this must incorporate some discussion of the variety – often very eclectic – of approaches to the study of services. An appreciation of the service sector will also be enhanced by an examination of the economic and social determinants of the growth and diversification of the supply/demand for services during this century, and particularly since 1945. The way in which this is translated into the contribution of service industries to employment, capital formation, or output is illustrated with empirical data at the global, national and regional scale.

A second major objective is to provide an outline of the theoretical bases for the analysis and interpretation of the location of service industries at different scales and in relation to different types of services. These theoretical concepts, many of which have their origins outside geography, provide a superstructure which is far from complete but is a prerequisite for any subsequent attempt to substantiate them by using evidence about the patterns and the processes which determine the location of services in the real world. The third major objective is, therefore, to offer a range of empirical material from studies undertaken in Western Europe, the United States, Canada and Australia to test, and to verify or contradict, some of the assumptions and conclusions derived from the theoretical approaches to service industry location. Such spatial selectivity does not represent a deliberate attempt to confine the book to service industries in developed economies; rather it reflects the pragmatic recognition of the more limited range of work undertaken in developing and centrally planned economies where the service sector remains either smaller or much less diversified than elsewhere.

Analysis of service industries is complicated by, among other things, the division between private and public sector provision, with the latter occupying an increasingly prominent role in many of the most advanced economic systems. In consequence, there are some fundamental differences in location patterns and priorities arising from the demand for, and supply of, services within these two broad subgroups. Hence the final broad objective is to try to convey the nature of these differences and how they have ultimately led to more critical appraisals of the role of service industries in public policies, with particular reference to the restructuring of regional economies and problems of overconcentration in major urban areas.

It was tempting, because of the diversity of the service industries, to write systematically about each of the major and generally recognized industries within the sector. But this surely would have lacked the cohesion, continuity and interest which the reader might reasonably expect. The structure finally adopted utilizes the broad objectives specified above, which essentially can be graded into those representing general or macroscale matters, gradually stepping down to finer levels of resolution. There seems more logic to this approach than one which relies on a step-by-step examination of separate service industries. Wherever possible, appropriate examples have been included with the underlying notion that these should be taken from across the whole range of the service industries. In practice, this may however also lead to coverage of too much territory and a lack of focus so that some activities (such as transport services) have undoubtedly received less attention than others. Some may perceive such selectivity as a weakness, but comprehensiveness might then have been attained at the expense of continuity. Perhaps it is also inevitable that my own particular interests have exerted some influence on the services selected but this is certainly not a book about office location within the guise of service industries. The two are often considered as synonymous; but this is far from accurate, and a conscious effort has been made to minimize the intrusion of the office location theme.

CHAPTER 1

Service industries: identity and delimitation

THE DEFINITION OF SERVICES

A distinctive feature of service activities is their heterogeneity; a characteristic which has undoubtedly dogged attempts at systematic analysis. Perhaps one should, therefore, begin by trying to define a 'service'? It is probably most easily expressed as the exchange of a commodity, which may either be marketable or provided by public agencies, and which often does not have a tangible form. Services seem to 'pass out of existence in the same instant as they come into it' (Greenfield, 1966, 7), or the term 'service' itself 'implies the existence of two parties, those rendering the service and those to which the service is rendered' (OECD, 1978a, 8). In other words, the industries which provide material and tangible commodities are agriculture or manufacturing or construction, while a bar, say, which stocks beer and lager under conditions designed to ensure that they are served at the right temperature to the paying customer or prepares 'bar snacks' at lunchtime is providing a service. Equally the preparation of an advertising portfolio for the lager manufacturer wishing to promote a new brand is a service but – and this is a constant source of confusion when trying to identify services – the firm that assembles the brewing vats and related equipment required to produce the new lager is classed as a secondary industry. Note that such a firm does not create the finished product which it assembles, that task may be performed by an engineering company.

Contrary to the popular view, service activities have always had a place in economic systems, although their attributes and significance for the operation and spatial organization of those systems have been subject to considerable change. Services have underpinned economic and urban development for the last 200 years; the exchange of raw materials between producers in the New World and consumers in Europe, the development of nationwide postal systems, improvements in urban health and mortality rates through better sanitation and cleaner water supplies, the production and distribution of coal gas for domestic/industrial lighting and heating, the distribution and consumption of durable and non-durable consumer goods, the collection and redistribution of taxes by various central or local government agencies and the exchange of legal documents permitting the purchase of a house or factory have all promoted more efficient operation of the economic and social system, or contributed to improvements in the quality of life. In this way services can be said to be pervasive in both developed and developing economies and, in common with the variety of political, religious, cultural and administrative systems which comprise these economies, there are spatial variations in the distribution, structure, productivity, or growth of service activities.

Such variations are facilitated by one very obvious difference between a manufacturing and service establishment: the former fabricates raw materials or components into products which are sold to many users, while the latter is often engaged in the sale of (or provision of) skill and knowledge over a period of time, the systems analyst in a computer/data processing firm or a consultant in a marketing or advertising agency. There may be limited scope for applying the use of time to solving or advising on identical problems but by and large the essence of many service industries is selling the time and talents of highly trained and experienced individuals. The implications for the location of service industries in which these aspects are important will be considered in Chapter 7.

In retrospect it is perhaps inevitable that industrialization, accompanied as it was by an exodus of population from rural to urban areas, would be conducive to the creation of an environment in which a legitimate concern for the production of goods would need to be supported by activities which sustained the competitiveness of enterprises, on the one hand, and enhanced the quality of life, especially in urban areas, on the other. In essence, this is the role which service industries performed in earlier periods and continue to perform in the contemporary world; the main difference now is that they are much more diversified and specialized than ever before. Symptomatic of this diversity is the way in which the production of services has become increasingly difficult to separate from the production of goods. Transport and distribution services exist in large part because there are goods to carry and sell; many research and development (R&D) services are concerned with identifying/producing new products for manufacturing firms; and advertising agencies assist with the sales promotion of manufactured goods, or financial services such as merchant banks with providing finance for investment in goods producing plant and machinery. Thus it has been shown that about 50 per cent of the employment in services in the United Kingdom (in both 1961 and 1971) is goods related (Gershuny, 1978; see also Gershuny and Miles, 1983). Why, then, should service industries be examined independently? The principal reason is that although many may well be goods related, this does not mean that they are geographically linked, so that by concentrating on the geography of manufacturing we cannot adequately explain the geography of services.

CLASSIFICATION OF SERVICE ACTIVITIES

Such is the diversity of service industries that it is also virtually impossible to make generalizations about the spatial patterns which they reveal without resorting at an early stage to an attempt to identify subgroups of activities most like one another but different from other groups. Such classifications provide the foundation upon which subsequent analysis can be based. But the task is a difficult one, as demonstrated by a range of possible alternative classifications (Table 1.1). These alternatives are not discussed in detail here since we are not concerned with the classification of service industries as an end in itself. For the interested reader some further points relating to the classification of service activities are provided in Appendix 1.

Table 1.1 Service industries: some alternative classifications

Basis of classification	*Usual labels and groups*	*Alternative labels and groups*
Input of capital and skill	Intensive/limited/ primary dependence on skill	—
Destination of output	Producer/consumer	Distributive/producer/ social/personal; productive/individual consumption/collective consumption; complementary/old/new
Occupation	White collar/blue collar	Tertiary/quaternary/ quinary
Origin	Public/private	Market/non-market
Location	Tied/foot-loose	Local/non-local
Premises	Office/non-office	—
Organization	Formal/informal	Modern/traditional

The problem is not just about how to devise an 'internal' grouping of services, but also in specifying the boundary between service and non-service activities. This is more difficult than it might seem because the location of the boundary – and the implications for statements about, for example, the size of the service sector – depends on the direction from which the task is approached. All the major studies of the service sector preface their analyses with statements about classification, and Stigler (1956, 47) is clear about the fact that there 'exists no authoritative consensus on either the boundaries or the classification of the service industries'; by 1978 the OECD could still only comment that 'it is accepted that there is no completely satisfactory delimitation'. Three kinds of classification are used throughout the remainder of this book and these will now be outlined in more detail.

Classification by industry sector

One of the most frequently used methods of classifying economic activities as a whole is by industry sector (Table 1.2). While it makes for a neat division into primary, secondary and tertiary sectors, the method has several limitations. Each major sector is subdivided into a number of industry orders (some examples are given in Table 1.2, n. 2). Essentially these orders are derived from a classification based either on the kind of raw materials which are used, such as non-ferrous metals or oil for the petroleum industry, or by reference to the nature of the final product, such as electrical goods, or cars, vans and lorries in the case of the vehicle industry. Such an approach is not very helpful when applied to services; many require a personal input for example, domestic or office cleaning, or a medical/legal practice, and this does not lead to any material output other than a cleaner home or office, or advice from the medical practitioner or legal adviser about how to resolve a problem or to complete a house purchase. There is, therefore, little to be gained by classifying services with reference to output in the way possible for primary and secondary (manufacturing) activities.

Table 1.2 Sectoral classification of industry orders

Sector	*Industry order*
Primary	Agriculture, hunting, forestry, fishing
Secondary	Mining and quarrying; manufacturing; electricity, gas, water;[1] construction[1]
Tertiary (service)	Commerce; transport, storage and communication;[1] finance, insurance, real estate and business services; community, social and personal services; activities not adequately defined (all included in the tertiary sector)[2]

Notes: 1 'Marginal' industry orders which, depending on individual studies, may or may not be included in the service sector or vice versa.

2 The NACE definition (see text) of the tertiary sector by industry is: wholesale and retail trade and restaurants and hotels; transport, storage and communication; financing, insurance, real estate and business services; community, social and personal services; the same classification is utilized in the ISIC definition.

Source: Derived from United Nations, 1948; rev. 1968.

When a list of industry orders has been agreed – and there is now some, although far from complete, uniformity between national censuses – it remains to decide which orders should be included in the service sector. Inevitably, perhaps, there is no internationally agreed definition although both the European Economic Community through its Nomenclature des Activités Économiques dans les Communautées Européennes (NACE) (see Statistical Office of the European Communities, 1970) and the United Nations (1948) through its International Standard Industrial Classification (ISIC) have endeavoured to encourage a standardized definition in the interests of more effective comparative analyses (see Table 1.2). Perhaps the definition of the service sector should vary according to the purpose of individual studies, but this only obviates the necessity to delimit it in a consistent way.

However, the UK for example has for some time deviated from these two schemes. The first Standard Industrial Classification (SIC) for the UK was issued in 1948, with subsequent revisions in 1958 and 1968 (Department of

Employment, 1983). The most recent revision (1980) not only takes into account the changing structure of British industry, but brings the classification closer into line with NACE and ISIC. There are substantial differences between the 1980 SIC and its predecessors both in the structure of the classification and the numbering system used. Most notable is the use of a four-tier structure with a hierarchical decimal numbering system rather than the twenty-seven orders in the 1968 SIC divided into minimum list headings (MLHs) with numbers unrelated to the roman numerals by which orders are identified. The 1980 SIC is divided into 10 divisions, 60 classes, 222 groups and 334 activity headings and a broad comparison, as it relates to service industries, is given in Table 1.3 (see also Central Statistical Office, 1979). As far as possible this new scheme distinguishes between agents and principals – i.e. between dealers who buy or sell on behalf of others and those who own the goods or carry the risks. Such a distinction is especially apposite to distributive and financial services and should lead to more homogeneous data for the headings concerned.

Table 1.3 General comparison of service industry classification using SIC 1980 and SIC 1968

Divisions, SIC 1980	*Orders, SIC 1968*
6 Distribution, hotels and catering, repairs	XXIII, XXVI (MLH884–888, 894, 895)
7 Transport and communication	XXII
8 Banking, finance, insurance, business services and leasing	XXIV, XXV (MLH871, 873)
9 Other services	XXV (remainder), XXVI (remainder), XXVII

Source: Department of Employment, *Employment Gazette*, 91, 1983, 118, table 1.

Some of the early studies by economists such as Fisher (1935) and Clark (1940) referred to services as the tertiary sector. There was no serious effort to define this sector, however, and this provoked those who thought that it simply encompassed all those economic activities which could not be conveniently grouped in the primary or secondary sectors to in fact use it that way. Once attempts were made to be more precise (see, for example, Kuznets, 1957), the debate revolved around inclusion or exclusion of particular industry orders. Most of the disagreements centre on construction, the utilities (gas, electricity and water), and transport and communication. Because the latter uses a large volume of physical capital, Fuchs (1965) proposed that it should be included with manufacturing. Similarly, the production processes used by the utilities have more in common with goods production, and construction also involves substantial physical capital. But both industries provide for the needs of other industry orders (e.g. office or factory premises, energy for goods production) as well as for 'social needs' in the form of housing, hospitals, schools and domestic heating, light and water.

Comparison of the industry-based definitions used in service industry studies over the last twenty years indicates that construction is invariably excluded but the other two 'marginal' industries may be subsumed into services. There seem to be no hard-and-fast rules; for the purposes of the analysis presented here, whenever it is necessary to present material in industry terms the service sector will, therefore, be taken to include: transport and communication; distribution (wholesale and retail trade); finance, insurance, real estate and business services; community, social and personal services (including hotels, entertainment, restaurants, recreation); and public administration and defence.

Consumer and producer services

Frequent reference will be made to services classified by industry but the exploration of location theory, empirical evidence about location, or the urban and regional development problems arising from the location behaviour of services (see Chapters 4–9) is structured around a useful distinction between consumer and producer services (Greenfield, 1966). This arises from a taxonomy of commodities originally proposed by Kuznets (1938) in which commodities are divided into consumer and producer goods and which can, in turn, be subdivided into those with perishable, semi-durable, or durable attributes. Durability is a difficult concept to invoke in the context of a service, but consumer services such as attendance at a football match, a visit to the hairdresser, or the use of a launderette, which yield utility over a short time, can be considered perishable. Semi-durable consumer services include advice provided by lawyers, assistance from accountants with the completion of tax forms, a course of dental treatment, or an agreement for regular maintenance and repair (if necessary) of a central heating boiler. The length of time over which the service yields utility is clearly an important principle here, so that durable consumer services must be those with long term value and might include architectural advice and assistance with the design of a new family house, or financial advice and assistance with house purchase or educational training which provides a skill necessary for career development or promotion.

Producer services can be similarly categorized; perishable producer services include daily cleaning of office and commercial premises (often undertaken by companies specializing in such work), window cleaning, waste disposal, and rapid document collection and delivery services. A solicitor dealing with a divorce case or an advertising agency providing copy for the promotion of a product is providing a semi-durable producer service. Examples of durable producer services are provided by management and business consultants advising on the way in which organizations should diversify their activities over the next five years, by computer consultants assisting with system selection, where it should be used within an organization and how it should be introduced, or by financial analysts on issues such as locations for profitable new investment or the prospects for long-term borrowing strategies. If this schema can be accepted, 'it is important to note the parallelism between goods and services, and that services can be analysed in terms which have heretofore been exclusively reserved for goods' (Greenfield, 1966, 9).

Identifying producer and consumer services

There remains the vexed question of whether it is possible to distinguish between consumer and producer services in a way which is helpful and reliable. Unfortunately the answer must be negative because very few published statistics allocate employment or income on the basis of whether they arise from, or create output for, intermediate or final demand. Indeed most of the industries classed as services have to satisfy demand from a highly variable amalgam of other businesses, individual consumers, government and non-profit organizations. Invariably it is necessary to rely on estimates of the proportion of all employees, for example, in a particular service industry engaged in meeting intermediate demand or making use of input–output tables to establish the ratio of intermediate to other kinds of output.

A good example of the first approach is provided by Greenfield (1966). He used detailed data on the revenues of service industries in the USA to estimate the proportion of producer service employment but found that revenue data was not available in comparable form for all industries and had to incorporate evidence from other sources to arrive at his estimates. Some 75 per cent of the US employees in transportation were allocated to producer services, 50 per cent of the employees in communications, and 50 per cent in finance, insurance and real estate. Greenfield found an almost equal division between business loans and consumer loans, both major sources of revenue for banks, although the number of employees required to handle the consumer loans is probably greater than one-half of the total. In relation to output, input–output tables showed that over 55 per cent of insurance and finance output was sold to intermediate users compared with just 35 per cent of the output from real estate activities. The other proportions for producer services are: legal (50 per cent); engineering and architectural (90 per cent); accounting, auditing and bookkeeping (90 per cent); miscellaneous professional services (75 per cent); and government (33 per cent). Although 'forced to make heroic asumptions' (Marquand, 1979, footnote 3, para. 1.22) about his data, Greenfield was then able to apply the above proportions to employment data for each service industry and show that between 1950 and 1960 employment in producer services in the USA increased by over 21 per cent compared with 15 per cent for all industries. Their share of total national employment also increased from 12.5 to 13.2 per cent.

According to a more recent estimate (Wood, 1983), some 18 per cent of employment, in Britain, was in producer services in 1981 (Table 1.4) and this might be a conservative figure, in that the proportion may be nearer 22 per cent (Daniels, 1985c). Almost 29 per cent of service sector jobs (excluding public administration) were producer related in 1975, a proportion which had actually declined marginally by 1981. Although more labour intensive than manufacturing, producer services have been able to generate a higher level of output without an equivalent increase in employment (e.g. by introducing the by-

Table 1.4 Estimates of producer and consumer employment in British service sector, 1975–81

Industry	*Producer*[1]			*Consumer*		
	1975	*1981*	*Change (%)*	*1975*	*1981*	*Change (%)*
Transport and communication	29.2	23.2	−15	4.1	3.7	−2
Distribution	19.1	18.8	6	29.1	2.1	−1
Insurance, banking, finance	24.1	27.4	22	5.2	5.7	23
Professional and scientific services	11.6	13.7	25	42.8	40.1	3
Miscellaneous	15.8	16.9	15	18.7	24.5	45
Totals (thousands)	2926 100.0	3150 100.0	8	7387 100.0	8160 100.0	10

Note: 1 Road haulage, dealing in industrial and building materials, advertising and business services have been allocated entirely to producer services; other services, such as rail, postal and telecommunications services, wholesale distribution, insurance, banking and finance, have been divided equally between producer and consumer markets.

Source: Wood, 1983, adapted from table 2.

products of information technology). Consumer services which are even more labour intensive have, until very recently at least, been less adept at substituting labour with technology, so that they have slightly increased their overall share of service employment and expanded at a higher rate between 1975 and 1981.

A similar pattern, over a longer time period, is evident in the employment structure of the USA (Table 1.5). It is difficult to make a direct comparison with the statistics for the UK because the industry groupings and definitions are different. However, producer services increased their share of employment from just over 8 per cent in 1959 to almost 12 per cent in 1977, a rate of growth only matched by non-profit services (education and health) and to a lesser degree government and government enterprises. Indeed non-profit services had a higher annual growth rate during the 1960s and 1970s but continue to comprise a smaller share of total employment. Generalizations must be made carefully here, but it can be concluded that producer services have been expanding more rapidly than other parts of the service sector whether measured through the medium of output, in relation to the average for total national product during the last two decades (Stanback, 1979), or as a share of total employment.

Greenfield (1966) used input–output matrices to verify or to substitute for his revenue data. Such tables are useful and are published annually in many countries, but service sector activities are rarely well covered. Marquand (1979), for example, has shown that input–output tables produced by member states of the European Economic Community (EEC) which indicate the consumption of producer services by other industries contain more 'gaps' than for the secondary or agricultural industry orders. Hence for 'all the most buoyant

Table 1.5 Changing employment shares and annual rate of growth in selected services: USA, 1959–77

Industry	*1959*	*Annual growth rate (%), 1960s*	*1969*	*Annual growth rate (%), 1970s*	*1977*
Transformative industries	38.3	1.6	35.1	0.1	31.6
Distributive services	12.2	1.5	11.0	1.8	11.4
Retail services[1]	12.7	2.8	13.0	2.4	14.2
Mainly consumer services	6.5	1.4	5.8	−0.4	5.0
Non-profit services	3.5	5.5	4.7	5.2	6.3
Producer services[2]	8.2	4.5	10.0	3.6	12.0
Government	18.6	3.5	20.5	0.7	19.6
All industries	100.0	2.5	100.0	1.3	100.0

Notes: 1 As a significant component of employment in the distributive services, retailing has been treated separately.
2 The producer services include finance, insurance and real estate (SIC 60–7 in US Census), business services (SIC 73), legal services (SIC 81), membership organizations (SIC 86), miscellaneous professional services (SIC 89) and social services (SIC 83, after 1974).

Source: Stanback and Noyelle, 1980, table 1, 9.

parts of the service sector – insurance, banking, business and financial services; professional services; and some of the expanding private consumer services such as entertainment and those concerned with tourism – the information is acknowledged to be grossly inadequate' (ibid., para. 1.22). It is necessary, however, to do the best possible with what is available, and this reveals that just under 36 per cent of the total output of UK industries in 1974 was destined for intermediate use (Table 1.6). There are wide variations between industries, not least among services with a range from 51 per cent for property-owning and managing to just over 11 per cent for lodging and catering.

It is interesting to note that the proportion of activity from all industries generating intermediate output in the UK is comparable with the average (40 per cent) for six European countries (Table 1.7). However, this disguises considerable variations between the countries in the proportion of such outputs from the same industry group and, therefore, in the overall importance of each industry to producer service output in each country. Germany has the highest proportion of intermediate output for wholesaling and retail trade; business

Table 1.6 Intermediate output as a proportion of total output: selected UK industries, 1974

Industry	*Total intermediate output (£m.)*	*Percentage of total output*	*Percentage of total intermediate output*
Agriculture, forestry and fishing	2,178.6	52.1	4.13
Extractive	367.2	78.7	0.70
Gas, electricity and water	1,957.0	41.2	3.71
Alcoholic drink	69.3	5.5	0.13
Fertilizers	279.3	73.9	0.53
Office machinery	26.5	14.9	0.05
Television, radio and sound-reproducing equipment	15.0	2.7	0.03
Clothing and footwear	104.0	5.9	0.20
Printing and publishing	1,231.4	57.7	2.34
Construction	1,177.2	10.9	2.23
Transport and communications	5,074.0	46.0	9.63
Distributive trades	3,238.7	25.5	6.12
Insurance, banking and finance	828.9	36.9	1.57
Property-owning and managing, etc.	811.1	51.2	1.54
Lodging and catering	439.2	11.2	0.83
Other services	5,583.3	48.2	10.60
Total intermediate output	23,380.7	33.6	44.34
Total, all industries[1]	52,673.0	35.7	100.00

Note: 1 Incorporates industries not included in table.
Source: Department of Trade and Industry (Business Statistics Office), *Input–Output Tables for the United Kingdom, 1974*, London, HMSO, 1980, table D.

services provided to enterprise are much lower (as a proportion) in the Netherlands and Belgium; education and health, provided as market services have a much higher level of intermediate output in France than elsewhere. Marquand concludes that:

> beyond the very broad similarities of structure of the tertiary sector between member states. . .the behaviour of individual sectors is shown to vary from country to country in respect to the nature of their markets, and probably (and consequently) in regard to the manner of operation of particular sectors too. It would be rash to generalize about the extent to which particular sectors supply producer services; any operational distinctions clearly have to be drawn at a much more detailed level which for practical purposes is that of the enterprise. However, one conclusion which is firmly established is that a substantial proportion of much service sector output throughout the market sector is sold to other producers rather than going direct to the final consumer. (ibid., para. 3.19)

There is one other useful piece of evidence about producer services in the input–output tables for the UK which is pertinent to the later analysis of location behaviour (see Chapter 7). This concerns the importance of intra-industry transactions relative to the total intermediate output of the major service industries. There are clear indications (Table 1.8) that some industries, 'other services' and insurance, and banking and finance in particular, are heavily engaged in intra-group transactions; over 88 per cent of the intermediate output of insurance, banking and finance is destined for establishments in the same group.

Table 1.7 Proportion of intermediate outputs from service industries: EEC countries (except UK), 1970

Service (NACE classification)[1]	*Belgium*	*France*	*Germany*	*Italy*	*Netherlands*
Recovery and repair	53.4	51.7	60.9	64.5	38.3
Wholesale and retail trade	21.3	19.6	41.7	21.0	23.9
Lodging and catering	5.8	16.6	38.4	12.2	20.0
Inland transport	42.6	71.1	55.0	45.5	6.5
Maritime and air transport	11.4	20.4	28.4	13.3	23.7
Auxiliary transport services	46.8	85.0	75.7	67.2	54.4
Communications	65.2	77.0	79.7	67.4	67.7
Services of credit and insurance institutions	76.8	81.3	78.7	89.0	72.2
Business services provided to enterprise	(41.3)[2]	77.5	91.4	86.0	56.0
Renting of immovable goods	1.8	11.2	5.0	19.6	0.0
Education and research: market services	n.a.	64.0	32.8	44.8	24.8
Health: market services	4.3	37.7	7.0	1.0	(6.4)[3]
Recreational, cultural, personal and other market services	n.a.	32.0	37.4	16.5	46.7
General public services	(0.0)	0.0	0.0	0.0	2.4
Non-market education and research	0.0	0.0	0.0	0.0	0.4
Non-market health services	n.a.	0.0	0.0	0.0	n.a.
Domestic services and other non-market services	n.a.	0.0	0.0	0.0	11.3
All industries and services	37.0	41.5	47.6	40.2	39.1

Notes: 1 Nomenclature des Activités Économiques dans les Communautées Européennes.
2 Includes education and research; market services; recreational, cultural, personal and other market services; domestic services and other non-market services.
3 Includes non-market health services.

Source: After Marquand, 1978, table 3.3.

Considerable space has been devoted to this discussion of producer services in the belief that, even though there are many problems of identification (which are not unique to the producer/consumer classification), they occupy an impor-

Table 1.8 Intra-industry transactions in relation to total intermediate output: UK service industries, 1974

Service industry	*Intra-industry transactions (£m.)*[1]	*%*	*Total intermediate output (£m.)*
Transport and communication	566.3	11.2	5,074.0
Distributive trades	108.2	3.3	3,238.7
Insurance, banking and finance	732.2	88.3	828.9
Property-owning and managing, etc.	7.0	0.9	811.1
Lodging and catering	11.8	2.7	439.2
Other services	13,032.0	233.4	5,583.3

Note: 1 Intra-industry transactions have not been included in the totals for intermediate output, so that the percentages only provide a rough comparison of the importance of intra-industry transactions relative to other kinds of intermediate output.

Source: Department of Trade and Industry (Business Statistics Office), *Input–Output Tables for the United Kingdom, 1974*, London, HMSO, table D.

tant place in a geographical analysis of services. Some compromise about which service industries to include in the producer group is clearly necessary but for the remainder of this book this will be taken to include insurance, banking and finance, professional and scientific services and some of the activities incorporated in the miscellaneous services group (but see Daniels, 1985c).

PRIVATE AND PUBLIC SECTOR SERVICES

The producer/consumer classification is based upon the destination of the output of service activities (see Table 1.1). It is an intuitively attractive dichotomy but, as we have seen, it is not easy to use because national employment and other statistics are rarely available in the appropriate form or detail. Therefore, it is sometimes helpful to examine services using an additional dimension (which can overlap with the producer/consumer division) related to the origin (or source) of a service in which the principal distinction is between private and public provision (see Table 1.1). One of the most notable trends since 1960 in the service sector has been the faster rate of employment expansion by public corporations, local authorities (in particular), central government and defence-related activities. The classification of services into private and public categories can also be characterized as a distinction between market and non-market services (Bacon and Eltis, 1976). Market (private) services are generally assumed to behave in response to a set of market forces, including demand factors, the behaviour of competitors, the availability of certain kinds of labour, or limitations in the kinds of service already available. Ultimately the arbiter in their birth, continuing existence, or death is profitability. Such is not necessarily the case with most public services, which while they may be subject to detailed scrutiny of their administrative overheads and manning levels (in the way typified by the reductions in civil service manpower and spending achieved by the Thatcher administration in Britain during the early 1980s) are more concerned with accessibility or availability to their clients than profitability. This has important consequences for the locational behaviour of such services, and this will be considered in more detail in Chapter 6. The main point to make here is that, however significant this taxonomy may be, it is not always easy to make a neat distinction between market and non-market public services. In the USA, for example, education and the medical/health industries are provided via market and non-market organizations; about 25 per cent of education and 75 per cent of medical services are market activities.

The difficulties of classifying public services have been widely noted (see, for example, Boulding, 1971; Lineberg, 1977; Teitz, 1968; Robson, 1976; Seeley, 1981). Seeley (1981, 2) notes the lack 'of accepted definitions, the ever-changing landscape of publicly sponsored and legislated services, the constant shift from public to private ownership and management (and vice versa) and the interdependence of public and private activities'. All this 'renders definition complex and often irrelevant or harmful, independent of place and time' (loc. cit.). The term 'public services' implies the involvement of the public (or its representatives) in paying for using and assessing the service, while funding is provided through national, regional, or local taxation and facilities are planned by a public agency or jurisdiction.

In developing countries it may be essential for the public sector, in the interests of providing suitable infrastructural services and utilities, to take responsibility for large areas of social and economic life (Ezeikel, 1976). In India, for example, the government provides various public services and utilities, together with a large variety of welfare activities and the promotion of social and economic changes; it conducts a large number of units engaged in trade and industrial production and regulates many social and economic activities. Like many developing countries, India has chosen to operate discretionary controls like industrial licensing, price controls; and so on, thus directly affecting vital sectors of the economy rather than limiting itself to generalized indirect controls such as monetary and fiscal policies. Hence the number of central government ministries and departments had increased from eighteen in 1947 to fifty-one by 1973; and all have increased the number of civil servants on the payroll. The Ministry of Agriculture employed 33,500 in 1965; even though it is a state rather than federal responsibility; the number of departmental committees alone increased from thirty-one in 1946 to more than 600 in 1976. The state administrations have also expanded. The twenty-one states each have between eleven and thirty-four departments as well as several technical departments set up to promote rural development. Specialized units dealing with health, education, industry, and so on, have also been set up at district level. Common to each of these publicly provided services is that all the tiers in the hierarchy have grown in size, depth and content. Employment in public administration has, therefore, doubled from 2.2 million in 1957 to 4.4 million in 1971.

A LITANY OF NEGLECT

Service industries, with the notable exception of retailing and transport, have largely been ignored by economic and urban geographers. The problems of definition and classification which have been briefly outlined may have played a part but there are other reasons and it is clear that geographers are not alone in this omission.

In his classic analysis of the relationship between economic growth and changes in the structure of the labourforce Clark (1940) was one of the first to observe that the economics of the tertiary sector had yet to be written and that many of his contemporaries were reluctant to admit that service industries even existed. Underappreciation of the job-creating potential of service industries has a long history which has been consolidated by theories of growth in which a distinction is made between basic and non-basic economic sectors of the local and regional economy. The basic industries have been viewed as the lifeblood of the economy, generating local or regional income through exporting their output. Non-basic activities are simply the result of growth that has already taken place in the basic sector industries and are, therefore, 'passive'. Services are considered part of the non-basic sector. Others such as Kaldor (1966) and Bacon and Eltis (1976) have advanced theoretical arguments suggesting that economic growth in advanced economies is adversely affected by the absorption of too much labour by service industries. Many of these arguments have been put forward on a very weak empirical foundation; hence Ginsburg, in a foreword to a book by Greenfield (1966, vii), noted that despite 'the growing significance of the service sector, it has been subjected only infrequently to detailed study'. Channon puts it even more strongly when he states that the

> service industries as a sector of the economy are extraordinarily under-researched. Despite the fact that in the major developed economies of the world they are rapidly replacing or have replaced the manufacturing sector as the dominant source of employment they are still the cinderella industries of academics and politicians alike. Indeed, in some strange way, many opinion leaders see them as positively non-productive and are endeavouring to reverse the tide toward the manufacturing sector where the making of things is seen as somehow more honourable. Quite why this ethic prevails is difficult to understand while those who endeavour to reverse what appears to be an inexorable tide are likely to suffer the same fate as Canute. (Channon, 1978, xv)

Studies of service industries in individual national economies are not easy to come by, especially for countries in the less developed world (some examples include Emi, 1978; Ezeikel, 1976; Frobel *et al.*, 1980; Lengelle, 1966a, 1966b; Lluch, 1975; Roggero, 1976).

The idea that the rapid increase in service employment since 1945 has been misunderstood is promulgated by Stanback (1979). He attributes this situation to three 'major failures of perception' (ibid., 1), comprising, first, an inadequate distinction between service sector output and employment which has created the mistaken view (see also Gershuny, 1978) that we are entering a new era characterized by increased demand for services and a levelling off of the demand for goods. There is in fact a high degree of complementarity between durable goods, in particular, and many services: the purchase of a television or a video-recorder is concomitant with increased demand for routine servicing and maintenance; the purchase of consumer goods is facilitated by an array of retail services; and the expansion of leisure and recreation activity generates demand for new and replacement sports equipment and facilities which allow participation. Secondly, the approach to the analysis of services has been insufficiently disaggregated, thus making an assessment of trends in growth difficult. This is a universal problem for the study of service industries which is a long way from a satisfactory resolution. Thirdly, Stanback suggests that misunderstanding about services is compounded by the inadequate attention devoted to the relationship between urbanization and service sector development, even though Stigler (1956) or Greenfield (1966) have written about the connection between the two phenomena. Consequently there is, for example, 'a lack of understanding of the extent to which the economic vitality of cities rests on their ability to compete as service centres within a national and international system of metropolitan places' (Stanback, 1979, 1–2).

Writing in the context of the work of industrial geographers Beyers (1983, 28–9) has recently commented on the geographer's 'pre-occupation with the philosophical context of manufacturing industrial change' which does 'little to gain the needed understanding of the process of more general structural change or the bases of growth of the services sector which have dominated recent employment expansion'. Moreover, 'it is clear that we must redirect our consideration of the role of services in industrial systems, if we are to also advance theory regarding their development in modern economies'.

In a series of seminars in 1983 on the emergence of the transactional city, which is seen by some as the ultimate product of the expansion of service activities in developed economies, Gottman puts forward the question 'is there a geography of services?'. He concludes that:

> the services, which are many indeed, may be differentiated, classified into categories and considered separately. Then for the various kinds of categories and services substantial spatial variation would appear for their distribution. This ought to be the first task of a geography of services. Linking specific services with certain locations and elucidating the reasons for the linkages observed could not fail to improve our understanding of human geography on the one hand and of the socio-economic functioning of society on the other. (Gottman, 1983, 63)

It is hoped that the approach used in the following chapters, and the examples cited, will elaborate some of the requirements for a geography of services. One question: 'what are service industries?', has already been addressed in this chapter, and this is followed logically by further questions such as: how have service industries changed over time?; what is their spatial relationship with other components of the economic system and with one another?; why do they exhibit these locational characteristics?; can their location patterns be explained by utilizing some of the models used elsewhere by geographers?; what have been the trends in the location of services?; and what is their significance in the unending task of coping with the processes of change in cities and regions? It will be demonstrated that these and other questions have been directed at selected service industries, in particular, retailing and warehousing, but the remaining, and larger, components of the sector have been largely overlooked by geographers.

SUMMARY

An effort has been made in this first chapter to lay the foundations for the remainder of the book. However unexciting, the need to define and classify services must be confronted early on because it provides guidance on how to structure the theoretical and empirical analyses presented in the subsequent chapters. In view of the heterogeneity of services it may never be possible to arrive at a universally agreed, comprehensive and unambiguous taxonomy. A notable feature of most existing classifications, however, is their simplicity. Many are dichotomous and quite at odds with the notion that services are diverse activities. More disaggregated classifications of the kind suggested by, for example, Browning and Singelmann (1975) simply increase the chance that individual activities will not fall exclusively into one category rather than another.

Everyone will not agree with the choices made in this chapter; they are not innovative, for instance, but the fact is that it is ultimately necessary to apply classifications to the available data. Simplicity has, therefore, been retained in order to make this task easier, and hopefuly to enable a more comprehensible analysis. As a result, there is a risk that nuances are missed but this is not a research monograph and we should not be overly concerned about the problem here. In the following chapters, therefore, service industries are mainly discussed with reference to the industry-based classification, according to whether they are producer or consumer services and whether they are provided by private or public sector organizations.

The chapter in conclusion has briefly outlined some of the reasons behind the now well-recognized neglect of service industries in the research of geographers and others. To some extent this provides a justification for the essentially descriptive approach used later. Finally, there remains the question as to whether there is a geography of service industries, and perhaps the queries signposted at the end of this chapter – and the contents of those that follow – will confirm that such a geography does indeed exist.

CHAPTER 2

Service industries in the economy: some preliminary empirical evidence

THE EARLY ROLE OF SERVICE INDUSTRIES

The service industries already comprised a large and growing sector of the British economy in 1861 (Lee, 1979, 1984). Over 31 per cent of the employed population were engaged in the service industries in 1861, and this had increased to more than 41 per cent by 1911 (Lee, 1979; see also Hartwell, 1973). With 3.3 million workers, the service industries already accounted for a larger proportion of total employment than agriculture (18.8 per cent) and more than one-half of the employees were in miscellaneous services. By 1911 the service industries had more than doubled the number of employees to 7.6 million, with transport and distribution growing more rapidly than miscellaneous services. Lee (1984) demonstrates that there were regional variations in service provision and his econometric results indicate that the incidence of manufacturing industry was not a precondition for the development of service industries. Income was the main determinant of growth in transport and distributive services employment, and since income was unevenly distributed, some regions had a larger share of service growth than others. The south-east and other regions in southern England took a relatively large share of the increases between 1861 and 1911, and 'economic growth was generated and sustained in those regions, by the combined effects of international trade and particularly investment, the consumer demand sustained by many centuries' accumulation of landed wealth and the self-sustaining capacity of affluence' (ibid., 154).

One consequence of this regional specialization (or perhaps contributing to it) was the pivotal role of the City of London for wealth-creating services in the eighteenth and nineteenth centuries. Although the balance of power and activity has changed, the principal services available in the City were already well established two centuries ago (Kay, 1985). Indeed the Bank of England had been overseer of flourishing markets in insurance and banking, commodities, shipping, foreign exchange and company stock since the last quarter of the sixteenth century. Informality was the order of the day in the mid-eighteenth century; coffee-houses were the venues for chance meetings or sometimes auctions which were advertised in advance. Lloyd's and Jonathan's, were two of the major venues with the former retaining its name as, in 1773, both the stock and insurance markets moved out of the coffee-shops and occupied their own premises as private clubs. The exchange of stock was centred on Jonathan's, which subsequently became the Stock Exchange, an institution largely concerned with floating and dealing in government debt.

The Bank of England, the East India Company and the South Seas Company were the principal dealers in stock in 1785.

Liverpool provides a more detailed example of the way in which services played a part in the economic development of the nineteenth-century city (Anderson, 1983; Chandler, 1968; see also Hartwell, 1973). During the eighteenth and nineteenth centuries the port of Liverpool became one of the foremost trading centres in the world, a process which was both facilitated by, and encouraged the appearance of, several service industries essential to the port's success and its ability to keep ahead of its rivals. According to Anderson:

> the development of the capital market and the emergence of financial institutions within it is of considerable intrinsic interest in a port-region such as Merseyside, where the links between finance, trade, and regional economic growth evolved over the classic period of the industrial revolution, and whose place in that wider process of change was of central importance to it....
>
> It was the regular and successful exploitation of new trade opportunities both overseas and in the home region – the African, West Indian, and domestic coal and salt trades in the eighteenth century, and those with the Americas, the Far East and in emigrants in the nineteenth – that produced the surpluses and the demand for financial machinery upon which the growth of specialist intermediaries could proceed. (Anderson, 1983, 26, 50)

Liverpool's overseas connections formed part of a £4,000 million investment by British companies between 1865 and 1914 (Cottrell, 1975). The expansion of Liverpool's business services during this period was also helped by the 'open' character of the merchant community (Anderson, 1983). New men and new ideas from other parts of the country could establish themselves and provide 'an important clue to the successful development of financial services over the period. For common to the growth of banking, insurance, and securities markets alike is the improvement in the quantity and quality of market information at lower costs' and 'banking access to the London money-market and the importance of close links with London correspondents, in insurance the key role of world-wide agencies, and in the stock exchange the regular listing of a widening range of company shares all served to relate the regional economy ever more strongly to that of the nation and beyond' (ibid., 50–1). This serves at the outset to scotch the notion that, even at this early juncture, service industries were essentially orientated towards and generated by demand from other economic activities in local markets; Liverpool service firms already had connections which extended well beyond their immediate hinterland. Such a feature has very likely prevailed since the nineteenth century and is relevant to later arguments (see Chapters 7 and 8) about the contribution of services to cities and regions with ailing economies.

But perhaps ports were in an especially favourable position to generate relatively high levels of service occupations. Data showing the occupational structure of three major British cities in 1871 (Lawton and Pooley, 1976) confirm that Liverpool had a thriving financial and insurance community by the middle of the nineteenth century and this, combined with its status as a major port and the related demand for transport services, provided employment for over 26 per cent of the male labourforce (Table 2.1). The learned professions and commercial occupations amounted to more than one-third of this proportion but in Manchester and Birmingham service occupations as a group were much less prominent than in Liverpool. These were primarily manufacturing cities which did not have a large group of transport occupations; by way of contrast, only 15.6 per cent of Liverpool's male labourforce was in manufacturing occupations compared with 32.9 and 42.4 per cent in Manchester and Birmingham respectively.

Table 2.1 Occupational structure of male labourforce (%) in selected British cities, 1871

Occupation, by industry	*City*		
	Liverpool	*Manchester*	*Birmingham*
Manufacturing			
Engineering	3.2	5.0	5.6
Textiles	6.8	17.4	8.4
Metals/minerals	3.2	5.3	18.8
Other manufacturing	2.4	5.2	9.6
Share of total male labourforce (%)	15.6	32.9	42.4
Services			
Learned professions	2.5	2.9	2.4
Domestic service	1.4	1.2	0.9
Transport	14.8	5.9	3.7
Commercial	7.5	8.0	4.6
Share of total male labourforce (%)	26.2	18.0	11.6

Sources: Anderson 1983, using data partly derived from Lawton and Pooley, 1976; the service occupations of nineteenth-century Liverpool are given in Anderson, 1983, table 4.1, 77–94.

But there has been a growing debate among economic historians and economists in recent years concerning the role of financial institutions in the process of industrialization and the degree to which banks, for example, have in fact played a direct or motivating role (Cameron, 1963). Indeed, 'there are a number of historical instances in which financial institutions constituted leading sectors in development; these institutions were "growth-inducing" through direct industrial promotion and finance' (Cameron, cited in Rudolph, 1976, 1). Gerschenkron (1962), for example, has suggested that banks played a leading role in impelling England along the path towards economic growth in the nineteenth century, while Rudolph (1976) points to the general agreement that in central Europe, especially in Germany and the Austro-Hungarian monarchy, the banks had intimate ties with industrial firms. After his detailed analysis of the nature of the actual relationships between banks and industry in the Czech Crownlands and Austro-Hungary, Rudolph (ibid., 184) finds, however, that the hypothesis that the 'great central European Banks *acted* as the entrepreneurial force which replaced government or individual activity in developing industry does not appear to be substantiated' and 'it is clear that the idea of banks as entrepreneurs, initiating industrial development and taking nascent firms through the dangerous years of youth and adolescence must be largely discarded' (ibid., 91).

It is also worth noting that by 1960 service industries already employed a larger share of the labourforce than manufacturing industry in both the more developed and less developed regions (defined in Table 2.3). Data on long-term changes in the shares of the three main sectors for both the developed and less developed countries assembled by Kuznets (1971) suggest that services were already significant in some economies by the mid-nineteenth century or even earlier (Table 2.2). The rise in the share of the service sector was generally widespread despite an absence of a consistent upward share in its product. By the 1850s and 1860s services already accounted for 17–25 per cent of the labourforce in several developed countries, including Britain, the Netherlands, Sweden and the United States (Table 2.2). Although some way behind Britain and the Netherlands in the mid-nineteenth century, the USA had easily surpassed them by the mid-1960s when over 56 per cent of its labourforce was in services, a change almost double that achieved over roughly comparable time periods by other developed countries. Even in some of the less developed countries – Egypt, for example – the share of the service sector rose more rapidly than the industry (manufacturing) sector during the early part of this century. India, on the other hand, achieved only a minor change in its service sector share of the labourforce between 1881 and 1961. In common with other observers Kuznets explains these early changes in terms of the difficulty of substituting capital for labour in the service sector of developed countries in the nineteenth century and its role as a refuge for an inadequately employed and, in more recent times, increasingly urban labourforce in the developing countries. As well as the indication of an early beginning, the speed of the structural changes revealed in Table 2.2 is notable. But these macroscale processes of change conceal substantial intra-sectoral shifts in service activities which have radically altered the character of the major industry orders over the last 100–150 years.

As will be shown later, this process of intra-sectoral change and adjustment is continuing. During the 100 years between 1850 and 1950 it took the form, initially, of expansion of finance, real estate and trade (especially retail trade). Domestic services, on the other hand, declined sharply, particularly during the period after the turn of the century since when they have been displaced in numerical significance by the expansion of professional services such as education and medicine or of non-market employment in public administra-

Table 2.2 Long-term trends in service sector share of labourforce, selected developed and less developed countries

Country and year	*Share in total labourforce (%)*		
	Agriculture	*Industry*	*Services*
Great Britain			
1851–61	21.6	56.9	21.5
1921	7.2	56.9	35.9
1961	3.7	55.0	41.3
Change 1851–1961	−16.0	0.0	+16.0
Netherlands			
1849	45.4	29.4	25.2
1960	11.0	50.5	38.5
Change 1849–1960	−34.4	+21.1	+13.3
Sweden			
1860	64.0	18.8	17.2
1910	48.3	32.2	19.5
1960	13.8	52.7	33.5
Change 1860–1960	−50.2	+33.9	+16.3
Japan			
1872	85.8	5.6	8.6
1920	54.6	25.4	20.0
1964	27.6	37.4	35.0
Change 1872–1964	−58.2	+31.8	+26.4
United States			
1839	64.3	16.2	19.5
1929	19.9	38.8	41.3
1965	5.7	38.0	56.3
Change 1839–1965	−55.9	+21.0	+34.9
India			
1881	74.4	14.6	11.0
1961	73.5	13.1	13.4
Change 1881–1961	−2.2	+0.6	+1.6
Egypt			
1907	71.2	14.1	14.7
1960	58.3	15.6	26.1
Change 1907–60	−12.9	+1.5	+11.4

Notes: Agriculture = agriculture, forestry, hunting and fishing.
Industry = mining and quarrying, manufacturing, construction, electricity, gas and water, transport and communication.
Services = Trade, banking, insurance, finance and real estate, ownership of dwellings, public administration and defence, and other services.

Source: Kuznets, 1971, extracted from table 38, 250–3.

tion and defence. Again, these characteristics are most clearly evident among those countries already classed as developed before the turn of the century.

The genesis of service industries can, then, be traced to the nineteenth century and earlier but most of the substantive expansion and specialization has only taken place during the recent past. This can best be demonstrated by turning to some further empirical evidence, beginning at the macroscale with intercontinental contrasts in the level of service activity and followed by an illustration of the relationship between economic development and the level of service activity, particularly the distinction between developed and less developed countries.

SERVICE INDUSTRIES AT THE MACROSCALE

It is useful to begin with an overview of the size of the service sector in relation to the two other major components of the economic system (Table 2.3). The estimates prepared by the International Labour Office show that as the share of the total labourforce in agriculture has decreased the proportion employed in services has expanded at a rate, in absolute terms, between 1950 and 1970 only just below that for manufacturing. While it is clear that at the world scale the labourforce in services was only one-half that in agriculture in 1970, the evidence in Table 2.3 suggests differentials in the expansion of services relative to the stage of economic development (see Chapter 3 for further discussion of this relationship). Hence in the more developed regions the agricultural labourforce has decreased by more than 40 per cent in twenty years (1950–70) as services and manufacturing industry have increased their share of total employment, with the former expanding most rapidly. In the less

Table 2.3 Labourforce in agriculture, industry and services, 1950–70

Region	*Year*	*Sector(%)*			*Total (thousands)*
		Agriculture	*Industry*	*Services*	
World	1950	64.3	16.3	19.3	1,100,150
	1960	57.7	20.1	22.2	1,297,400
	1970	51.0	22.9	26.1	1,508,613
Change, 1950–70 (%)		+8.5	+92.6	+85.1	
More developed regions[1]	1950	37.6	30.4	32.0	397,436
	1960	28.1	34.6	37.3	441,798
	1970	18.3	37.6	44.1	487,930
Change, 1950–70 (%)		−40.2	+51.9	+68.9	
Less developed regions[2]	1950	79.5	8.4	12.2	199,714
	1960	72.9	12.7	14.4	855,602
	1970	66.6	16.0	17.5	1,020,684
Change 1950–70 (%)		+21.6	+175.9	+109.3	

Notes: 1 More developed regions: Japan; Southern Africa; temperate South America; North America; Eastern Europe, Northern Europe, Southern Europe and Western Europe; Australia and New Zealand; and Soviet Union.

2 Less developed regions: China and other East Asia; eastern South Asia; middle South Asia; western South Asia; Eastern, Middle, Northern and Western Africa; Caribbean; Middle America mainland; tropical South America; Melanesia; Polynesia; and Micronesia.

Source: Adapted from International Labour Office 1977, table 3, 40.

developed regions the absolute number of workers in agriculture was still expanding in 1970 but at a rate much lower than for industry or services. The change between 1950 and 1970 is larger for both these groups but it has, of course, occurred against a much lower base of wage-labour than in the more developed regions.

Nevertheless, it is the magnitude of these changes which has fuelled the idea that economic systems are making a fundamental transition from an industrial to a post-industrial state (Bell, 1974; Gershuny, 1978; Kumar, 1978; Gershuny and Miles, 1983; Kellerman, 1985), a process also sometimes described as 'de-industrialization' (Blackaby, 1978; Bluestone and Harrison, 1982). Bell posits that post-industrialism, among other things, is accompanied by a smaller and smaller proportion of disposable income being used for buying goods as 'a third sector, that of personal services begins to grow: restaurants, hotels, auto services, travel, entertainment, sports, as people's horizons expand and new wants and tastes develop' (1974, 128).

The pace of this change varies, an expectation created by the very basic macroscale evidence in Table 2.1, but it is very significant because where the process is most advanced, such as North America, services clearly occupy a dominant position in the labourforce (Table 2.4). Almost 62 per cent of the North American labourforce was in service industries by 1970 compared with

Table 2.4 Labourforce in agriculture, industry and services: continental comparisons, 1970

Continent	*Sector (%)*			*Total (thousands)*
	Agriculture	*Industry*	*Services*	
Africa	71.5	11.3	17.2	136,653
Latin America	40.8	21.8	37.4	89,166
North America	4.1	34.2	61.7	95,764
Asia	64.8	17.5	17.8	859,730
Europe	20.8	40.5	38.8	201,428
Eastern Europe	34.9	37.8	27.6	53,231
Western Europe	9.8	44.5	45.7	62,560
Oceania	23.7	30.4	46.0	8,184
USSR	25.7	37.7	36.7	117,688

Source: Kuznets, 1971, table 3, 41–8.

only just over 17 per cent in Africa, where agriculture is by far the largest sector. As a general rule, in those continents where it is not the principal source of employment, the service sector is larger than manufacturing but the most notable exceptions arc Eastern Europe and the USSR, where the service

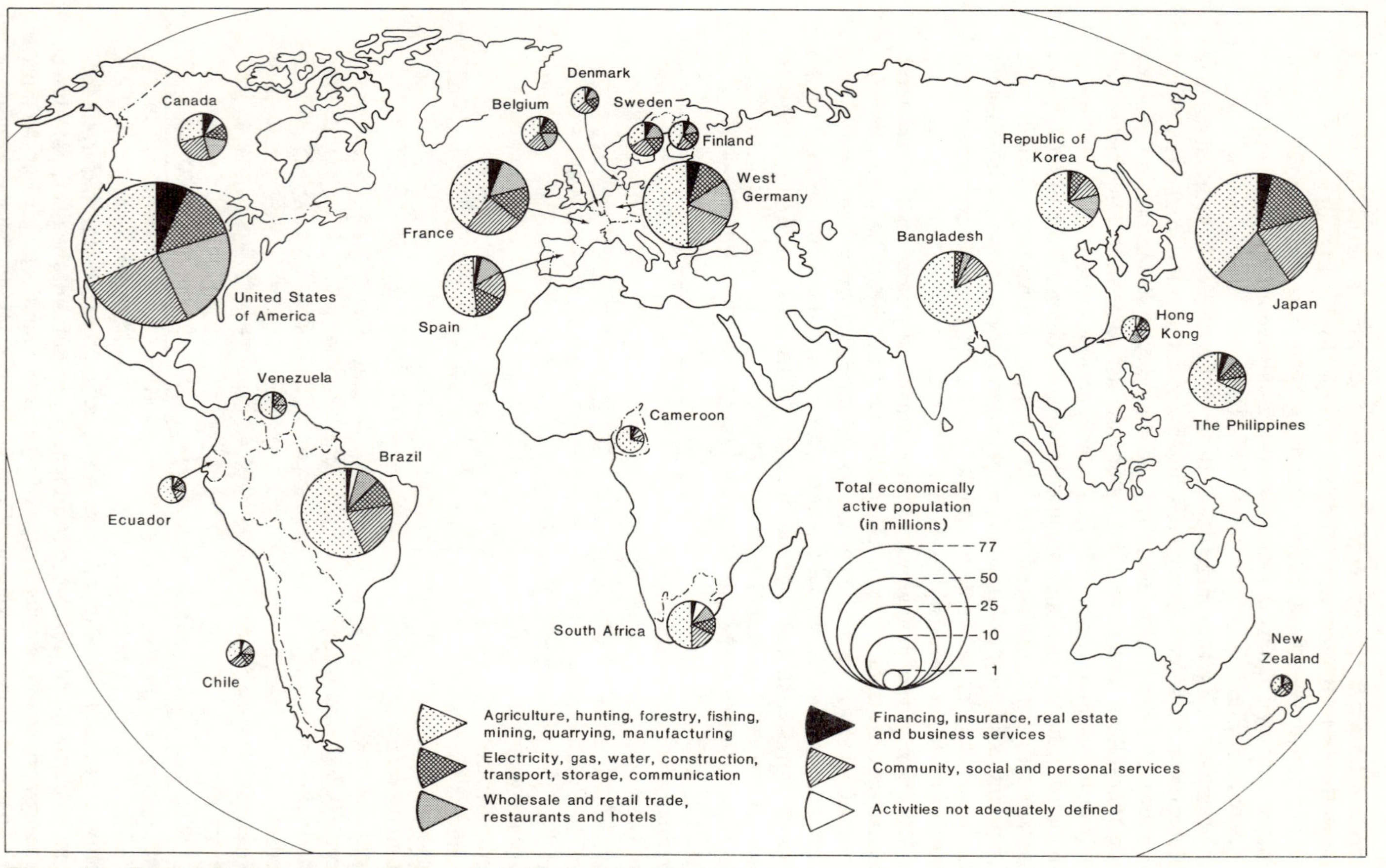

Figure 2.1 Economically active population, by industry, selected countries
Source: Compiled from data in United Nations, *Demographic Yearbook, 1979*, New York, UN, 1980.

sector historically has occupied a comparatively minor position (Ofer, 1973; Bergson, 1964). In 1897 some 11 per cent of the labourforce in Russia was in services; by 1940 this had only increased to 18 per cent and to approximately 37 per cent by 1970 (Ofer, 1973; International Labour Office, 1977).

Although the data is far from complete and not necessarily available for the same census year, a more detailed comparison of the size of the service sector in a range of countries is attempted in Figure 2.1. Note that the economically active populations in primary and secondary sectors are grouped together but the data for the service sector is disaggregated from the smallest to the largest category (reading each pie-chart clockwise). This helps to distinguish between those countries, mainly in the developed world, where finance, insurance, real estate and business services along with transport and communication comprise a significant proportion of the total economically active. In the less developed countries (LDCs) not only is the service sector smaller, it is also less diverse and usually dominated by transport and communications, the distributive trades and community, social and personal services (see also Bennett and Tucker, 1979).

Bangladesh (1974) and the USA (1970)* are clearly at opposite ends of the spectrum with respect to the level of service industry representation. Only 16 per cent of the economically active population of the former are in services, and of these just 1.7 per cent are employed in financial services, 10.7 per cent in transport and communications and the majority of 63.6 per cent in community and related services. Many of the latter are, of course, provided by the public sector. Community and related services comprise just over 37 per cent of the service sector in the USA, with financial and business services accounting for almost 10 per cent and transport and communications over 18 per cent of the economically active population. Even in Sweden (1975), where the concept of the welfare state is much further advanced than in most other countries, community and personal services represent only 46 per cent of total employment in services (the overall proportion in this subsector is much the same as the USA), with transport and communications employing as many as wholesale and retail distribution (22.3 per cent). In accord with the distinction between market and non-market economies made earlier, the level of financial and related services in Bulgaria (1975) is the same as Bangladesh (1.7 per cent) but, and this underlines the problem of trying to make generalizations, a smaller proportion of the economically active employed in services are classified as in community and related activities (43.8 per cent). Transport and communications, at just over 35 per cent, is much larger than in the three other examples cited.

Some of these differences can be attributed to variations in national census definitions which, in turn, have been regrouped to fit the United Nations classification of industry orders used in Figure 2.1. But even allowing for this, there are substantial differences between individual national economies. These can be partially explained by reference to economic development variables but the proportion of the variation explained leaves a large residual which reflects

*Data from the last national census available for each country.

individual national circumstances. This must be borne in mind whenever an attempt is made in subsequent chapters to make general statements about the location or organizational structure of services; their very heterogeneity as a group seems to persist in the context of variations in their status and structure in individual countries.

OTHER MACROSCALE VARIATIONS IN SERVICE ACTIVITY

Apart from comparing the relative importance of the service sector in different parts of the world and its functional structure, it is also useful to make some comparisons based on more selective criteria. With information exchange being important in service economies, telephones provide a crude indication of the level of specialization reached in different countries (Figure 2.2). There are clearly very large differences between nations as well as a large gap between the communications of north and south. Of the 500 million telephones in the world about 90 per cent are concentrated in just 15 per cent of the countries in the International Telecommunications Union (a United Nations agency created in 1932). It has been suggested that this discrepancy operates to the

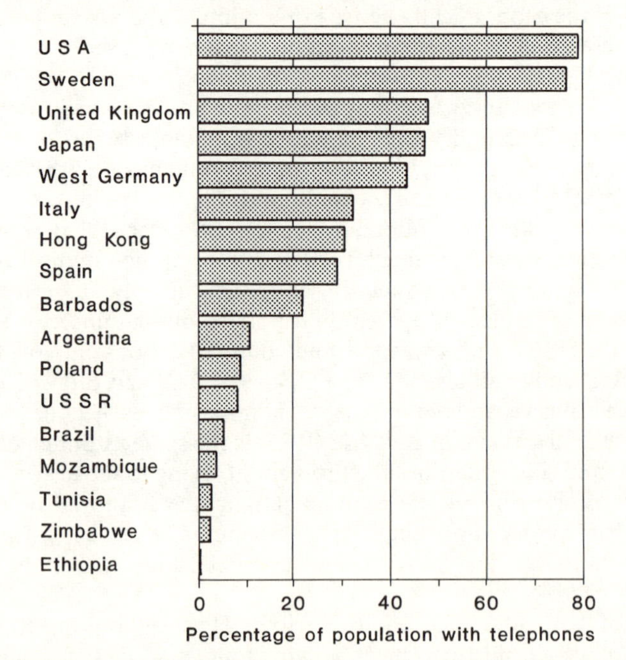

Figure 2.2 Access to telephones in developed and less developed countries
Source: AT&T data, *The Times*, 27 June 1982.

disadvantage of the developing world because the benefits accruing to the developed nations through advances in telecommunications, reflected in improvements in industry, education, health and transport, are missed. The pace of technological change is likely to aggravate the situation, especially if the developing countries continue to be relatively unaware of the importance of telecommunications technology in the fight against disease, ignorance, hunger and poverty. Less than 2 per cent of the aid provided by the United Nations, for example, is used to assist the south's improvements in telecommunications. Meanwhile there is unprecedented demand among the developing countries for more advanced facilities such as satellite-based radio, television and business communications and the 'gap' illustrated in Figure 2.2 could become even wider.

National assessments of the significance of telecommunications may affect the degree to which service industries in general contribute to internal economic development as well as to earnings and benefits from international trade. Another service industry, banking, has an even more important role in this respect. The commercial banks have come to play a crucial role in channelling capital flows around the world (OECD, 1983). Between 1973 and 1981 new international lending expanded at an annual rate of 20 per cent. This was partly a response to the needs of multinational corporations and the result, on the one hand, of the massive deposits placed with commercial banks following the sharp increases in oil prices imposed by the Organization of Oil-Exporting Countries (OPEC) in the early 1970s, and on the other, the demand from oil-importing countries for finance to pay for the more expensive oil. Massive payment imbalances have arisen in the developing countries in particular where debts to the banks increased at 23 per cent per annum during the 1970s.

The beneficiaries of this increased international flow of capital have been a relatively small number of banking centres (Figure 2.3), mainly in Europe, the USA, and south-eastern Asia. London, New York and Paris banks have long standing historical roles as centres for international deposits and lending and their main concern, in recent years, has been the shift in their share of international business as a result of the emergence of some competing centres. Most of these have adopted conscious policies for expansion. Singapore, for example, now has higher gross lending than Hong Kong. It emerged as a financial centre at the end of the 1960s because of the establishment of the Asian dollar market. Rapid development of the service sector of the economy has been one of the cornerstone's of the government's strategy for growth aided by Singapore's role as a major entrepôt; the number of offshore banks increased from fifty to seventy between 1980 and 1982. Similarly, Bahrain has become one of the world's major bank lending centres within a relatively short time; 65 offshore banks operated there in 1982, taking in deposits from the governments and central banks of surrounding rich Arab countries and making loans.

The power base of the banking industry remains heavily concentrated in Western Europe, North America and Japan (Table 2.5). Of the assets held by the world's top 500 banks in 1982 almost 44 per cent were controlled by banks with headquarters in Western Europe and almost 27 per cent by those headquartered in Japan and the Far East. The banks with headquarters in less developed countries in Latin America, or Africa and the Middle East (especially Africa), have only just over 8 per cent of the assets but over 26 per cent of

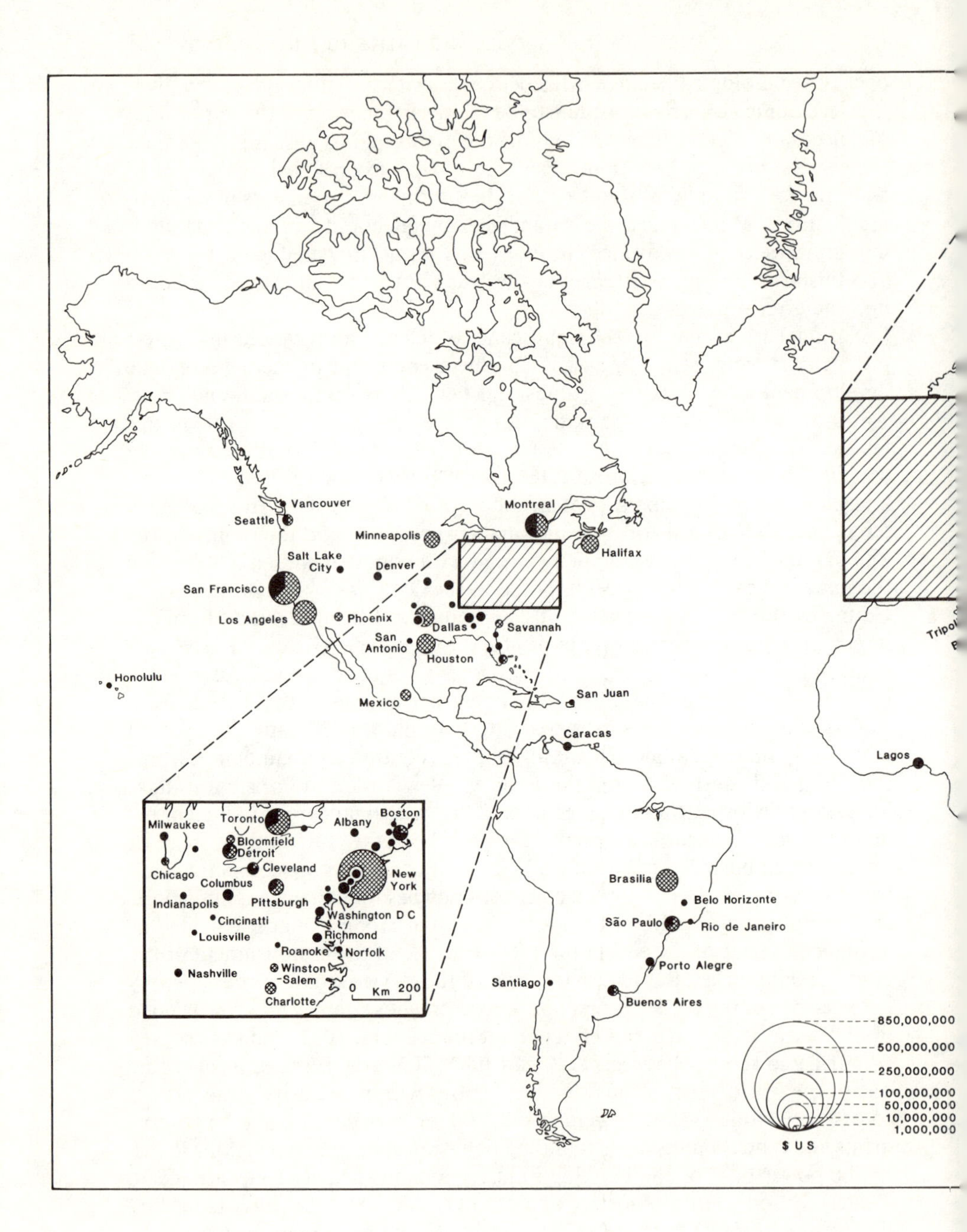

Figure 2.3 Location of the headquarters of the world's top 500 banks
Source: Compiled from data in the *Banker*, November 1983.

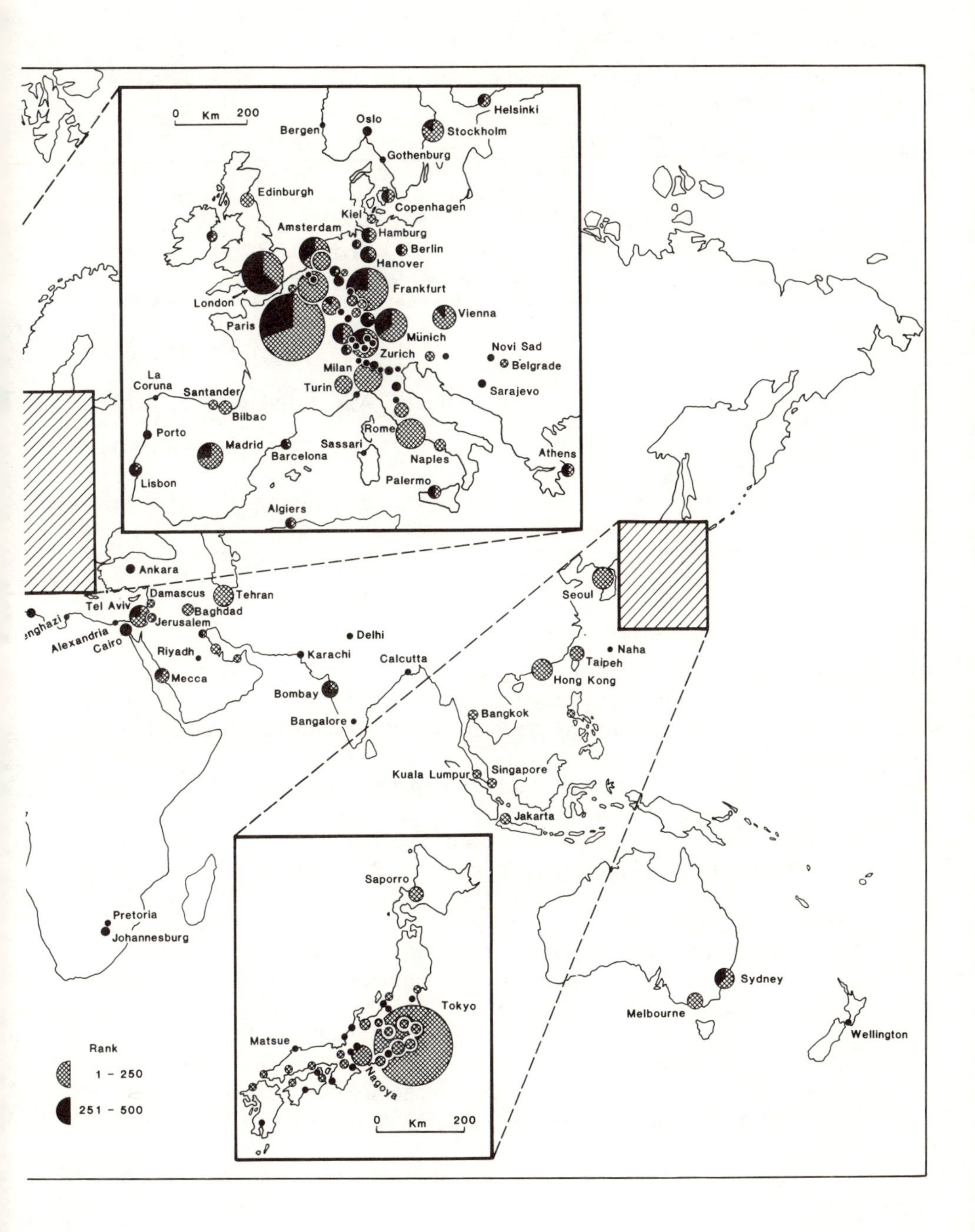

0 Km 200
Bergen
Oslo
Helsinki
Stockholm
Gothenburg
Edinburgh
Copenhagen
Kiel
Amsterdam
Hamburg
Berlin
Hanover
London
Frankfurt
Paris
Vienna
Münich
Novi Sad
Zurich
Belgrade
La Coruna
Milan
Sarajevo
Santander
Turin
Bilbao
Porto
Rome
Madrid
Sassari
Barcelona
Naples
Athens
Lisbon
Palermo
Algiers
Ankara
Damascus
Tehran
Tel Aviv
Baghdad
Jerusalem
Alexandria
Cairo
Delhi
Riyadh
Karachi
Calcutta
Mecca
Bombay
Bangalore
Seoul
Naha
Taipeh
Hong Kong
Bangkok
Kuala Lumpur
Singapore
Jakarta
Pretoria
Johannesburg
Saporro
Tokyo
Matsue
Nagoya
0 Km 200
Sydney
Melbourne
Wellington
Rank
1 - 250
251 - 500

Table 2.5 Summary of distribution of world's top 500 banks, by headquarters location, 1982

Area	*No. of headquarters and rank*		*Total assets*		*Total employees*		*Assets ratio*[1]
	1–250	*251–500*	*(US $m.)*	*%*	*(thousands)*	*%*	*(US $)*
Latin America	4	12	69,000	1.1	253,525	5.9	0.27
Africa and Middle East	22	24	306,865	4.9	273,340	6.4	1.12
South East Asia, Aust.	7	10	143,781	2.3	609,509	14.2	0.24
Europe	106	91	2,710,118	43.6	1,649,387	38.4	1.64
United States, Canada	54	79	1,389,528	22.4	948,835	22.1	1.46
Japan and Far East	57	34	1,594,822	25.7	557,203	13.0	2.86
Totals	250	250	6,214,514	100.0	4,291,799	100.0	1.45

Note: 1 Assets per employee.
Source: Calculated from data published in the *Banker*, July 1983.

the employees (Table 2.5). The assets ratio provides a further guide to the efficiency and potential strengths of the major world banks; Japan has the highest ratio, which is eleven times greater than the ratio for Latin American and south-east Asian banks.

Two final examples provide evidence for spatial disparities in service activities at the international level. The first example, access to health personnel and hospital services, reveals wide variations in a way which is disproportionate to need (Figure 2.4). Not only does this reflect the well-known fact that health services are more widely accessible (although not necessarily more available, for reasons of cost) in the more developed countries, especially Western Europe, North America and Australasia, but also means that the role of community services as a source of employment is much reduced. Certain less developed countries have a better level of provision than average, such as Libya, Zimbabwe, or Venezuela, where some of the high revenues from oil or copper mining have been expended on improving health and other services. In general, however, the majority of the LDCs have between four and five times as many persons per hospital bed and ten times the population per physician as most of the developed countries.

International tourism provides a second, rather different, example because of its potential for generating national income through 'invisible' trade. Invisible earnings arise from the provision of services to people living abroad (invisible exports) in contrast to 'visible' exports which comprise the sale of tangible goods abroad (Committee on Invisible Exports, 1981). Profits from investment in foreign enterprises, the insurance of a foreign factory, the raising

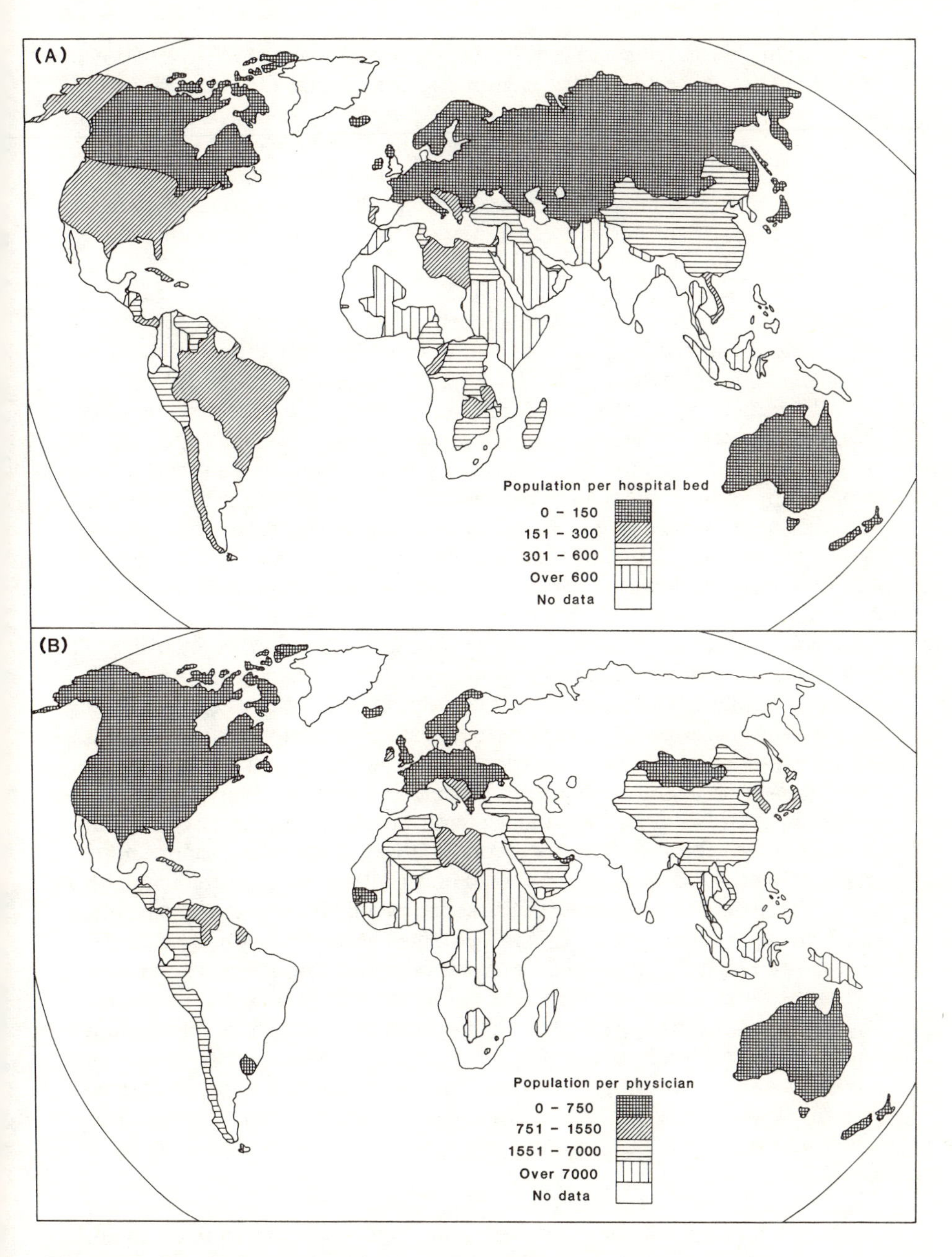

Figure 2.4 Population per hospital bed (A) and per physician (B), selected countries, 1977
Source: Compiled from data in United Nations, *Statistical Yearbook, 1979–80*, New York, UN, 1981.

of capital by a foreign borrower, the transport of foreign goods in a nation's shipping fleets, accounting advice given to a foreign client, or expenditure on services by foreign tourists all contribute to earnings from invisible exports. The significance of tourism is that it is seen in many developing countries as a panacea, particularly with reference to its indirect rather than direct effects. Its direct effects on employment are very limited but the indirect employment multiplier can be substantial, depending on the employment component of the various activities connected with the supply of tourist services (Wanhill, 1983). International tourism is more important to developing countries than domestic tourism because the latter depends on the domestic standard of living, while the former is exogenous and independent of the level of development of the host country. Statistics for the structure of tourist expenditure in India (Government of India, 1969), for example, show that accommodation (hotels, etc.) only accounts for 30 per cent of the total, internal travel by air, rail and road comprises 26 per cent, restaurants and entertainment 12 per cent and shopping more than 28 per cent. Sabolo (1975) argues, however, that the direct employment effects of shopping expenditures, for example, should not be overestimated; their main effect is simply to increase the total number of hours worked. Nevertheless, tourism has so many connections with other economic activities, including agriculture, food and clothing industries, that its overall impact on employment in particular is best assessed by using the techniques of macroeconomic analysis.

Because final demand expenditure by international tourists constitutes an item for which changes are exogenous to the economy in question, Sabolo (ibid.) uses input–output analysis (for which the same exogenous demand is presupposed) to assess the demand for labour. The method involves adding an exogenous increase of final demand to the various sectors connected with tourist demand; this is followed by a calculation of the output of each sector directly required to meet the intermediate demand from all the other sectors. This information is then used to estimate the resulting overall demand for labour using coefficients of productivity (for a full explanation see ibid., appendix 3, 157–66; see also Bureau of Industry Economics, 1980; Leontief, 1966). It is necessary to make several assumptions when using this method of analysis; that tourist demand and the resulting expenditures emanate exclusively from international tourists; that demand arises from a new group of tourists added to existing flows; that tourist investment will lead to new investment in tourist industries; that the flow of tourists is uniform throughout the year; and that the relationships within the input–output table (such as import coefficients or productivity) are stable.

After applying the model to three case studies, Japan, Greece and Spain, Sabolo concludes that tourism does appear to be a signficant creator of employment, although its impact will vary between countries according to their structural and economic characteristics. For the developing countries this is an important conclusion because the performance of the international tourist industry is largely independent of the size and other attributes of the national market, a factor which can have adverse effects on investment in manufacturing.

Taking a hypothetical investment of US $1 million in tourism and an associated US $3 million in tourist consumption in one year (1960), the former creates 800 jobs in Japan, 550 in Spain and 670 in Greece with almost all the increase created indirectly. By far the largest number of jobs follow the rise in tourist consumption, ranging from 3900 in Japan to 2700 in Greece and 1500 in Spain. Almost one-half of these jobs are created indirectly in Japan and Greece but 60 per cent in Spain. Sabolo suggests that the differences in the number of jobs created mainly arise from the degree of integration of the respective national economies and variations in the use of labour for identical production tasks. The ratio of indirectly to directly created jobs also depends on the structure of the economy; the induced relationships tend to be greatest in the more complex economies especially if local production is substituted for imports. In the less developed economies investment leakage may cause problems because the demand experienced by those sectors linked to tourism may only be fulfilled through the purchase of raw materials, equipment, or professional guidance from outside the country.

The pattern of actual tourist arrivals in 1977 is shown Figure 2.5(A) as a proportion of all the arrivals in the countries for which data are available. Western Europe and North America are the principal destinations with Spain (13 per cent), the USA (7 per cent), the UK (4 per cent), Canada (5 per cent), Austria (4.5 per cent), France (10.0 per cent) and Italy (5.7 per cent) absorbing a large proportion of all arrivals. While these seven countries receive over 49 per cent of the arrivals, they attract almost 52 per cent of the receipts. Hence the kind of tourist, the length of stay, or the demand for ancillary services such as golf courses, supplementary excursions to places of interest, or special courses in a wide range of recreational pursuits will determine the level of tourist-related invisible earnings. Therefore, although Spain has the largest number of arrivals, they only generate 7.5 per cent of the total receipts by the countries shown in Figure 2.5(B), while Italy (9 per cent), the USA (11.6 per cent) and the UK (6.4 per cent) attract a disproportionate share of tourist expenditures.

Some 11.7 million visitors from overseas came to the UK in 1982, one of the few countries not only to retain, but to increase its share of the international tourist market. These visitors spent £3229 million, an increase of 9 per cent on 1981. This makes tourism Britain's top invisible export and one of the biggest job-creating industries. Certain countries, such as the UK, must rely on international tourism rather than short-term (even one-day) visits across international boundaries of the kind which can take place in mainland Europe (for example, between West Germany and the North Sea coastal recreation areas in the Netherlands), and this could be significant for the arrivals–income ratio (Mathieson and Wall, 1982). This means that although many of the less developed countries do not figure prominently in the arrivals statistics, they may still be able to attract significant receipts. Saudia Arabia, Libya and Morocco are some current examples and there must be potential for others to benefit in the same way.

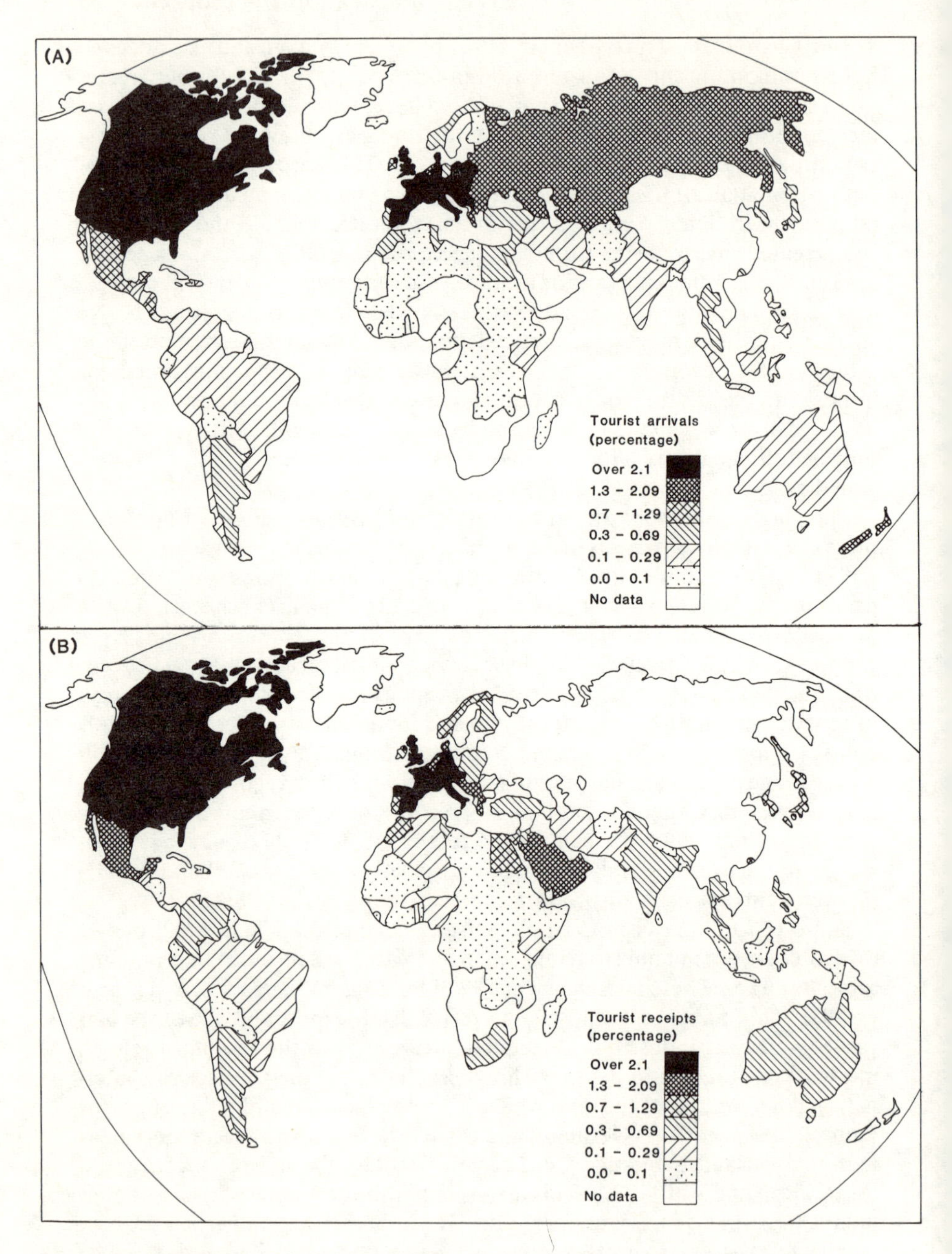

Figure 2.5 Tourist arrivals (in thousands) (A) and tourist receipts (US $m.) (B), selected countries, 1977
Source: As Figure 2.4.

OTHER 'INVISIBLE' TRADE IN SERVICES

Tourism is only one source of invisible trade in services. Transport, investment income and a wide range of other services such as royalties, commission, management and agents' fees, insurance, consultancy work and personal income are also included (Committee on Invisible Exports, 1981). In a country like Britain where gains from the oil sector are now flattening out and manufacturing exports restrained by reductions in competitive and innovative capabilities, overseas earnings generated by business services such as consultancy are likely to become more important. This involves the export of skills and experience by both private and public sector industries. Public corporations such as Britoil, British Rail and British Steel earned £342 million from overseas consulting work in 1982 (*The Times*, 20 June 1983), an increase of 57 per cent (£124 million) over 1981. Such growth can have benefits for the export activity of private sector consultants who are used to provide advice for some of the public sector projects.

Travel is an important element of invisible receipts in the low-income countries (Table 2.6) along with other services (36.8 per cent), whereas investment income and transport contribute to more than 56 per cent of invisible receipts by the high-income countries. The low-income countries must expend a large part of their invisible payments on the purchase of specialist services connected with investment activity, while expenditure on travel is only one-half of that found for the high-income group. In common with national service employment and its contribution to GDP, trade in services as a proportion of receipts or payments has expanded between 1964 and 1973 (Table 2.6). One notable feature, however, is the way in which the proportion of invisible payments as a proportion of total payments has increased more rapidly in the low-income countries (indeed it represents a higher overall proportion in 1973), thus increasing their dependence upon externally provided services with consequent losses for wealth-creating activities in the national economy. The gap between receipts and payments is also significantly larger for the low-income countries.

During the last decade trade in services has become one of the most rapidly growing and changing sections of international trade (Moroz, 1983; Clarke, 1965; Cowell, 1983). It has been estimated that services now account for one-third of world trade amounting to a massive $550,000 million in 1980 (Committee on Invisible Exports, 1981). About 20 per cent of the world total is accounted for by the USA, with France and Britain in second and third place respectively (9.4 per cent and 9.2 per cent of world trade in invisibles). Until recently most of the services crossing international boundaries were ancillary to, or in support of, the trade of goods (such as the financial services which permit trade in goods, transportation, or consulting services). Now services are traded in their own right: advertising, market research and management consulting, telecommunications services, computer services, specialized financial services in banking and insurance and business services. The spread of multinational enterprises has encouraged the growth of international trade in services (see Chapter 7), even though some of the transactions will be between

Table 2.6 Trade in invisible services (receipts and payments), by sector and changing dependence on invisible trade (1964–73), low- and high-income countries, 1973

	Percentage of total invisibles				*Percentage of total receipts*	
	Transport	*Travel*	*Investment income*	*Other services*	*1964*	*1973*
Receipts						
High-income countries[1]	30.0	20.7	26.1	21.4	22.0	24.2
Low-income countries[2]	19.7	30.3	18.5	36.8	20.0	20.5
Payments						
High-income countries	26.1	22.2	32.1	24.1	23.4	24.7
Low-income countries	25.2	14.1	39.3	21.5	22.5	25.0

Notes: 1 High-income countries are the USA, UK, Denmark, Belgium/Luxembourg, France, Denmark, Netherlands, West Germany, Norway, Finland, Austria, Sweden, Switzerland, Yugoslavia, Saudi Arabia, Libya, Canada, Australia, Japan and Israel.
2 Low income countries are Greece, Italy, Portugal, Spain, Argentina, South Africa, Brazil, Mexico, Venezuela, Iran, Egypt, Indonesia, Turkey, Republic of Korea, Singapore and Ireland. The division between high- and low-income countries is $3000 per capita for 1973.

Source: Committee on Invisible Exports, 1976, tables 4a, 6a and 6b.

the constituent parts of the same enterprise. Unfortunately there is a shortage of reliable statistics on international trade in services; it remains difficult to measure and analyse trade volumes, patterns and composition.

In 1981 the current account of the UK balance of payments showed an overall surplus of £6013 million (seasonally adjusted), with more than 52 per cent of this positive balance attributable to invisibles (services, interest, profits and dividends and transfers) (Department of Trade and Industry, 1983). Over 64 per cent of the balance arising from invisibles was derived from services, 16 per cent from interest, profits and dividends, and 20 per cent from transfers. The most recent estimates for 1982 (first three-quarters) show that the positive balance has been maintained overall, but with the contribution of services increased to 55 per cent. Most of the invisible surplus is derived from the activities of private sector services and public corporations, which has helped to cover deficits in government transactions. Trade in invisibles, such as telecommunications and postal services, advertising, films and television, royalties and licences, airlines, shipping, construction work overseas (i.e. fees to architects, quantity surveyors and civil engineers) and the services provided by the City, is vital to Britain's balance of payments since in only twelve of the last 200 years has the export of goods exceeded the import of goods (Cowell, 1983). The surplus on invisible earnings has usually made up the loss on visible trade.

It is also worth noting that the service industries are not just an important source of invisible earnings. Some of these earnings (Table 2.7) will eventually find their way back to the consumption of inputs. Hence between 1976 and 1982 the proportion of fixed capital expenditure (at 1975 prices) attributed to distributive and service industries (except for shipping) increased from 54 per cent (£3959 million) to 72 per cent (£6658 million) (Department of Trade and Industry, 1983). Almost 27 per cent of the expenditure (at 1975 prices) in 1976 was by financial services (other than leasing). Some 38 per cent of the total expenditure by all services was on new building work, 37 per cent on plant and machinery and the remainder on vehicles. By 1981 the pressure to invest in new office technology is reflected in 53 per cent of total expenditure devoted to plant and machinery and 31 per cent on new building work. Leasing and other service industries now represent 58 per cent of total expenditure by industry group.

THE IMPORTANCE OF PUBLIC SECTOR SERVICES

Whether measured in expenditure terms, as a broad aggregate, or as capital expenditure on goods and services and capital expenditures and transfer payments to firms and households, there has been a dramatic expansion of public sector employment since 1945 (see, for example, Peston, 1978; Gartner *et al.*, 1973). Demand for non-market or public services has arisen for a number of reasons. First, from demographic changes in the population – for example, more people are living longer after retirement and, therefore, require a wide range of health services which market activities may not want to provide at a cost which the retired can afford. Secondly, the number of children eligible for schooling and the number of years for which education is compulsory has also increased, thus ensuring sustained demand for education-related service employment. Thirdly, there are national, state, or local minimum standards for student–teacher ratios and many countries have introduced policies for the provision of more post-school and adult education services as a way of generally improving knowledge and skills. Fourthly, most social and community services must also be provided through publicly funded channels; the acceptance by society of an obligation to look after the interests of disadvantaged children, to assist families with social and related problems, to provide probationary services, or to have a penal system (to name but a few) has also ensured that employment in these services has grown steadily. Fifthly, public sector employment is generated by the need for advanced societies to regulate their internal activities or to have agencies responsible for their interaction with other states or participation in trade blocs such as the EEC, OPEC and Comecon, or in defence arrangements such as NATO. Various departments of state have long existed to fulfil a variety of roles from foreign affairs to town planning and have become major white-collar employers. In consequence,

Table 2.7 Fixed capital expenditure by manufacturing, distributive and service industries: UK, 1976–81

Industrial sectors and type of asset	*Year(%)*[1]		
	1976	*1978*	*1981*
Manufacturing	43.7	41.6	31.5
Distributive and service industries	56.3	58.4	68.5
Total (£m.)	7615	9054	9337
Of which, by industry:			
Wholesale distribution	8.5	8.9	6.1
Retail distribution	16.4	15.1	14.6
Leasing	12.4	19.4	28.0
Other finance	26.6	20.9	20.0
Other industries	36.1[2]	35.6	29.9
By type of asset:			
New building work	38.5	30.8	26.1
Vehicles	24.3	28.5	21.0
Plant and machinery	37.2[2]	40.7	52.8

Notes: 1 At 1975 prices seasonally adjusted.
2 Columns sum to 100.

Source: Department of Trade and Industry, 1983, tables 1 and 3, 395–6.

public expenditure has increased as a proportion of total expenditure and the income generated in the public sector has risen as a ratio of national income.

Some comparative data for public sector employment in the member states of the EEC is shown in Table 2.8 (Bannon *et al.*, 1977). Almost 25 per cent of total employment in Denmark in 1973–4 was in local and central government and there was more than one public sector employee per 100 population. But there appears to be no relationship between the size of a nation's population and the proportion of total national employment in the public sector. West Germany has only 0.5 public sector employees per 100 population, although its population is twelve times larger than that of Denmark. The UK is almost as large as West Germany but has a marginally larger proportion of public sector workers and 0.8 per 100 population. Explanations for the range shown in Table 2.8 would seem to arise from contrasts in political and bureaucratic philosophies and the degree of public, as opposed to private, provision of services such as health and education. In 1970 in the USA almost 16 per cent of the labourforce was in general government (Hiestand, 1977), suggesting that even economic systems based upon a strong capitalist ethic still require a wide range of public agencies to administer defence, trade, regulation of privately provided services such as air transport or environmental protection programmes. All of the individual states also require their own administrative and regulatory bureaucracies to manage their affairs.

Table 2.8 Public sector employment in the EEC, 1973–4

Country	*Population (thousands)*	*Employment (thousands)*	*Total employment in public sector*	*Employees per thousand population*	*Percentage of total employment*
West Germany	61,973	26,126	3,430[1]	55.35	13
UK	56,021	24,010	4,338	77.44	18
Italy	54,901	18,140	2,336	43.10	12
France	52,133	20,663	3,528	67.67	17
Netherlands	13,349	4,563	617	45.91	13
Belgium	9,742	3,783	510	52.35	13
Denmark	5,022	2,355	563	112.23	23
Ireland	3,051	1,037	180	59.00	17
Luxembourg	353	151	14	39.09	9

Note: 1 Estimates produced by the EEC Information Office.
Source: Bannon *et al.*, 1977, table 2.21, 70.

Some of the statistical attributes of a burgeoning public sector in the UK are illustrated in Figure 2.6. Public sector employment in service industries rose from 4.5 million in 1961 to 6 million in 1979 (+33 per cent) compared with an overall increase in public sector employment of 13 per cent (7.5 million to 8.5 million). The chart (A) is also included in Figure 2.6 for private sector services to provide a comparison of their scale and broad trends. The public sector data is divided into three groups: publicly owned corporations (B), central government (C) and local authorities (D). It is clear where the growth has occurred, primarily in central government and local authorities rather than the public corporations where the broad trends are similar to those for private sector services, even though the shares of the major industry sectors are very different. Although the expansion of local authority employment has slowed down – and even decreased overall – since 1976 when it comprised some 13.2 per cent of total employment, education continues to occupy a prominent place (almost one-half of the total in 1982). Over 43 per cent of the jobs in education are classified as part-time and, along with social services, this accounts for a large proportion of all part-time employees (32.7 per cent) in local authorities. Other significant sectors within local authorities include construction (5 per cent), recreation, parks and baths (3.8 per cent), and refuse collection and disposal (2 per cent). Other services provided by local authorities include transport, public libraries and museums, environmental health, town

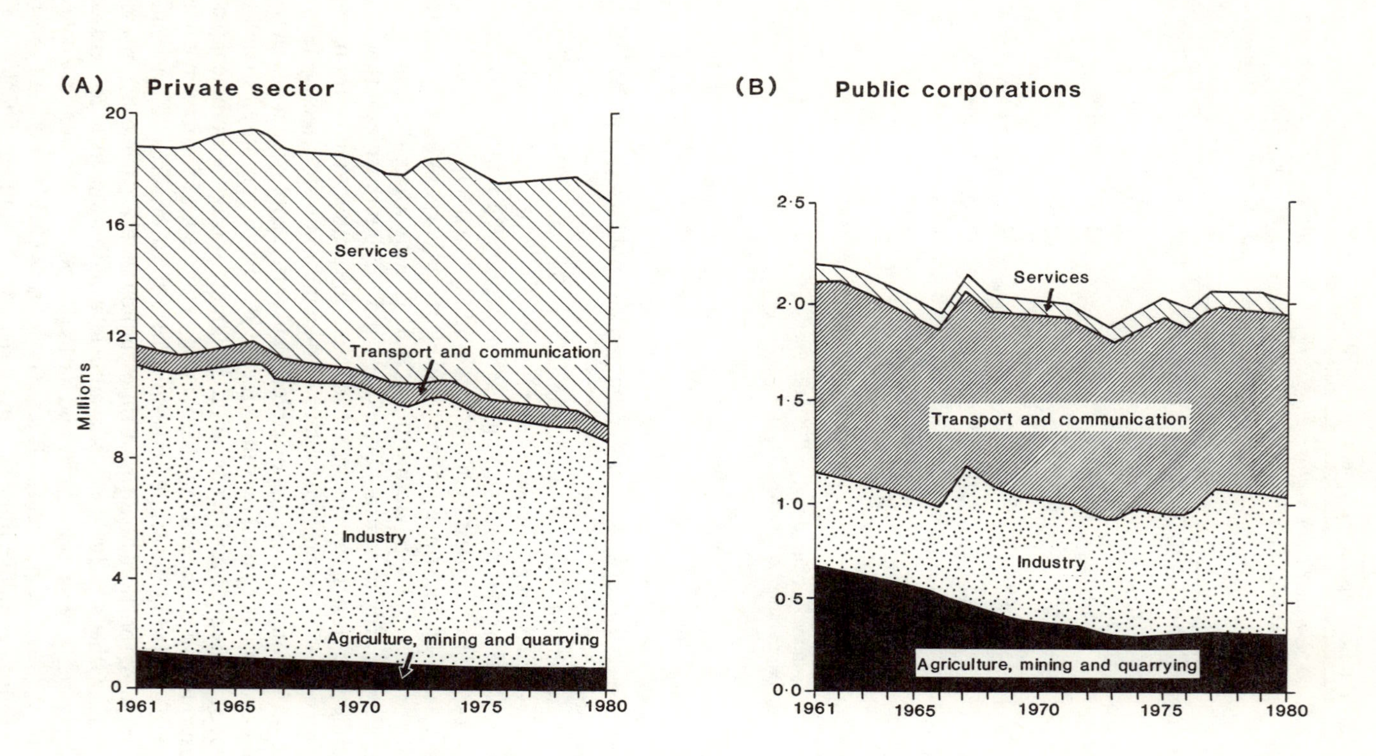
(A) Private sector
Millions
20
16
12
8
4
0
1961
1965
1970
1975
1980
Services
Transport and communication
Industry
Agriculture, mining and quarrying
(B) Public corporations
2·5
2·0
1·5
1·0
0·5
0·0
1961
1965
1970
1975
1980
Services
Transport and communication
Industry
Agriculture, mining and quarrying

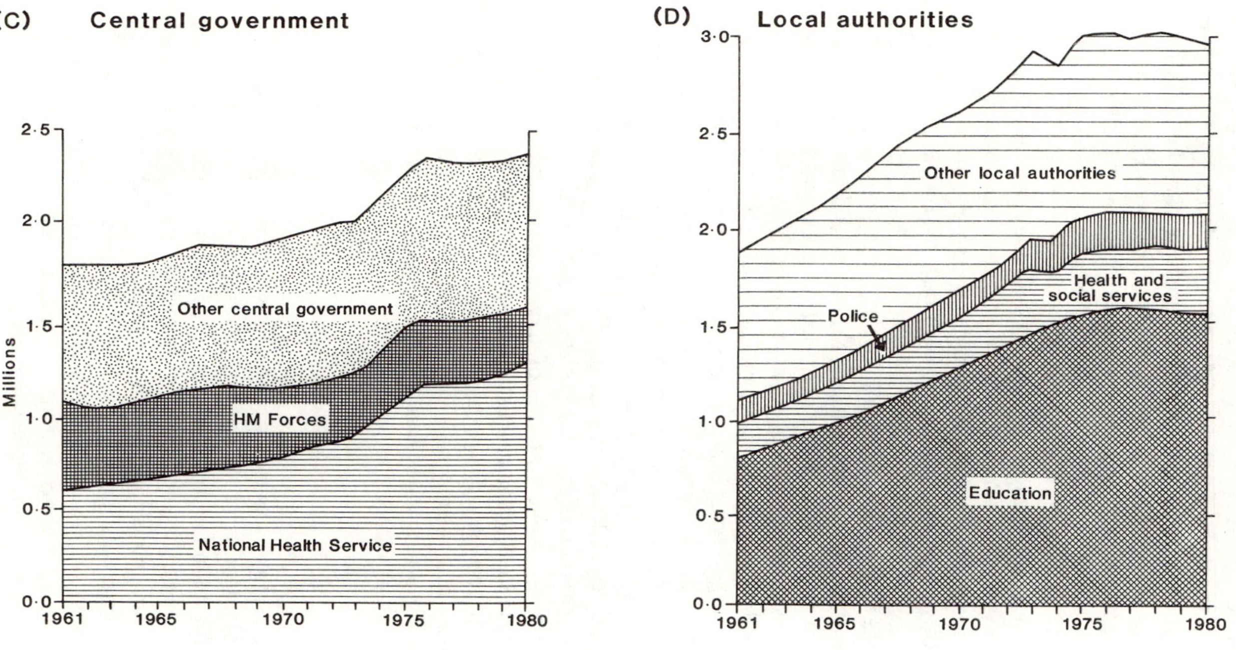

Figure 2.6 Employment in British public corporations (B), central government (C) and local authorities (D) 1961–80; the sectoral distribution of private sector employment (A) is included for comparative purposes
Source: Central Statistical Office, *Social Trends 12*, London, HMSO, chart 4.8, 66.

and country planning, fire services, police and probation, magistrates' courts and agency staff.

The expansion of educational services, for instance, has been influenced by continuous changes in the age structure of the population caused by peaks and troughs in births such as that following the postwar baby boom which created a heavy demand for primary and elementary schools in the mid and late 1950s, secondary schools in the late 1950s and early 1960s and, eventually, higher education establishments during the late 1960s. Concurrent with this the 'value' of education was reassessed, particularly with reference to the range of recipients, which has been steadily extended to include those without formal qualifications arising from their school years, those wishing to further their general education as 'mature' students and those able to attend courses offered by community colleges in the USA or colleges of further education in the UK.

The rapid increase in health service employment has been promoted by demographic forces as well as a recognition of the merits of illness prevention and cure. The administrative–bureaucratic role of the public sector may also grow because the market economy does not always give rise to the most effective or to safe competition – the automobile industry has only adopted exhaust emission measures, energy-saving engines and safety devices such as seat belts after public regulations have required it. Pharmaceutical companies are required to undertake careful product tests before marketing drugs, overseen by public agencies such as the Food and Drug Administration in the USA or the Health and Safety Executive in the UK. Control of additives used in food processing and colouring and the annual or triennial testing of motor vehicles for structural and other defects are some examples of service jobs created by a need to protect the public interest. Urbanization has created a need for public services such as refuse collection, sewage disposal and police services – which the private sector may not wish to provide even though the public are consumers of these services in much the same way as they are consumers of the goods produced in the private sector.

SUMMARY

The significance of service industries for the economic development of cities and regions in the eighteenth and nineteenth centuries, together with some temporal statistics for a number of countries over the period 1850–1960, has been outlined. However, the most rapid expansion, diversification, and specialization of service industries has taken place since 1945, and this has been illustrated with some data for the period 1950–70. There are some notable variations in the proportion of service activities in different countries and between developed and less developed regions. These differences extend to structural contrasts. The new, more developed and specialized service activities, such as finance and banking, marketing and advertising, are much more prominent in the developed regions, while in the less developed regions the old or more 'traditional' services, such as transport or retail trading and distribution, comprise a large part of the service sector.

The spatial variations evident at the industry level are also shown to exist for some more specific examples such as telecommunications, the distribution of the headquarters of the world's largest banks, and access to hospital and physician services. International tourism is an example of a service which in a wide range of countries, irrespective of their development status, can be used to generate direct, but especially indirect, benefits for the economy. The potential and actual significance of tourism as a mechanism for relieving spatial inequalities consequent upon the distribution of other types of services has also been discussed.

Services provided via the public sector give national and local government an opportunity to exert some influence on the distribution of service sector growth and the chapter, therefore, concludes with some UK statistics demonstrating the size and importance of public sector service employment.

In Chapter 3 some of the possible explanations for variations in the level and range of service industries will be explored with particular reference to economic considerations. There is also some discussion of alternative perspectives which the disciplines of sociology and political economy can provide, allowing us an opportunity to incorporate some additional empirical material.

CHAPTER 3

Some causes and further consequences of the emergence of services

In Chapters 1 and 2 the emphasis on definition, classification and macroscale variations in the distribution of service activities has, of course, overlooked so far any attempt at explanation. It is not intended that this chapter should completely fulfil this role – the explanation of spatial differences in distribution will come later. But this will not be comprehensible without recourse to aspatial factors which also contribute to our understanding of the spatial patterns of service activity. There is a general consensus that economic systems pass through stages of development for which one symptom is an increasing proportion of employment in service industries. There is less agreement about how far these changes will proceed; or whether they are part of an ongoing evolutionary process rather than representative of a new kind of 'post-industrial' economy and society; or the role which technology will play; or whether they presage detrimental effects on spatial patterns of economic development within individual countries as well as between them. Much seems to depend on the disciplinary viewpoint and three examples – from economics, sociology and political economy – will be considered here.

SERVICE INDUSTRIES AND ECONOMIC DEVELOPMENT

That the growth of service industries and the expense of other kinds of industry is in some way linked with the level of economic development has been examined by several economists. Neither Fisher (1935) nor Clark (1940), however, in their widely cited references to a progression, as economic growth advances, in the allocation of labour from primary to secondary and finally to service employment (Figure 3.1) provide a systematic analysis of the factors responsible for this restructuring (Fuchs, 1965; Oberai, 1978). Gershuny provides a concise summary of standard accounts of the stages of economic development and their implications for the growth of services:

> Rostow's account of sectoral development is that at the early stage of civilization, with rudimentary agricultural techniques, the majority of any population must work on the land; as techniques and productivity improve they are not so constrained and may move on to different categories of production – secondary and then tertiary. Rostow asks where we go from the stage of high consumption of material goods; Bell answers that we pass on to the next category of consumption, the consumption of services; Dahrendorf similarly though in different terms, that we pass to the public

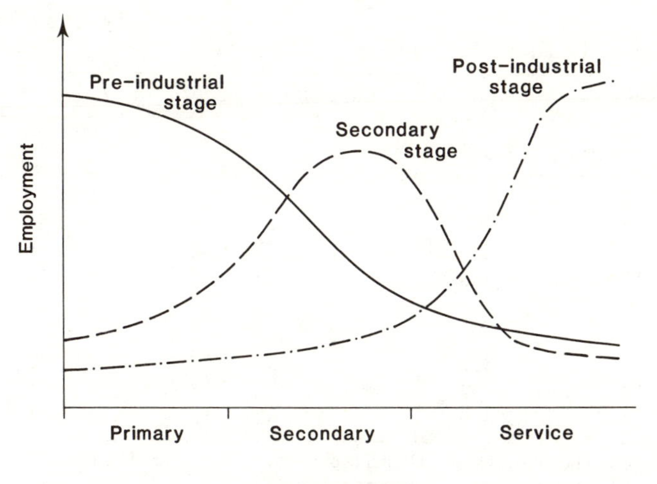

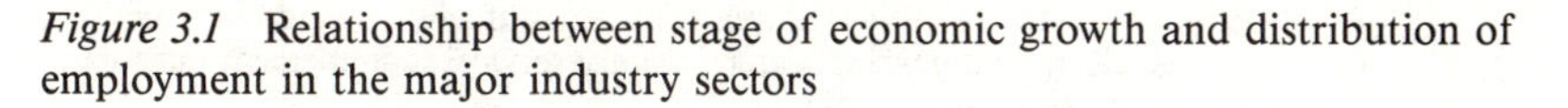
Figure 3.1 Relationship between stage of economic growth and distribution of employment in the major industry sectors

> provision of non-material products, education and leisure activities; Schumacher, that we turn our attention to social and spiritual values. Galbraith, rather differently, sees the trend as ever-increasing material consumption, but only as a result of the machinations of the great post-capitalist corporations of the 'planning system' for whom economic growth is a requisite of survival. Once these corporations are controlled, then the pattern of development may be trimmed to the public purpose, which is assuredly not the continued growth in consumption of material goods. The view of the future of employment for all these writers is simply a continuation of Rostow's drift through sectors classified by the nature of their final product; they see post-industrial employment in a humanized form of manufacturing industry and increasingly in tertiary sector employment. (Gershuny, 1978, 141–2)

Although this tells us something about the variety of interpretations of the outcome of the economic development process on the part of several distinguished observers, it is less illuminating about the causes of the uncertainty which it conveys so effectively. Gershuny uses it to argue that there is certainly a clear shift towards tertiary patterns of consumption in the sense of more abstract, largely recreational or intellectual, activities but he argues that this need not be equated with any necessary growth of tertiary employment on an equivalent scale (see also Gershuny and Miles, 1983).

Oberai (1978) shows that the sectoral distribution of wage employment also changes in association with the process of economic development (Table 3.1) (see also Galenson, 1963). In those countries with the lowest average income (group I) in 1960 wage-labour absorption is mainly attributable to agriculture with services not far behind. For the countries with intermediate per capita

Table 3.1 Comparison of labourforce and wage employment, by sector in low-, intermediate- and high-income countries, 1960 and 1970

Group[1]	*No. of countries*	*GDP per capita US $*	*Wage employment (%)*			*Labour force (%)*		
			A	*B*	*C*	*A*	*B*	*C*[2]
1960								
I	19	196	37	24	35[3]	60	14	21[3]
II	10	447	14	33	50	29	26	41
III	13	1,656	5	45	48	14	39	45
1970								
I	19	322	31	26	39	49	18	28
II	10	992	9	36	53	22	29	45
III	13	2,890	4	42	52	11	35	51

Notes: 1 Group I countries have GDP per capita below $300; group II between $300 and $700; and group III over $700 per capita.
2 A = primary sector (agriculture, hunting, forestry and fishing).
B = secondary sector (mining and quarrying, manufacturing and construction).
C = services (electricity, gas and water, sanitary services, commerce, transport, and storage and communications).
3 Row totals do not sum to 100 because unemployed and activities not adequately described are excluded.

Source: Oberai, 1978, tables 1 and 2, 7 and 9.

incomes (group II) services were already the main source of wage employment in 1960 with manufacturing occupying a much higher position than in group I countries. Services are even more clearly the dominant sector for wage employment in group III countries in 1960, and the main change by 1970 is the one expected: services are even further forging ahead as the construction and manufacturing sectors remain largely unchanged (Bhalla, 1970).

One interesting question arising from Table 3.1 is whether there is a systematic pattern of change in the sectoral distribution of employment. The changes which have taken place between 1960 and 1970 for group I/group II countries show that the share of the services sector rises much faster (in relation to wage employment) than does the share of the manufacturing sector. Increases of between 9 and 15 per cent compare with a contraction in agriculture of the order of 23 per cent. The change between group II/group III suggests that we cannot assume that the service sector share of wage employment will continue to rise, indeed it reveals a moderate decline with the secondary sector as the principal beneficiary. The service sector's share of the overall labourforce does continue to rise, but again the rise is less dramatic than the increase in the secondary sector.

There is evidently, then, a link between income and services; as per capita income increases the service sector share of total employment increases. This has figured prominently in explanations for the differential growth of such activities between less developed and developed countries but the distinction becomes blurred when an alternative measure of the service sector contribution, gross domestic product (GDP), is used (Table 3.2). There is only limited variation in the proportion of GDP attributable to service activities among the countries in the top three groups in Table 3.2. In the lowest-income countries the contribution of services to GDP is much smaller but the range is also greater than for countries with higher per capita incomes. It is also apparent that the average share of service sector output in all countries is greater than the average service sector share of employment, except for countries in the highest-income group where the differential is narrower. This raises questions about the relative productivity of services in low- and high-income economies; the wider differential between output and employment in the former could be interpreted as indicative of higher productivity until it is recalled that agriculture is more important in the low-income countries and this may significantly affect the productivity level of the service sector (Bureau of Industry Economics, 1980).

Table 3.2 Size of service sector in relation to GDP by annual per capita income, 1963–73

Per capita annual GDP in US $, 1973	*Number of countries*	*Average service sector share of domestic product (%)*	
		1963	*1973*
5000 or more[1]	6	61	62
		(5066)[5]	(4966)
3000–4999[2]	6	62	62
		(5470)	(5367)
1000–2999[3]	4	63	62
		(4584)	(4677)
Under 1000[4]	9	49	52
		(3668)	(4172)

Notes: 1 USA, Sweden, West Germany, Canada, Australia, Denmark.
2 Norway, France, Belgium, Finland, Austria, UK.
3 Italy, Singapore, Greece, Portugal.
4 Mexico, Chile, Turkey, Tunisia, South Korea, Philippines, Thailand, India, Indonesia.
5 Figures in parentheses show the range.

Source: Bureau of Industry Economics, 1980, table 2.2, 9.

Recent data (Hopkins, 1983; see also Gemmell, 1982) do, however, show that service industries are steadily increasing their contribution to GDP and share of the labourforce in developing countries (Table 3.3). The trend is particularly clear in those groups of countries which include areas dependent on tourism, for example, middle and low Latin America and the Caribbean (Table 3.3). By 1980 over 50 per cent of GDP was derived from services which also occupied a larger proportion of the labourforce than in any of the other groups of developing countries. In China and Africa (low) service industries increased their share of total employment only slowly between 1960 and 1980 but this disguises a substantial contribution (over 40 per cent in 1980) to GDP in Africa (low). By 1980 the service sector was the largest source of GDP and a larger

Table 3.3 Contribution of services to GDP and labourforce in developing countries, 1960–80

Developing countries	*1960*		*1970*		*1980*	
	GDP[1]	*LF*[2]	*GDP*	*LF*	*GDP*	*LF*
Latin America and Caribbean (middle)	48.1	33.6	48.8	38.6	51.0	42.4
Latin America and Caribbean (low)	46.5	21.7	47.5	27.1	51.3	30.9
Asia (middle)	43.9	19.8	44.4	24.2	45.7	28.1
Asia (lower)	32.2	15.5	34.1	19.1	39.8	22.6
India	39.6	14.6	32.6	17.2	38.1	20.6
Africa and Middle East (oil)	23.1	20.0	18.8	25.3	29.6	29.9
Africa and Middle East (middle)	37.5	17.9	36.7	21.6	44.4	25.4
Africa (low)	32.1	7.3	38.2	9.6	42.4	11.4
China	—	9.8	—	11.9	22.0	14.1
All developing countries	39.1	14.5	39.0	17.7	44.2	21.0

Notes: 1 Gross domestic product.
2 Proportion of total labour force in service industries which are defined as commerce, transport and communications, public and private services.

Source: Hopkins, 1983, 461–78.

source of jobs than the manufacturing sector. This has led Hopkins to note that

> those who see agricultural development as the key to overall development are likely to be running against the course of history. Hence it might be profitable to undertake new research on the dynamics of both the industrial and service sectors as major absorbers of manpower in the future and, in particular to give more weight than hitherto to the relatively neglected service sector in developing countries. (Hopkins, 1983, 465)

In the specifc case of Britain the contribution of services to GDP during the period 1960–80 conforms with the overall trends (Figure 3.2). Well over one-third (41.8 per cent) of Britain's GDP was created by manufacturing industry in 1960 compared with just over 37 per cent from all the service industries combined (excluding the 'other' category in Figure 3.2). By 1980 this pattern had been reversed, so that service industries accounted for more than 48 per cent of GDP and manufacturing for only 31.2 per cent. Oil revenues (included in mining and quarrying) have risen slowly since 1975 (from 2 to 7 per cent of GDP) and transport and communication has reversed the downward trend since 1960. Most notable, however, has been the steadily increasing contribution to GDP of insurance and banking (3.4–10.5 per cent between 1960 and

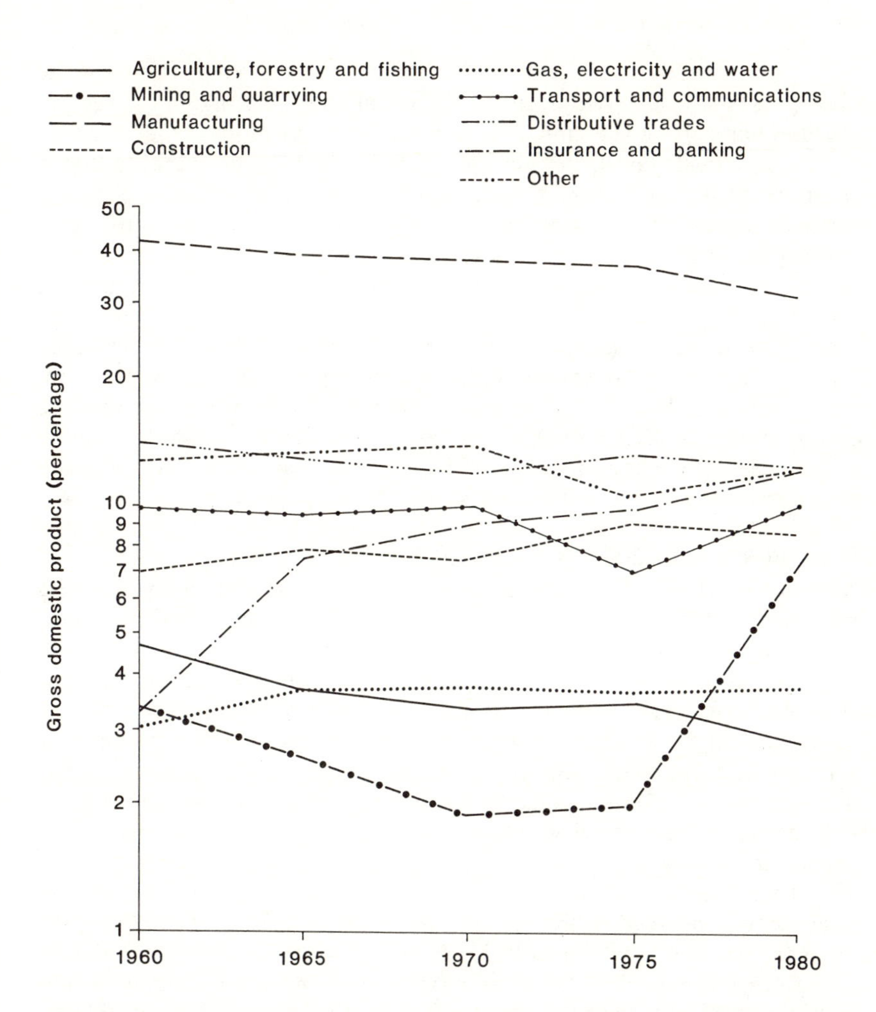

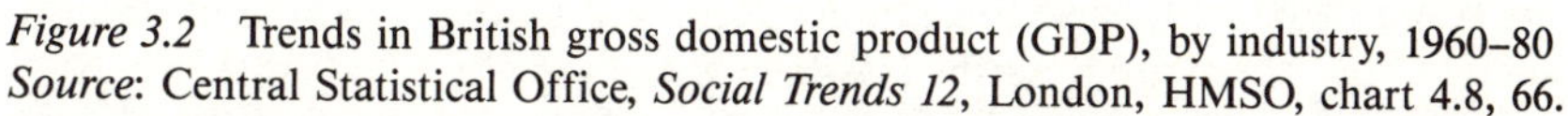

Figure 3.2 Trends in British gross domestic product (GDP), by industry, 1960–80
Source: Central Statistical Office, *Social Trends 12*, London, HMSO, chart 4.8, 66.

1980). Simultaneously the growth of public administration and bureaucracy, much of which is a service function, has reduced the combined contribution of production and trade to GDP from 87.8 to 74.9 per cent. The main losses of manufacturing industry output have been in the mature/capital-intensive and/or undifferentiated sectors such as textiles, mechanical engineering and basic metals and materials. Newer industries, such as motor vehicles, chemicals, electrical engineering and instrument engineering, have tended to increase output.

It is, therefore, misleading to think that the rapid expansion and increasing dominance of services in economic systems is primarily a symbol of maturity in their development. In practice, services make an early appearance whether measured in terms of the labourforce or wage employment. This confirms an observation already made in Chapter 2 and a number of reasons can be advanced. First, the development of the social and economic infrastructure of group I and group II countries is invariably dependent upon the provision of facilities made available through public agencies; hospitals, schools, transport services, postal and telecommunications. Many of these are labour-intensive creators of wage employment. Secondly, the low elasticity of factor substitution in the service sector means that as the economy grows, accompanied perhaps by urbanization, there is more scope for entry into unskilled but labour-intensive services. Finally, Oberai (1978) suggests that where manufacturing output is expanding, there is an employment multiplier which requires the expansion of central/local government services as well as of commercial services. This analysis confirms Sabolo's (1975) conclusion that the view of service industry growth in developing countries as essentially residual is superficial. There 'is no fundamental difference between the developing and developed countries, and Colin Clark's analysis is valid for both' (see ibid., 143).

Some causes of the growing demand for services

In order to account for the shifts in national economic structure which they had observed, both Clark and Fisher tend to stress the importance of differences in income elasticity between the major sectors of the economy (as well as the influence of changes in productivity). They, therefore, viewed expansion of employment in services as demand centred and largely reliant on income effects.

First, per capita income will determine the patterns of consumption in an economy in accordance with sectoral income elasticities of demand. A rise in the income of an individual, a household, or a nation is usually associated with increased demand for goods and services. The ratio of the percentage increase in income and the percentage increase in demand gives the income elasticity; if the increase in each is the same, the income elasticity will be zero. Some items consumed by households, for example, exhibit elastic demand, that is the demand for them increases disproportionately with a unit increase in income (transport, health services, or recreation are examples); equally there are some items with elasticities below zero, that is we need them in order to survive, and these are described as inelastic (food, fuel, or light).

Income elasticities vary widely from one country to another, and although some care must be exercised when making comparisons because of wide differences between national surveys some examples are given in Table 3.4 (Sabolo, 1975). The statistics relate to the income elasticities for consumption of 'new' services, such as health care, education and tourism, and the ranges in value shown for Japan and Malawi refer to low/high income levels in the former and small towns/Blantyre (the capital city) in the latter. For the less developed countries, or individuals at low income levels, increases in income are reflected in a more than proportionate change in expenditure on education, while for the highest income levels the increase in resources leads to a less than proportionate increase in expenditure on education. The needs of most of the recipients in this category have already been met. It, therefore, seems that the demand for 'new' services emerges only after a minimum income threshold has been crossed (see ibid.).

Table 3.4 Differences in the income elasticity of consumption of 'new' services, selected countries

Type of service	*Country*				
	Japan	*India*	*Malawi*	*Sudan*	*Tunisia*
Medical care	0.4–0.8	0.6	1.0–1.2	3.2	1.5–2.1
Education	1.3–1.0	1.6	1.2–0.8	2.6	—
Hotels	2.2–1.1	—	—	—	—
Cinemas	0.4–0.8	—	—	—	—
Travel	1.3–1.0	—	—	—	—
Recreation/ tourism	—	1.2	0.5–2.8	2.4	1.5–1.8

Note: For the ranges of elasticities the first value refers to the lowest incomes, and the second value to the highest.
Source: Sabolo, 1975, table 6, 42.

But there are several difficulties attached to the use of elasticities or income to explain, or to forecast, the demand for services and the associated expansion of employment (Fuchs, 1965). Elasticities are measured in terms of real output or consumption but for many services accurate measures of output are still not available. Adjustment patterns to changes in income do not occur quickly, so that observed patterns of consumption may depend upon past as well as present levels. Related to this is the point that elasticities are subject to change; those which apply now may not be applicable retrospectively or in the future. An important characteristic of services which has already been noted (see Chapter 1) as significant for spatial analysis is that many of them produce intermediate outputs, sold to other firms, as well as final outputs, sold directly to consumers. In some cases individual service activities may be solely concerned with one or the other but changes in income will affect both intermediate and final demand. The extent of its effect will depend upon the relative proportions of services and goods used in production.

Some other difficulties mentioned by Fuchs include the fact that the demand for some services seems to depend upon the distribution of income in the economy as well as its average level (for instance, domestic services used by the wealthier households) or that demand is subject to the vagaries of taste, technological change and relative prices. Substantial changes in income may also be associated with changes in the level of urbanization and this makes it difficult to determine whether the observed change in demand is related to income or urbanization.

The technical arguments are complex and will not be considered further. Suffice to say that some historical and contemporary evidence does demonstrate an association between increased expenditure and the sectoral distribution of employment. A historical example is provided by Stigler (1956), who used data from a very early study of workingmen's family budgets (dating from 1797) to make a comparison with similar information for 1937–8 (Table 3.5), subdivided between agricultural and non-agricultural workers. Although such a comparison must be made cautiously because of the size of the samples or the bias towards lower-income households, it is clear that relative to total expenditure food costs diminished sharply, by almost one-half. Apart from the 'miscellaneous' category, the other expenditure categories show moderate rises. This leaves the miscellaneous group, which was almost non-existent in 1797, to reveal the largest relative increase to three times its level at that time. Included under the miscellaneous heading in 1937–8 were household furnishings and tobacco, and services such as insurance premiums, travel, health and employment insurance, medical care, entertainment and union dues. Stigler (1956) also provides evidence for a similar shift in household expenditure as incomes rose in Massachusetts between 1874–5 and 1935–6.

Table 3.5 Composition (%) of expenditures of English working families, 1794 and 1937–8

Category of expenditure	*Agricultural workers*			*Non-agricultural workers*		
	1794	*1937–8*	*Difference*	*1794*	*1937–8*	*Difference*
Housing	4.6	8.3	3.7	6.0	12.7	6.7
Food	74.5	48.4	−26.1	73.9	40.1	−32.9
Clothing	9.0	9.1	0.1	5.0	9.5	4.5
Fuel and light	4.4	8.6	4.2	5.4	7.6	2.2
Miscellaneous	7.6	25.6	18.0	9.6	30.1	20.5
Total	100.0	100.0	—	99.9	100.0	—
Number of families	60	1491		26	8905	
Persons per family	5.9	3.8		6.2	3.8	
Average annual expenditure	£39.6	£150.6		£40.3	£224.3 [1]	

Note: 1 Annual expenditure has been rounded to the nearest value in decimal currency.
Source: Stigler, 1956, table 1, 3 (after Eden, 1797, and *Ministry of Labour Gazette*, 1940).

More contemporary data for household expenditure patterns in Britain show a continuation of the trend away from 'essential' to 'discretionary' expenditures; alcoholic drink, durable household goods (which incorporate a service element), transport, and vehicles and other services have increased their share of total expenditure from 31.2 per cent in 1963 to 36.6 per cent in 1979 (Table 3.6). 'Other services' includes expenditure on items such as leisure. Either through working fewer hours or 'enforced leisure' (i.e. unemployment), expenditure on active recreation provided by private and public clubs has increased, along with a growing number of theme parks, open-air museums, wildlife parks and stately homes. Similar trends are apparent in the USA (Muller, 1981), where personal consumption increased by 2.5 per cent per annum between 1972 and 1977, while outlays for housing and gasoline rose by 4.1 per cent and 6.5 per

Table 3.6 Structure of British household expenditure (%), 1963–79

Item of expenditure	*1963*	*1979*	*Difference*
Housing	10.6	14.6	4.0
Fuel, light and power	6.6	5.6	−1.0
Food	29.2	23.2	−6.0
Alcoholic drink	3.9	4.8	0.9
Tobacco	5.7	3.0	−2.7
Clothing and footwear	9.2	8.3	−0.9
Durable household goods	5.8	7.5	1.7
Other goods	7.1	7.7	0.6
Transport and vehicles	11.8	13.9	2.1
Services	9.7	10.4	0.7
Miscellaneous	0.4	1.0	0.6

Note: Columns sum to 100.
Source: *Annual Abstract of Statistics, Great Britain*, 1963, 1979, London, HMSO.

cent respectively. As a result, shopper goods sales as a percentage of personal consumption declined by 13.2 per cent as housing increased by 8 per cent. It is thought unlikely that a higher share of personal income will be devoted to shopper goods during the 1980s. Food, a major 'essential' expenditure item, accounted for almost 30 per cent of the 1963 total but had decreased to almost 23 per cent by 1979.

The demand for producer services

One of the principal consequences of the changing structure of expenditure on services has been an increase in the demand (and supply) for producer services. A number of reasons can be enumerated. First, there is a desire on the part of other producers to arrive at the existing level of output at lower cost; external services will be sought if a specific function can be performed at lower cost, and with no loss of quality, outside the firm. Since the external labourforce used in providing such a service will specialize in providing the required input, it represents a saving in terms of better productivity. Secondly, a company may want to improve the quality or quantity of its output by using its own resources; it may find it easier to achieve this goal by engaging a computer analyst, a market research consultant or an advertising agency specializing in making the company's product attractive to final consumers. A third determinant of demand for producer services is the desire on the part of some companies to 'hive off' unpopular tasks such as those jobs involving unsociable working hours or low status and repetitive work. Fourthly, many small and medium-sized enterprises cannot justify the cost of retaining specialized staff on a full-time basis and will, therefore, look to outside agencies which, fifthly, can also respond to erratic labour requirements, both specialized and routine, such as temporary secretarial or clerical assistance. Producer services can also thrive because, according to Greenfield (1966), many firms believe that they should retain small, compact and relatively homogeneous labourforces. This keeps down training and retraining overheads and reduces the chances of costly labour disputes. In these circumstances it may also be easier to introduce technological and organizational innovations, of the kind supplied by producer services which may of course reduce the demand for them in the long run. The need to imitate competitors, the need to cope with risk and uncertainty, especially in relation to technological change and obsolescence, also generate demand for producer services. Finally, most firms require, at some time, independent guidance on their growth prospects, thus creating demand for financial auditors and consultants, management consultants, or market research firms.

The supply of services

However strong the demand for services, their availability ultimately depends upon the assembly of the factor inputs necessary to create supply (Kuznets, 1971). The supply of labour is crucially important for the growth of existing, and development of new, service activities. The main factors usually used as surrogate variables for this are population growth, rur-urban migration and educational attainment of the population. Although these were very likely influential in the nineteenth century in developed countries, the empirical evidence is now best documented for less developed countries. Hence, using data from a sample of 27 LDCs, Sabolo (1975) shows that the elasticity of labour employed in the tertiary sector in relation to total population is 1.4 ($r = 0.76$); in other words, the labourforce employed in services grows more rapidly than the population. In addition, the faster the population grows, the higher the rate of increase of employment in the tertiary sector. This leads to the situation in many LDCs where a proportion of service employment is 'redundant' or residual as soon as the total population begins to increase rapidly, a phenomenon which is accelerated by the inability of the manufacturing sector to absorb the growing supply of labour. For rur-urban migration the elasticity is 0.8 ($r = 0.86$) because of the way in which, for example, in the main cities of, say, Brazil, 80–85 per cent of immigrant males acquire remunerated employment within one month of arrival and much of this is in traditional activities in the service sector (Santos, 1968, quoted in Sabolo, 1975). The importance of incorporating supply-side factors in an assessment of the differential expansion of services in different economic conditions is that the surprisingly large volume of services in some LDCs seems to be the product of the combined effect of the pressure of labour supply in urban areas and a strong private demand for services.

The supply of services in developed economies also relies to a degree upon the availability of labour with the appropriate skills, but the innovative and adaptive abilities of the smallest through to the largest organizations is also important in a sector which is increasingly vulnerable to rapid technological change and growing competition at the international scale. In regions experiencing a decline in their traditional manufacturing industries the supply of new services which substitute for job shedding by other industries or as a mechanism for other economic activities to remain viable and competitive depends on the diffusion of knowledge of the latest technology or innovations and the ability or willingness of local entrepreneurs, as well as larger established companies, to take risks.

Access to the latest technology has quickly come to occupy a central role in the range and quality of the services available. This began with the development, initially, of the computer and more recently a wide range of information-handling equipment linked by telecommunications (Barron and Curnow, 1979; Sleigh *et al.*, 1979; Hills, 1982; Otway and Peltu, 1983). The service industry applications of information technology are very varied but such applications of telecommunications and machine technology are very diverse and depend not just upon the ability of manufacturers to produce new and more powerful or efficient equipment, but upon the demonstration of its value to individual activities; by designing suitable computer software, developing data networks or document reproduction and transfer systems for general as well as very specific and specialized purposes, and providing ongoing support, including updating, for the equipment once it has been installed. This has not only diversified the supply of services, but also created changes in the type of labour or occupational skills required – a trend that is likely to continue as the decline in the cost of computers, for example, continues (following the introduction of mini- and microcomputers) so increasing the accessibility of computing power to medium and small users who in many cases have no detailed knowledge of the technology or the equipment and will, therefore, require the assistance of specialist support services. In addition, the trend towards comprehensive solutions for user needs, including decision-making models, has sustained and diversified demand for highly skilled personnel (the consequences of technological change for the geography of services are discussed in more detail in Chapter 10).

The welfare function of services

The limited scope for substituting other factors of production, the demand for services and the high level of labour input have all contributed to the recent emergence of service industries. But to these another consideration may be added. According to Sabolo (1975), individuals will tend to seek jobs in services, so that they can achieve maximization of their welfare functions. Two main factors, one monetary and the other psychological, are involved; income from the job is clearly significant but it may also hold particular interest and prestige and offer the individual an opportunity to fulfil certain aspirations. For any given level of educational qualification Sabolo (ibid.) shows that in the LDCs, for example, income prospects in the service sector are much better (especially for unskilled and semi-skilled individuals) than in the primary and secondary sectors. Since the majority of the labourforce in the developing countries is unskilled, excessive growth of service employment in recent times is not surprising. To this can be added the general preference for service sector employment, possibly reflecting the content of education and the increasing participation of females in the labour market. Trends in personal tastes have, it seems, 'led the machine to be preferred to the plough and now the pen to the machine' (see ibid., 121).

SOCIOLOGICAL VIEWS ON THE EMERGENT SERVICE ECONOMY

Sociologists equate the appearance of the service economy with a new type of society which is commonly labelled post-industrial (Touraine, 1971; Bell, 1974; Gersuny and Rosengren, 1973). It has also been variously referred to as 'high tech' or 'sunrise', or a 'knowledge based' or 'information based' economy. Touraine describes it as a

> new type of society now being formed. These new societies can be labelled post-industrial to stress how different they are from the industrial societies that preceded them – although in both capitalist and socialist nations – they retain some characteristics of these earlier societies. They may also be called technocratic because of the power that dominates them. Or one can call them programmed societies to define them according to the nature of their production methods and economic organization. This last term seems to me the most useful because it most accurately indicates the nature of these societies' inner workings and economic activity. (Touraine, 1971, 3)

The term 'post-industrial' has been chosen in order to underline or stress the difference between this society and its industrial predecessor although, as we have seen, not everyone is convinced that there really has been a significant change. Bell (1974) stresses that the concept of a post-industrial society is very broad and suggests that its meaning is best explained with reference to a set of dimensions or components which he enumerates as follows:

(1) Economic sector: the change from a goods-producing to a service economy.
(2) Occupational distribution: the pre-eminence of the professional and technical class.
(3) Axial principle: the centrality of theoretical knowledge as the source of innovation and of policy formulation for the society.
(4) Future orientation: the control of technology and technological assessment.
(5) Decision-making: the creation of a new 'intellectual technology'.

The social changes accompanying the above-mentioned changes are summarized in Table 3.7, where a comparison is made with pre-industrial and industrial societies.

In the post-industrial period the processing and recycling services are the main mode of production. Survival, or the strategic knowledge in Bell's terminology, depends on access to knowledge which is an organized set of statements of facts or ideas presenting a reasoned judgement or experimental result that is transmitted to others through some communication medium in some systematic form. Information technology, which allows the accumulation, processing and exchange of records, data and other material, allows knowledge to be obtained quickly as well as comprehensively. Computers, microprocessors, facsimile devices, teletext systems, or the more conventional

Table 3.7 Contrasting attributes of pre-industrial, industrial and post-industrial societies (after Bell)

	Pre-industrial	*Industrial*	*Post-industrial*	
Principal economic sector	*Primary* Extractive Agriculture Mining Fishing Timber	*Secondary* Goods producing Manufacturing Processing	*Tertiary* Transportation Utilities *Quinary* Health, education, research government, recreation	*Quaternary* Trade Finance Insurance Real estate
Technology	Raw materials	Energy	Information	
Occupations	Farmer Miner Fisherman Unskilled worker	Semi-skilled worker Engineer	Professional and technical Scientists	
Methodology	Common-sense Experience	Empiricism Experimentation	Abstract theory, models, simulation, decision theory, systems analysis	
Time perspective	Orientation to the past, *ad hoc* responses	*Ad hoc* adaptiveness Projections	Future orientation Forecasting	
Axial principle	Traditionalism: land/resource limitation	Economic growth State or private control of investment decisions	Centrality of and codification of theoretical knowledge	

Source: Bell, 1974, table 1-1, 117.

telephone are all central to this kind of economic system and its related society, whereas industrial society – suggests Bell – relied on created energy in the form of oil, gas, coal and electricity.

The enhanced significance of intellectual technology in the post-industrial world brings about important changes in the occupational structure of the labourforce. Blue-collar workers become numerically less significant, although they may still make a disproportionate contribution to the national product, and are replaced by workers in white-collar occupations. This process has been characterized, first, by an expansion of long-established occupations such as clerical work and numerous professional skills in accounting, banking, teaching, or medicine; and secondly, by the appearance of new occupations created either as by-products of new technology (computer programmers and engineers, microprocessor and copying machine salespersons and consultants of various kinds) or as a result of the demands generated by societies with more excess income to expend on leisure (ranging from holidays abroad to the regular use of both private and public sports clubs and related facilities). Bell also foresees 'de-skilling' of some occupations to create low-skill service jobs by stressing that expertise will be the basis for continued expansion of service occupations in post-industrial societies. It remains to be seen whether this replacement will completely compensate for jobs displaced in other sectors and whether the value of service sector output to economies in terms of their contribution to the creation of national wealth will match that of the primary and secondary sectors. Bacon and Eltis (1976), for example, are not convinced that this is happening and suggest that this may run counter to the long term national economic interests of a country like Britain.

SOME EMPIRICAL EVIDENCE PERTAINING TO THE DEVELOPMENT OF WHITE-COLLAR OCCUPATIONS

The emphasis on occupational change in the model of post-industrial societies can be tested by citing some data for the distribution of white-collar occupations, the ratio of male to female workers in employment, the incidence of full- and part-time employment and the growth of self-employment.

The growth and distribution of white-collar occupations

There are clear differences in the proportion and type of white-collar occupations in developed and less developed countries (Figure 3.3) or between countries classified according to GDP per capita (Table 3.8). Occupations classed as professional and technical or as clerical are among the largest of the white-collar groups in the high income countries of Western Europe and North America. These are the countries at the leading edge of the transition to a post-industrial economy; West Germany, France, Sweden and Canada are typical examples with a range from 10.6 per cent (West Germany, 1970) to 22.1 per cent (Sweden, 1975) in professional and related occupations and between 11.6 per cent (France, 1975) and 18.9 per cent (West Germany, 1970) in clerical and related tasks. Surprisingly, perhaps, Japan does not belong to this group; its professional workers, which are making a major contribution, for example, to the research, development and marketing of products for its external trade, is the next smallest group after administrators and managers. It could well be that Japan is a special case, in that the proportion of the economically active performing clerical jobs is sustained by the difficulties, so far unresolved, of converting the complexities of the Japanese alphabet into a code easily handled by modern office technology. Many routine tasks which can elsewhere be undertaken by electronic typewriters, word processors and computers must still be completed using more traditional and, by implication, labour-intensive methods. The devolved structure of administration in states organized on a federal basis, such as West Germany, may also affect the aggregate level of white-collar employment, its occupational structure and, as we shall see later, its location within the state or region.

In a second group of countries (classified as less developed in Appendix 2),

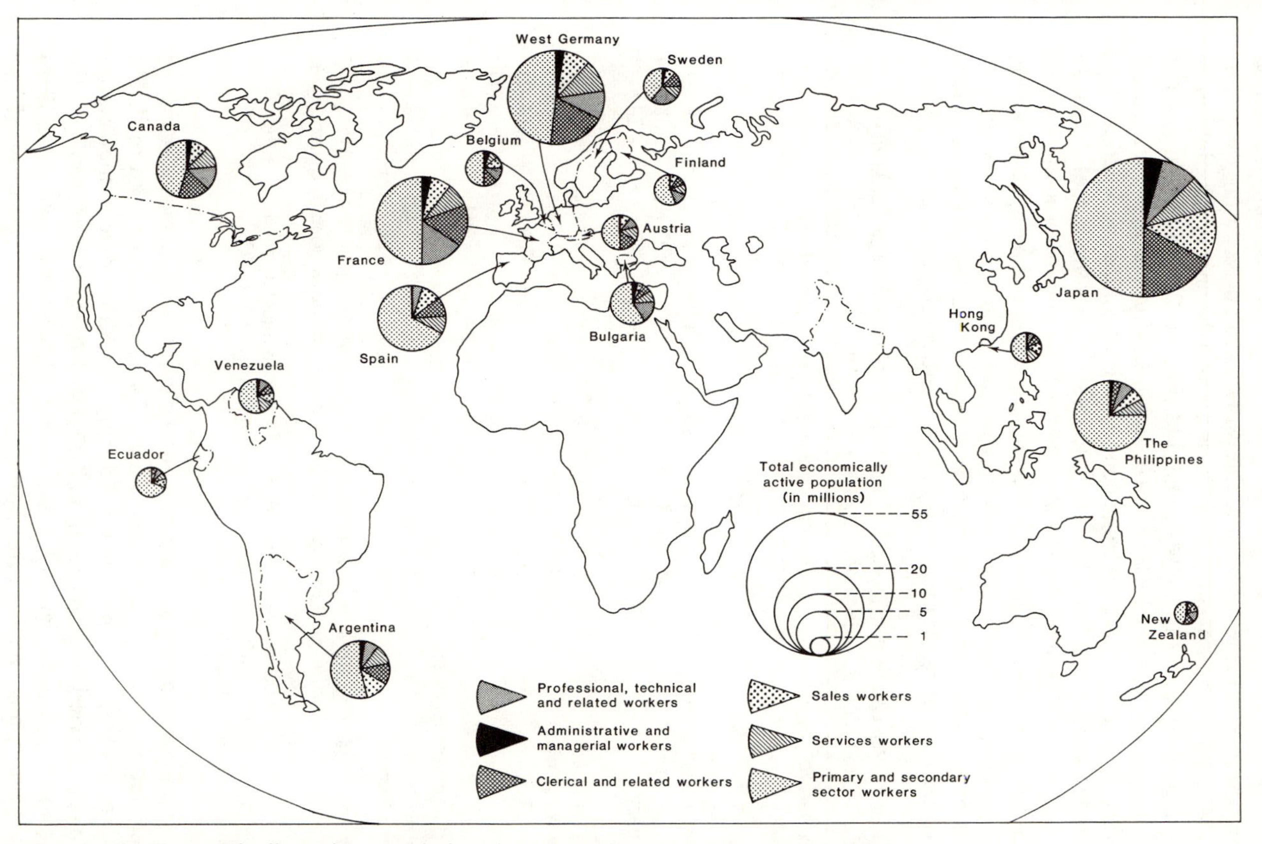

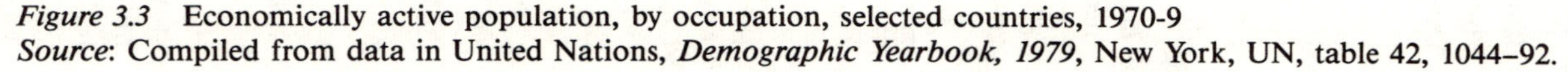

Figure 3.3 Economically active population, by occupation, selected countries, 1970-9
Source: Compiled from data in United Nations, *Demographic Yearbook, 1979*, New York, UN, table 42, 1044–92.

and shown in Figure 3.3, service workers (those engaged in transport and communications or in the utilities in white-collar occupations other than administrative or professional) and salesworkers comprise the largest proportion of tertiary sector occupations. These are also the only occupations in the low-income countries to show a small but positive change in their share of the labourforce between 1960 and 1970 (Table 3.8) but the difference is much smaller than that for high-income countries. Professional occupations have expanded their share in a uniform way, reflecting the import of externally trained workers to assist the development of the countries in the low and intermediate groups. This emphasizes the importance of occupations arising from traditional and informal service activities in these economies and confirms expectations created by earlier observations about the intra-sectoral characteristics of service industries in less developed countries.

Table 3.8 Percentage distribution of the labourforce, by occupation in low-, medium- and high-income countries, 1960 and 1970

Occupation group	*Year and group (by GDP per capita)*[1]					
	1960			*1970*		
	I	*II*	*III*	*I*	*II*	*III*
White collar	12	23	30	16 (4)	28 (5)	38 (8)
Blue collar	18	31	42	21 (3)	33 (2)	38 (−4)
Primary	65[2]	39	24	56 (−9)	33 (−6)	20 (−4)
Professional	2	5	9	4 (2)	7 (2)	11 (2)
Administrative	1	2	3	1 (—)	3 (1)	4 (1)
Clerical	3	7	10	4 (1)	9 (2)	14 (4)
Sales	6	9	8	7 (1)	9 (—)	9 (1)
Service	6	11	10	6 (—)	12 (1)	10 (—)
Craftsmen, production process, transport	18	31	42	21 (3)	33 (2)	38 (−4)
Agricultural	59[3]	30	14	50 (−9)	21 (−9)	10 (−4)

Notes: 1 Based on nineteen group I countries ($196 and $322 GDP in 1960 and 1970 respectively); ten group II countries ($447 and $992); and thirteen group III countries ($1656 and $2890).
2 Most column totals do not sum to 100 due to rounding and other errors.
3 Absolute difference between proportions for 1960 and 1970.

Source: Oberei, 1978, tables 5 and 6, 16–17.

It is important to note that white-collar occupations are by no means exclusively connected with service industries (Figure 3.4). Almost one-third of the 5.5 million workers in British manufacturing industries were administrative, technical and clerical (AT&C) staff in 1982 (Department of Employment, 1982; see also Crum and Gudgin, 1977; Delehanty, 1968; Gujerati and Dors, 1972). The majority are male workers, especially in mechanical engineering (16 per cent of male AT&C workers), shipbuilding and marine engineering (16 per cent), vehicles (12 per cent), and paper, printing and publishing (8 per cent). Female AT&C workers are mainly employed in paper, printing and publishing (13 per

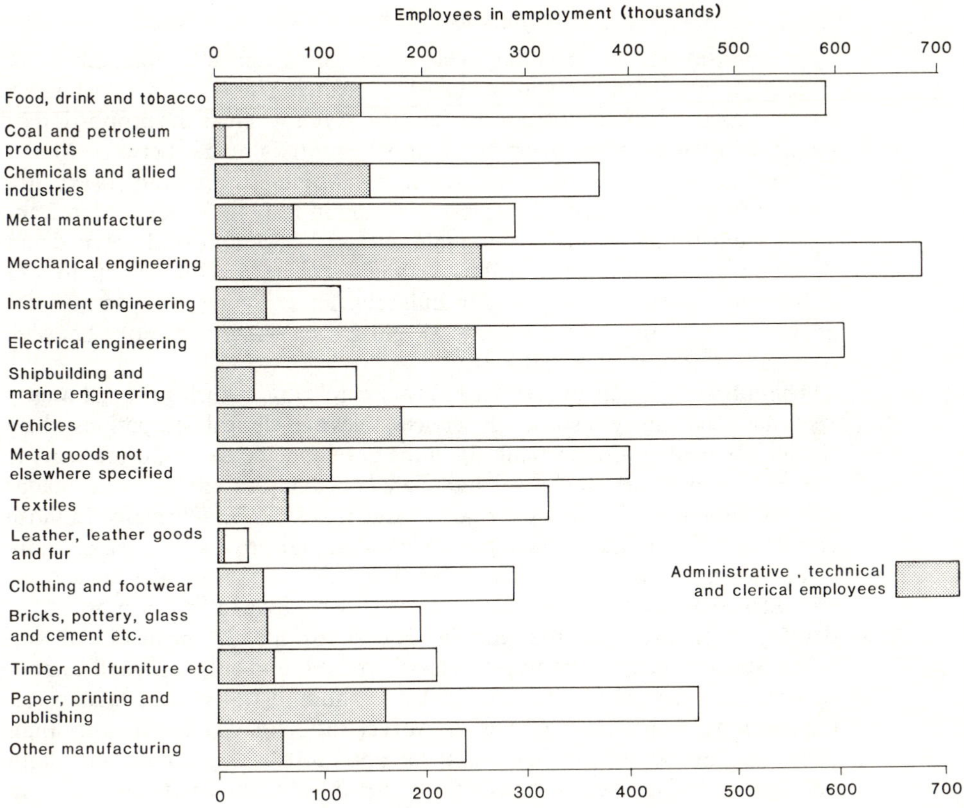

Employees in employment (thousands)

0 100 200 300 400 500 600 700

Food, drink and tobacco
Coal and petroleum products
Chemicals and allied industries
Metal manufacture
Mechanical engineering
Instrument engineering
Electrical engineering
Shipbuilding and marine engineering
Vehicles
Metal goods not elsewhere specified
Textiles
Leather, leather goods and fur
Clothing and footwear
Bricks, pottery, glass and cement etc.
Timber and furniture etc
Paper, printing and publishing
Other manufacturing

Administrative, technical and clerical employees

0 100 200 300 400 500 600 700

Figure 3.4 White-collar employees in British manufacturing industries, 1982
Source: Compiled from data in *Department of Employment Gazette*, 90, table 1.10, s.14.

cent of total AT&C workers), mechanical engineering (13 per cent), electrical engineering (12 per cent), and food, drink and tobacco (10 per cent). The proportion of total female workers in individual industries in AT&C jobs varies much more – from 63 per cent in shipbuilding and marine engineering to 11 per cent in clothing and footwear – than the male proportion, which has a range between 46 per cent and 18 per cent for all the industries shown. Many of these white-collar jobs are not attached to manufacturing plants, as such; they may be located in separate regional or headquarters establishments or in research and development facilities. Almost 32 per cent of the employees in US manufacturing industries were also engaged in AT&C occupations in 1978 (Browne, 1983), and in selected industries the proportion is much higher: examples include computers (69.5 per cent), office accounting machines (51 per cent), and radio and televisual communications (46.1 per cent).

Ratio of male to female employment

The expansion of white-collar occupations has been accompanied by an increase in the number of women in the labourforce. This is an important index for the development of post-industrial society, but whereas the proportion of employed men engaged in services in most countries ranges between 40 and 50 per cent, the values vary more widely for women. Marquand (1978) cites a range from 50 per cent of employed women in Italy to 76 per cent in Denmark, while the ratio of female to male workers in a sample of developed and developing countries (see Appendix 2) ranges from 0.16 in Guatemala to 0.73 in Sweden and Finland, and 0.88 in Bulgaria. These figures apply to the total economically active and, therefore, disguise even wider variations between occupation groups (Appendix 2).

The entry of women into service industry employment during the last twenty years has been largely absorbed by services, sales and clerical occupations where the female–male ratio, particularly in the developed countries, often exceeds 1.0. Hence the ratios for clerical and related occupations are 1.90 for Canada, 2.60 for Finland, 1.94 for New Zealand and 1.20 for West Germany. In those countries where female emancipation is not yet far advanced, such as India or Ecuador, the ratios are much nearer to or below the national average (0.04 and 0.56 respectively). Salesworkers also show positive ratios for most of the developed countries in the sample. The ratios for the professional and administrative occupations are much less favourable and only in a small number of cases do they exceed 1.0, especially for administrative and managerial occupations. Thus the high clerical worker ratio for Canada is more than counterbalanced by ratios of 0.75 and 0.18 for professional and administrative/managerial occupations respectively. The pattern is similar in West Germany and New Zealand but in Sweden and Finland the values range around unity for professionals and are much higher than elsewhere for administrators and managers, although still well below unity. The picture is more confused for the developing countries; the ratio exceeds unity for professional workers in the Philippines and is as low as 0.01 for administrators and managers in India. Overall Marquand (ibid. para. 2.5) notes that the concentration of female employees in service sector occupations is even more pronounced than their concentration in service sector industries and comments that in the UK 'women are nearly five times as likely to be employed in a tertiary sector occupation as in a secondary sector one, but less than three times as likely to be employed in a tertiary sector industry'. There is little doubt that the high probability that clerical workers will be women, combined with the high growth rates for this kind of work since 1945, account for these ratios.

In the USA females occupied 49 per cent of service jobs in 1970 compared with 23 per cent in non-service activities (Stanback, 1979). But as already noted, women are found to a disproportionate degree in those service jobs which attract the lowest wages; fully 74 per cent are concentrated in low-income segments according to Stanback (see Table 3.9). Clerical, service, salesworkers and semi-professionals/technicians comprise most of these low-income jobs.

Table 3.9 Male and female service employment, by occupation and income group: USA, 1970

Occupation group	*Income*							
	Male				*Female*			
	Total	*High*	*Medium*	*Low*	*Total*	*High*	*Medium*	*Low*
Professionals	17.1	8.7	8.3	0.0	13.9	2.1	11.8	0.0
Semi-professionals	7.9	4.4	0.8	2.8	7.5	0.4	0.5	6.6
Managers/administrators	15.0	7.2	7.8	—	3.9	2.0	1.9	—
Salesworkers	11.8	5.5	2.4	3.9	9.6	0.9	0.4	8.3
Clerical workers[1]	9.4	—	4.2	5.2	34.3	—	4.8	29.5
Service workers	12.5	—	—	12.5	26.2	—	—	26.2
Craftworkers/operatives	21.3	—	10.2	9.1	4.0	—	0.9	2.9
Labourers	5.3	—	0.5	4.7	0.7	—	0.0	0.7
Total	100.0	27.7	34.1	38.2	100.0	5.5	20.4	74.1

Note: 1 Office and non-office clerical workers combined.
Source: Stanback, 1979, compiled from table 11, 55; originally computed from microdata for a one-in 1000 sample of 1970 US Census of Population.

Male workers comprise a much smaller proportion of the low-income segment (38 per cent), with medium- and high-income segments especially significant for professionals, semi-professionals and managers/administrators.

While the number of males employed in UK service industries increased from 5.6 million to 6.2 million, an increase of 11 per cent, between 1961 and 1979 (Figure 3.5), the number of female service workers grew by almost 50 per cent, from 4.7 million to 7.1 million. The changes have been uneven; male workers decreased by 16.5 and 7.5 per cent respectively in transport and communications and the distributive trades but this was counterbalanced by 16.2 and 9.6 per cent increases in the number of female employees. In the four other industrial sectors shown in Figure 3.5 the growth of female employment has exceeded by at least one-half as much again the increases for males, especially in insurance, banking and finance (51.1 and 105.5 per cent respectively) and in professional and scientific services (57.6 and 84.6 per cent respectively). The latter is especially significant since female workers in this sector accounted for 36 per cent of the total in 1979 compared with 29 per cent in 1961 when the principal employing industry was the distributive trades.

The growth of part-time employment

The extent to which the changing female–male ratio in service industries represents 'real' expansion of labour inputs should not be overestimated. Another clear trend in the employment of service sector workers, which has been associated with the increase in the proportion of female clerical workers, has been the growth of part-time employment. For example, in view of the routine nature of many job specifications in the clerical group and the kind of work involved in sales activity the opportunities for part-time working are considerable and well suited to those individuals in the working population

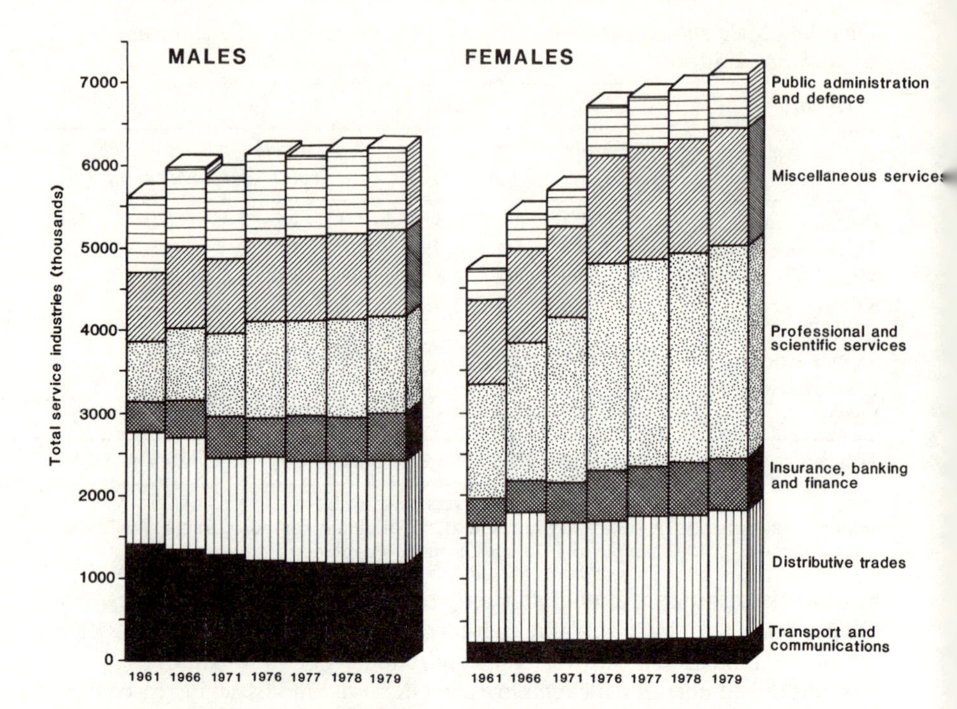

Figure 3.5 Employees in British service industries, by sex, 1961–79
Source: Central Statistical Office, *Social Trends 11*, London, HMSO, 75.

who need to divide their time between supplementation of household income and fulfilling other responsibilities, such as taking and collecting their young children to and from school, because the principal wage-earner is engaged in full-time employment. Part-time working is more prevalent in the UK service sector for both males and females; almost 47 per cent of the women employed in service industries were engaged part-time in 1981 compared with 39 per cent in 1971 (Central Statistical Office, 1983). Almost 42 per cent of female workers in all industries were part-time compared with less than 6 per cent of male employees (Department of Employment, 1983).

Temporary help is a variation on part-time employment which has recently become important in the US labour market. Between 1978 and 1982 the number of temporary-help workers increased by 25 per cent to 2.5 million at a time when the entire US labourforce only expanded by 5.5 per cent. Although originally confined to clerical tasks, work on this basis is now available in accounting, word processing and nursing. In a depressed economy the use of temporaries is attractive to employers lacking the confidence to increase their permanent workforce, and it helps to keep payrolls low during the peaks and troughs of the business cycle.

The incidence of part-time employment is significant because it places limitations on the distances which service sector workers (as well as part-time workers in other industries) are likely to be prepared to travel to work. Not only does

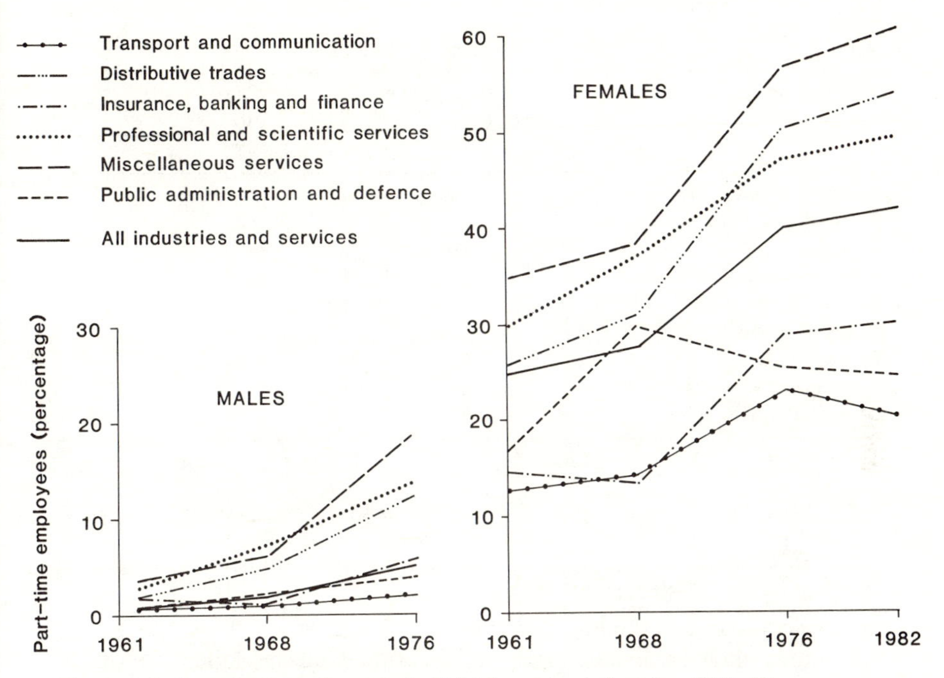

Figure 3.6 Part-time employment in British service industries, 1961–82
Sources: Compiled from Department of Employment, *British Labour Statistics, 1886–1968*, London, HMSO; *Department of Employment Gazette*, November 1977, February 1983.

this lead to compact, fragmented labour markets, but it also suggests that there will be certain demands on the locational behaviour of services attempting to gain access to suitable labour supplies. In the context of Stanback's (1979) observations on the relationship between occupation and high/low income status it is also likely that the association betwen clerical and related work and low-wage status is compounded by the tendency towards part-time employment in the same subsectors. Miscellaneous services and the distributive trades in Britain employ a large proportion of part-time workers, 49.2 and 60.8 per cent of females in these industries in September 1982 (Figure 3.6). Part-time work in miscellaneous services is especially notable in restaurants and cafés (65 per cent), public houses (86 per cent) and clubs (84 per cent) and in retail distribution (57 per cent). Consumer services are most likely to generate demand for part-time workers but some producer sevices, such as business services, had 64 per cent of their female workers (female–male ratio 1.3) in part-time jobs in 1981. Indeed these labour markets may be dominated by the availability of part-time staff, which may lead to lower wages than otherwise might be the case for full-time workers.

Self-employment

Reference was made in Chapter 1 to the ease with which individuals can enter

Table 3.10 British self-employed, by industry, 1971–81

Industry sector	*No. of self-employed (thousands)*			*Change (%) 1971–81*
	1971	*1976*	*1981*	
Transport and communication	74	92	125	68.9
Distributive trades	497	463	467	−6.1
Insurance, banking and finance	52	53	84	61.5
Professional and scientific services	197	210	213	8.1
Miscellaneous services	382	368	390	2.1
Total services	1202	1186	1279	6.4
All industries and services	1954	1933	2057	5.3

Source: Central Statistical Office, 1983, table 4.8, 62.

the service sector and thereby contribute to its expansion in the less developed economies and to its diversity and specialization in the developed economies. The informal sector is the principal product of this flexibility in the less developed countries, but in developed countries it is manifest as self-employment which tends to be concentrated primarily in the construction and transport industries. While this has contributed to the expansion of self-employment in some countries, the overall trend during recent years has been for the proportion to decline. The absolute number has, however, increased between 1960 and 1975 in a way not replicated in the secondary or primary sectors. This is confirmed by more specific statistics for Britain (Table 3.10), which show that there were more than 2 million self-employed people in 1981. Although still the largest in proportional terms, self-employment declined in the distributive trades by more than 6 per cent but increased by more than 60 per cent in insurance, banking and finance, and transport and communications. In some personal services, such as those provided by dentists, hoteliers, watch and shoe repairers and shopkeepers, employers and the self-employed outnumber employees. Self-employment is, therefore, concentrated in small businesses which continue to be formed because they are labour intensive, represent a response to markets which are continuously variable with respect to demand, preference and fashion, and provide opportunities for entrepreneurs to acquire the skills or accumulate the capital to allow survival as larger enterprises. Overall, the self-employed now represent almost 9 per cent of the employed labourforce in Britain.

POLITICAL ECONOMY AND SERVICES

Hirschorn (1974, 16) interprets the transition from industrial to post-industrial

America in the context of the view that 'all historical developments have historical roots as far back in time as we care to trace them'. He concedes, however, that while it is possible to pinpoint the Industrial Revolution at the period approximately 1750–1830 when 'men began to re-orient their conceptual frameworks, their value systems, and ultimately their political institutions, to cope with and to ultimately shape the new processes of development' (ibid., 17), it is rather more difficult to find a similar turning-point for the emergence of a society in which service industries are a major kind of economic activity. The period 1890–1920 was one of expansion and consolidation of an urban proletariat, largely in industrial centres, associated with increasing capitalization of industry. This prepared the way for a significant change during the 1920s in the USA; there was zero growth in the absolute number of manufacturing production workers, manufacturing output increased by 60 per cent and the annual hours worked fell by 6 per cent. Hirschorn (ibid., 24), therefore, concludes that 'there was a new "structure of productivity" unleashed in the 1920s which was both labour and capital saving and which no longer required the rapid accumulation of capital and labour to increase both productivity and production' and 'the decade of the twenties can be viewed as a technological, economic and ultimately socio-political turning point'.

While it may have been a turning-point in a market economy such as the USA, which promoted the subsequent expansion of service industries, events have not followed the same timescale in centrally planned economies. There are few easily accessible studies, but Ofer does provide an interesting insight to the place of the service sector in Soviet economic growth:

> [an] outstanding feature of the Soviet economy is the relatively small volume of services produced. The place of most types of services in the economy is notably small when the USSR is compared with other countries of the same or even lower levels of development. To the student of modern economic growth such a limited service share is rather paradoxical. (Ofer, 1973, 1).

One explanation might be that the Russians follow Marx in his conviction that the service industries are unproductive and create no material value. The result is that services are assumed to have no valid existence and tend to be condemned (Lewis, 1973; Bergson, 1964). Because it has full control of the economy, the government has been able to see to it that services are not allowed to absorb too large a part of the scarce factor resources available. Hence in the USSR and other communist bloc countries services are excluded from calculations of the national income.

But Ofer (1973) argues that all this is to oversimplify the degree to which ideological frameworks interfere with national economic needs. He suggests that in practice the small size of the service sector in the USSR is a by-product of the novel social and economic system and, in particular, the strategies for economic growth implemented by the Soviet authorities during the forty-five years preceding the early 1970s. The targets for economic growth, sectoral priorities created by these targets and the means employed to achieve them have together caused the production of services to be curtailed, especially those connected with final consumption and urbanization (retail trade, personal and public services). The abolition of private ownership of productive assets and the organization of the state using the principles of central planning has also dampened the production of services. The government-run planning apparatus substitutes for wholesale trade the financial and other business services.

Thus in the financial sector the collection and direction of resources for investment is handled by public agencies with the assistance of a very small banking system. In a market economy a great many transactions are needed: first, to solicit funds from the public; secondly, to distribute these among investors; and thirdly, a series of transactions to redistribute the investment proceeds back to the individuals or institutions. The first and last of these stages is almost entirely absent in the Soviet economy because investment funds never reach individuals and thus do not have to be solicited back. It is also likely that the actual allocation of investment funds in a centralized system is less expensive in resources and manpower because all the targets and information are gathered in one place and there is none of the duplication of, for example, financial services or investment studies and portfolios by competing institutions in market economies. A crude indication of this difference is that south-east London has between nine and twelve times more banks per 100,000 population as Moscow or Warsaw (Hamilton, 1979).

Urbanization has contributed to the expansion of services in market economies, especially in developing countries, but in the Soviet system it is practically impossible for those who migrate to the cities to set themselves up upon arrival as independent retailers, providers of personal services not already available, or as craftsmen. By using a combination of permanent shortages of agricultural labour, prohibition of private enterprise and, if necessary, the withholding of permits to move to cities (especially the large cities where miscellaneous services have the best chance of being viable), the socialist economic system is able to economize on services arising from urban lifestyles, including various administrative and municipal services required by urbanization.

Hamilton (ibid.) also shows how the view of services in centrally planned economies produces fundamental contrasts in the spatial structure of cities. The East European city is dominated by services mainly in the public consumption sector (welfare goods and services), while the capitalist city is overwhelmingly dominated by retailing and personal services for private consumption. Some of the reasons for this will be apparent from Ofer's observations in relation to cities specifically:

> Education, science, culture, recreation, and medical care are well developed and relatively labour-intensive in the socialist city because these activities are: free rights, necessary for the full sociocultural development of every citizen; investments in the future to foster a healthy, highly productive, and enlightened society; and major attractions to female labour. Not surprisingly, many socialist cities have doctor to patients, hospital beds

> to patients, and teacher to pupils ratios that are equal to, or significantly higher than, equivalent ratios in any capitalist cities. Retailing is far more restricted in socialist cities, whether measured in floorspace, turnover, employees, or numbers of establishments per thousand inhabitants because: State ownership has largely eliminated competition; because planning has imposed norms on the ratio of shops and services to the population and arranged these hierarchically within large cities; and because, in many cases, family budgets are spent more on necessities, alcohol, and cigarettes and less on scarce consumer durables. Indeed, supply, not income, is the critical factor, for while wages are low, welfare goods and services are extremely low priced or free. Family budget 'surpluses' feed per capita bank deposits which are amongst the highest in the world. (ibid., 217)

One other interesting observation about services in the Soviet economy is the conflict between the conscious effort to make employment available for women and the very limited provision of domestic and personal services which would ease the burden for those females in work. The substitution of household tasks with services such as child care or laundries available outside the household is very limited and most other household services such as domestic appliances or private transport are not generally available at a cost which most households could afford.

SUMMARY

While service industries have been neglected in academic research or in studies by international and national agencies, it is clear that the work which has been completed reveals a diversity of views regarding the causal factors in their growth and diversification. The statistical rigour sought by economists contrasts with the rather 'softer' interpretations used by political economists and sociologists. This is not to say that the latter do not employ empirical evidence, but rather that they surround this evidence with more general conceptual frameworks which are harder to substantiate but essential stimuli for further debate and identification of research directions. Academic observers disagree about whether a post-industrial economy can be viable. Keynesian and monetarist economists are highly sceptical, while at the other end of the scale Stonier (1983) suggests that by the end of the first decade of the next century only 10 per cent of the labourforce will be producing all society's material needs, leaving services and the information which they use as the wealth of nations.

Yet the transition from manufacturing to services is neither straightforward nor inexorable. Indeed it may be inappropriate to think of it as a transition involving clearly labelled phases along a trend line; rather there is a blurring of the distinction between different kinds of production process such that blue- and white-collar tasks become increasingly difficult to differentiate, or that may take place at similar locations, and give rise to less easily indentified industrial complexes than in the past. The fact is, however, that most of this is speculation because the contribution of geographical enquiry to the general issues posed by the appearance of new services has at best been sporadic. As we shall see, there have been some significant contributions but the obvious spatial questions arising from the variety of explanations for the appearance of service industries have yet to be adequately explored.

CHAPTER 4

Service industry location: the central place model

INTRODUCTION

An exploration of the rationale for the emergence of the service sector, and of the broad characteristics of the changes within it, and some of the explanations for these changes provides a necessary context for the remainder of the book. We now turn to the theories used to account for location and spatial distribution of services, beginning with the central place model and moving on (in Chapter 5) to extensions of this model and some other approaches to location modelling. Some reference has been made to the fact that many services cannot be stored and subsequently transferred to the point of consumption; there is often a substantial dependence on the provision of information, advice and expertise upon demand, and in infinitely variable contexts and complexity. This means that while manufacturing and service industries may have in common some factors of production such as labour, capital or land, others such as technology or knowledge are more central to services than to manufacturing.

For the production of goods it is necessary to take into account the cost and availability of labour, the level of access to raw materials or suppliers, and the relationship between the value of the good and the cost of transporting it to markets. These and other factors are extensively used by location theorists to arrive at optimal location solutions for goods-producing activities (Weber, 1909; Hoover, 1948; Greenhut, 1956; Isard, 1960; Smith, 1982). While the solutions produced are certainly not irrelevant to service industries, such 'modelling requires a greater degree of specification of input and output occurrence, usage and value than is feasible in a consideration of service sector establishments' (Edwards, 1983, 1328).

In industrial location theory transport is one of the key determinants of location choice, although the availability of a labour pool is also considered – especially if, for example, additional transport costs can be offset by labour savings. The latter would lead to the choice of a labour-oriented rather than a transport-oriented location. Weber (1909) also incorporates agglomeration of several plants in one place as a third determinant of location because the advantages conferred by such proximity may outweigh the higher transport costs resulting from choosing a location other than the minimum transport cost location (the classic Weberian model). Although he relegated agglomeration economies to third place as a possible determinant of locational choice (after transport or labour orientation), Weber had identified a factor which is much nearer the top of the list of locational priorities for many service industries.

AGGLOMERATION ECONOMIES

Agglomeration confers economies, which are external to a service production unit, and which result from locational association between either a number of similar or totally different production units. The development of spatial clusters of service and other economic activities may result from a sequence of decisions, in which those made after the initial location choice by one firm will determine the choices of its competitors, especially if it is assumed that the first firm selects a location as near as possible to its minimum transport cost location (Pascal and McCall, 1980). But Isard (1956) and Smith (1966) suggest that this is too simplistic: the agglomeration of all firms at one location assumes that they will all benefit in a similar way from that agglomeration. In practice, the response of firms will depend on their own strategies, the procedures employed in their locational decision-making, and the number and kinds of concession they are willing to make in order to reap the benefits of sharing a location with competitors.

The classic illustration of this problem is the linear market duopoly model (Hotelling, 1929), where an attempt is made to establish the profit-maximizing locations for two producers selling an identical product in circumstances where consumers are evenly distributed and one unit of output is purchased by each consumer in any given time period, and demand is inelastic. In addition, Hotelling assumes that production costs are zero, that each producer could supply the total market if allowed, relocation can occur at zero cost and each producer can locate anywhere, and that transport costs are constant and equivalent to a value per unit of output per unit of distance. Within this framework, the two producers will 'jostle' for the best location until the stage is reached when they are both located at the centre of the market. At this shared location each producer will not be able to increase profits by relocation, and their prices will be the same in order to ensure their share of the market. It is interesting that in terms of transport costs to the consumer this solution is the least satisfactory and illustrates how market forces and the search for profit maximization through agglomeration can displace social considerations. For the latter the minimum transport cost location would lead to each producer occupying a dispersed location within the market area (at each quartile within the linear market).

The assumptions incorporated in this model are of course very restrictive, not least the notion that each producer has no knowledge of the other's behaviour or likely reaction to his own decision, or that the demand for the product is inelastic. These and other objections could be seen to reduce the credibility of the agglomeration hypothesis but this would be to interpret the concept in a rather narrow way. Agglomeration is also motivated by the dependence of producers – within the same industry or in different industries – upon each other, external economies, variations in population density and the opportunity which is offered to reduce uncertainty. By choosing locations near to competitors or to external services, new producers minimize the risks involved in starting up. Such risks will be greatest for small firms and will be higher as the distance from their markets increases. Large firms will require to be in locations offering a large enough market, so that the combined effect is the promotion of agglomeration, especially in cities. Uncertainty may also be reduced in the case of relocating firms by a choice of location as near as possible to the original location (see, for example, Logan, 1966; Keeble, 1968); by establishment near to the entrepreneur's home in the case of new firms; or by following the locational example of firms already established.

Linkages are the key to agglomeration economies. A single-establishment service firm, such as an advertising agency, will have a number of different types of links. First, it will have links with the suppliers of the services (or inputs) that it needs to assemble its own 'product', such as printers, market researchers, graphic materials suppliers and financial services. Secondly, it will have links with the purchasers of its advertising copy and strategic advice about where or how frequently to advertise (usually other manufacturing and service firms). Thirdly, it is likely that the advertising agency will have links with other agencies either as a subcontractor or through inviting other advertising agencies to work for it on behalf of clients. Fourthly, it will have links with the media which carry advertising: the press, radio and television stations, magazine and journal publishers, and public transport operators whose vehicles and rolling stock carry advertising. Finally, the advertising will probably have links with government and other agencies which regulate the location, quality, timing and content of advertising, especially for certain products such as drugs or cigarettes.

There are, then, service linkages, marketing linkages and, in some instances, but less frequently than for manufacturing, production linkages. The relative importance of each will depend upon the type of service activity, together with its size, or sometimes the value of its turnover (or 'portfolio' for an advertising agency), and its organizational structure. The potential to use some or all of these linkages is clearly greatest when service firms are close together and can then increase their revenues or reduce their overheads or, ideally, both.

The value to the service firm of being part of a geographically limited agglomeration also endows it with other external economies, which arise from the wider environment of which it is a part. Agglomerations create demand for labour (both skilled and unskilled), public and private transport facilities and a whole range of other infrastructural facilities such as telephone lines, postal services, waste disposal, and electricity and gas supplies. The economies accruing to individual service activities will increase as the size of an agglomeration increases. Isard (1956) has shown, however, that the curves for different types of scale economy associated with, for example, labour or energy are variable and beyond a certain size of agglomeration (defined as the total population of an urban area) economies of scale diminish and diseconomies begin to occur. The greater the demand for labour, for instance, the larger the area from which it is recruited; this makes for longer journeys to work at higher cost to the individual worker, to the community which provides the transportation infrastructure and ultimately the firm which has to pay higher wages to retain or attract staff with the appropriate skills and quality. The firm's overheads will, therefore, be reduced unless the cost of services is increased but this might reduce its competitiveness. This greatly oversimplifies the way in which diseconomies begin to arise but, as we shall see in Chapters 8 and 9, these have been an important factor in the promotion of change in the location of some service activities.

THE CENTRAL PLACE MODEL

Christaller (1933) developed central place theory to account for some regularities in the size and spacing of settlements in southern Germany. He found that these regularities were clearest among those settlements which provided goods and services for their own residents as well as for those in the surrounding rural areas. Such settlements were defined as 'central places' and Christaller identified their relationship to one another, that is their spacing and size, by reference to the tertiary activities, in particular retailing, which each settlement contained. Indeed Christaller viewed his theory as having an equivalent standing to Von Thunen's theory of 1826 of agricultural location and Weber's (1909) industrial location theory.

Since retail and related tertiary activities make a major contribution to Christaller's theory, we can assume that they demonstrate certain regularities in the way that they go about locating within the settlement system. It is assumed that the suppliers of goods and services make their location decisions on an isotropic plain over which a rural population, with equal levels of purchasing power, is uniformly distributed. Within the context of these assumptions it is then possible to consider the economic determinants of the size, location and ultimately the spatial organization of tertiary services. In order to flourish each firm must be able to find a niche (or market) within which it can supply its good or service, and the size of this market will be determined by a number of factors. First, the volume of demand; there must clearly be a minimum number of potential consumers from among whom sufficient individuals will wish to obtain a service or good at sufficiently frequent intervals to make the firm viable. Where population densities are very low, market areas will thus need to be much more extensive than in those areas where population density (and perhaps income) is high. Secondly, each firm must take advantage as far as possible of economies of scale in the production process; some firms will need to be much larger than others in order to be able to compete effectively for customers and on price. A greengrocer or gas station has a much smaller minimum efficient size for its staffing, physical facilities, turnover, or inventory than a department store or hypermarket which, by definition, must provide a wide range of merchandise and therefore need to command much more extensive market areas in terms of minimum numbers of potential clients. Thirdly, market areas will be determined by transportation costs, which comprise delivery costs for the seller and travel costs to the place of consumption of a good or service for the buyer. For any particular good there will be a maximum distance which an individual will be prepared to travel to obtain it, and likewise there are are usually maximum distances prescribed by firms for the delivery of a good such as furniture or bulky electrical equipment.

The location of tertiary services is, therefore, the result of a minimum size of market or threshold for each firm with the size of the market determined by the above factors. This means that because some tertiary activities have much lower thresholds than others, more of them can be fitted into Christaller's hypothetical landscape, a feature encouraged by the further constraint that each firm cannot make excess profits. Hence the market areas for firms providing similar services are of equal size which, in turn, means we can expect regularities in the spacing of the locations at which individual services of the same type are available (Figure 4.1). Services with low market thresholds, such as bars, grocers' or chemists' shops, or post offices, will not only occur frequently, but tend to be associated with the smallest (or lowest-order) settlements. Those with much larger market thresholds, such as hotels, department stores, furniture stores, or jewellers' shops, will occur less frequently and will be located, therefore, in fewer settlements (Figure 4.1). Clearly, some of the lower-order settlements will also act as locations for these higher-order tertiary services, thus creating a hierarchy of settlements in the central place system in which there is a limited number of centres with the full range of retail and similar services and a very large number of centres with a small number of selected tertiary services.

As a general rule, the more complex and specialized the service, the more likely it is to be found in only the very largest central places where it can gain access to a market of sufficient size. All these larger central places will incorporate all of the tertiary services found in the lower-order central places since these will also be required by the residents of the higher-order centres. Thus the hierarchy of retail activities provides a functional basis for the hierarchy of settlements, or conversely, each central place exports services to its hinterland and imports goods and services from central places higher up in the urban hierarchy or from places outside the immediate confines of the national economy.

It should be apparent from the examples cited that central place theory is especially relevant to an interpretation of the location and spatial organization of consumer services. Empirical studies which confirm the basic tenets of central place theory are numerous (see, for example, Berry, 1967; King, 1962; Everson and Fitzgerald, 1969; Johnston, 1964; Marshall, 1969). The theory seems to work particularly well in rural areas where the population is distributed in such a way that individual tertiary services may have a monopoly within their market areas. The opportunities for differentiation through specialization and competition in products and services is much more limited than it is in urban, or more densely populated, areas. In other words, the organization of trade areas according to Christaller's marketing principle ($k = 3$) works well in rural areas. It is assumed that consumers converge on individual nodes to obtain their requirements, and if they cannot meet them at one level in the

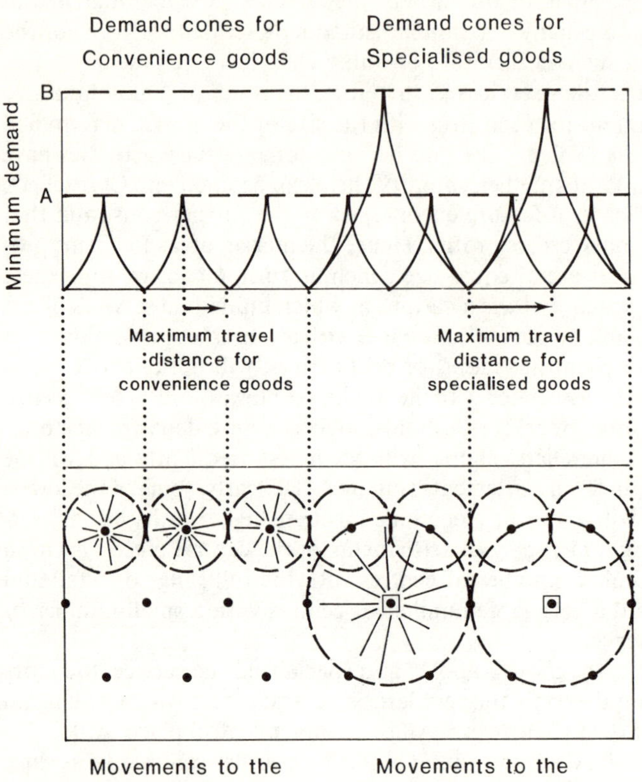

General pattern of the Trade areas

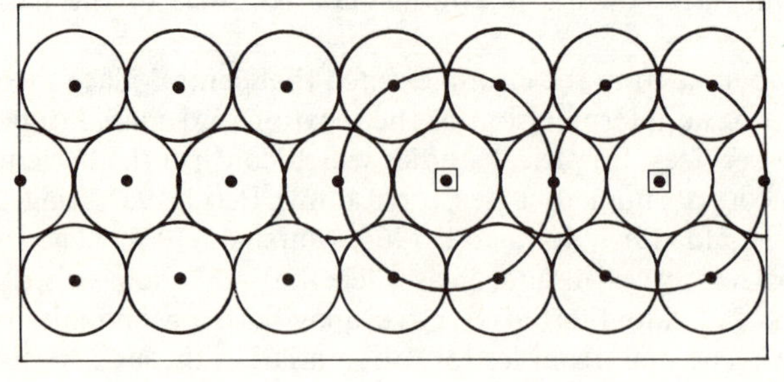

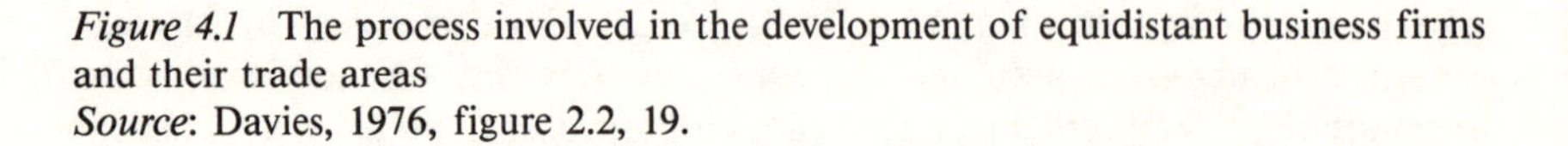

Figure 4.1 The process involved in the development of equidistant business firms and their trade areas
Source: Davies, 1976, figure 2.2, 19.

hierarchy, they will move up to the next level in the settlement hierarchy. At the very top is the largest and only central place which has a market area embracing all of the region or country (Figure 4.2). The interdependencies between the elements of the settlement system, then, are essentially hierarchical whether the Christaller model is used, as in Figure 4.1, or the more flexible Lösch model, as in Figure 4.3, which does allow some interaction between centres at the same level in the hierarchy but only within the 'city rich' and 'city poor' wedges that are characteristic of the central place landscape generated.

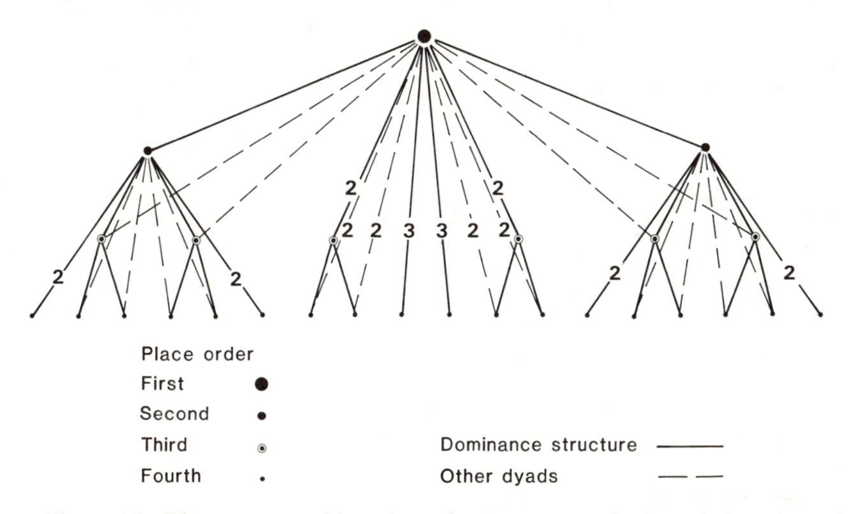

Figure 4.2 The structure of interdependencies, or growth transmission channels, in a Christallerian central place model organized according to the marketing principle
Source: Pred, 1977, figure 1.1, 19.

Modelling the location of services within urban areas is more complicated because of the ability of firms to differentiate the services they provide relative to those of their competitors while, at the same time, being able to choose locations directly adjacent to their competitors and so benefiting from agglomeration economies. This permits the consumer to make comparisons between firms offering similar products or services at different prices, on preferential credit terms, or with longer term or more comprehensive servicing arrangements. Some examples include automobile sales and servicing; the availability of various types of credit accounts on different terms in competing department stores which offer a similar range of products; or retail outlets specializing in electrical appliances, such as televisions or hi-fi equipment, but competing on price and duration or speed of back-up repair services. In these circumstances consumer services are sharing markets in which there is considerable overlap. Market structures are, therefore, different in urban areas; 'nesting' may occur, for example, and it may be more appropriate to use Christaller's modified marketing model in order to comprehend the resulting location patterns.

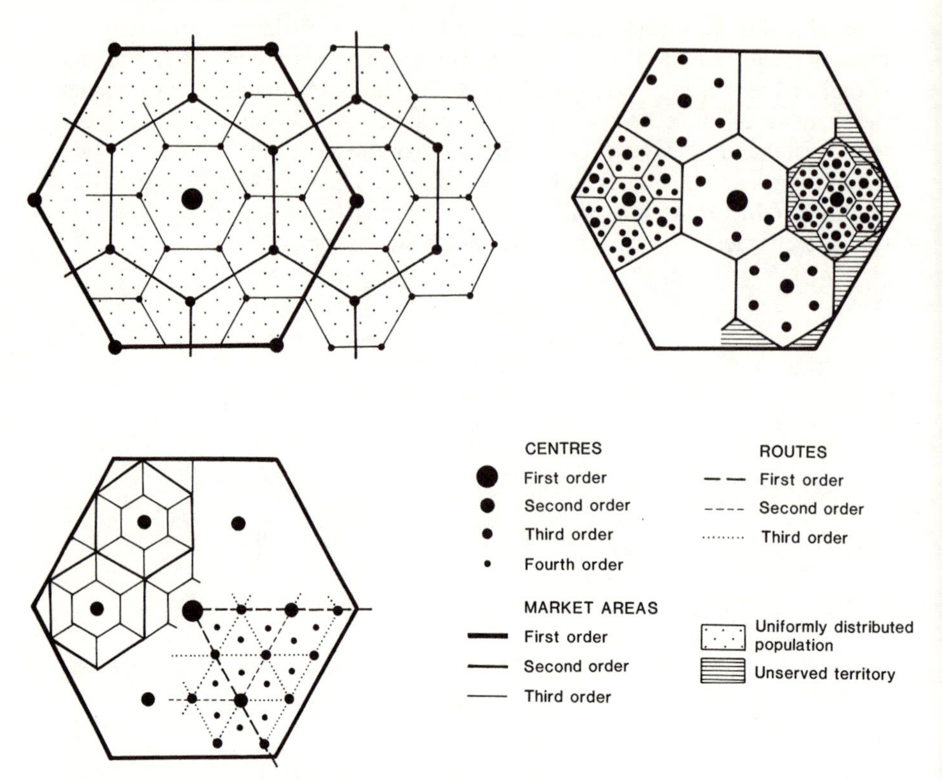

Figure 4.3 Further trade area arrangements according to Lösch
Source: Haggett, P. *Locational Analysis in Human Geography*, London, Edward Arnold, 1965, figure 5.4, 119.

In the marketing model it is assumed that consumers can travel freely in all directions but such freedom may in practice be modified by the transport facilities which link selected central places in rural areas rather better than others, or which confer superior opportunities for movement along particular segments (such as intra-urban rail or freeway routes) within urban areas. Together with the intra-urban distribution of population (an inverse linear exponential relationship, see Warnes, 1975), this distorts the spatial pattern of central places (shopping centres within urban areas) and the organization of the hierarchy may also be modified (Warnes and Daniels, 1979). Perhaps a more effective application of central place theory to the intra-urban location of retail services can be achieved by relaxing the fixed-k assumption employed by Christaller in such a way that the market principle ($k = 3$) applies to the location of fourth-order centres or individual or small groups of convenience stores (see ibid.). The administrative principle ($k = 7$) can be used for the location of second order centres (Figure 4.4) and the transport principle ($k = 4$) for third-order centres. Using this and modifications to the shape of the associated market areas and the location of retail centres within them to

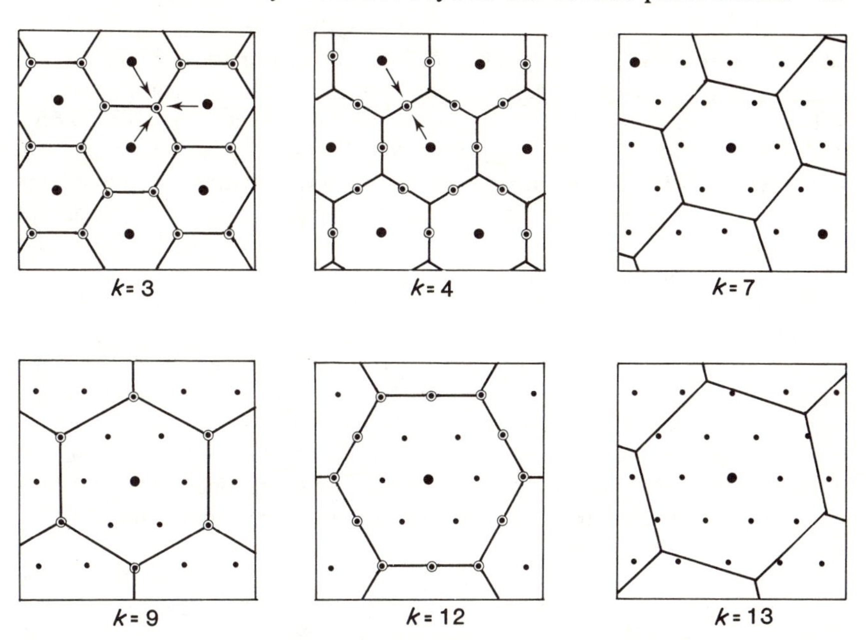

Figure 4.4 A variable *k* central place model for shopping centre location within an urban area
Source: Warnes and Daniels, 1979, figure 3, 393.

reflect the typical pattern of urban population density, a pattern of shopping centres is produced which – at least with reference to British cities – contains features similar to those present in observed distributions. The most widely repeated feature in the observed pattern of intra-urban retail centres is the tendency for second-order centres to be located closer to the centre than to the periphery; the density of centres to be much higher in inner than outer areas, especially with reference to lower-order centres; market areas are asymmetrical and tend to be elongated radially; and commercial ribbons along arterial transport routes are common and invariably associated with the older inner areas. These features are broadly replicated in the patterns generated using the modified theoretical assumptions (Figure 4.4).

Christaller also suggested a second modification to the basic marketing model which recognizes that consumer movements to services may be constrained by administrative boundaries. The relevance of this modification will be seen later when the location and access to certain public services such as libraries, schools and hospitals is considered. The opportunities for specialization and sharing of market areas in the larger central places are also conducive to the generation of demand for services which need only be infrequently provided in rural areas; fire protection for a much wider range of buildings and land uses, parking facilities such as multi-storey car parks and parking meters, or public transport.

Central place theory provides us with some important clues about the distribution of service activities. It can be hypothesized, for example, that all service industries will not be evenly distributed in relation to population either within a city or between regions. It can be expected that consumer services will be more evenly distributed or possibly 'over-represented' in regions at the national scale than industry as a whole (Keeble *et al.* 1982). Conversely, producer services will be especially prominent in central regions which enjoy the comparative advantage endowed by their accessibility from within the nation as a whole; distance will inhibit the location of producer services in peripheral regions. This will leave the latter with residual 'specialization' in consumer services of the kind provided by the public sector, by tourism, or by the distributive trades. The relative specialization of the central regions in manufacturing and producer services will also generate higher incomes and, therefore, superior spending power which stimulates a higher level of consumer service provision. Consequently the theoretical dichotomy in the location of certain types of service between centre and periphery is complicated by other factors. Keeble *et al.* (ibid.) suggest that intermediate locations or regions also exist where it might be expected that the representation of consumer and producer services would be less polarized.

Other hypotheses then suggest themselves. For example, it can be hypothesized that some services will therefore be more readily accessible to users than others (in terms of time taken to reach them, distance travelled, or frequency of use), or that some services will attach more importance to agglomeration economies than others. Such hypotheses can, of course, be put to the test by using some empirical evidence about the location of service activities in the real world.

Initially it is necessary to consider how best to approach the task of verification: should specific types of service industries be used as a basis for exploring their distribution? Should groups of similar/related services be examined to see whether they behave, with reference to location, in some common way? Should either of these alternatives be examined at different spatial scales to establish which provides the best 'fit' with the theoretical constructs? The answer to such questions is to some extent determined by the kind of data which is available. Perhaps the most convenient way to proceed is to explore the location of groups of services at a relatively high level of aggregation, followed by a more disaggregated approach guided by the conclusions from the preliminary stage. Information is most readily available for employment by industry and occupation and the following analyses will use one, or both of these, as appropriate.

THE LOCATION OF SERVICE INDUSTRIES AT THE REGIONAL LEVEL WITHIN THE EEC

In general, the first hypothesis is borne out by the empirical evidence relating to subregions of the member states of the European Economic Community (EEC) (Table 4.1). Consumer services are most evenly spread with an ill-defined contrast between the centre and the periphery, although the latter's dependence

Table 4.1 Mean regional percentage employment, by industry sector in central, peripheral and intermediate regions of the EEC, 1973–9

Regions	*Year*	*Agriculture*	*Manufacturing*	*Services*		*All services*
				Producer	*Consumer*	
Central	1973	5.3[1]	33.7	11.9	37.6	49.5
regions (35)[2]	1979	4.0	31.0	12.7	41.6	54.4
Intermediate	1973	10.2	33.6	9.9	35.6	45.5
regions (39)	1979	8.5	31.3	10.8	39.0	49.7
Peripheral	1973	18.8	23.8	8.3	36.9	45.2
regions (31)	1979	14.6	20.6	8.7	43.2	51.9
EEC regions:	1973	11.1	30.7	10.1	36.6	46.7
Total	1979	8.8	28.0	10.8	41.1	51.9

Notes: 1 Percentage.
2 Number of regions grouped into each category.
Source: Keeble *et al.*, 1982, table 4.11, 96.

on these services between 1973 and 1979 increased at a greater rate (+17.1 per cent) than in the central regions (+10.6 per cent). On the other hand, the 'gradient' from central to peripheral regions in the location of producer services is clearly apparent. The change in the proportion of these services between 1973 and 1979 represents a smaller increase in specialization in peripheral regions (+4.8 per cent) than in the central regions (+6.7 per cent).

South-east England, central Belgium, the western provinces of the Netherlands, the Paris region, and the Rheinland-Pfalz are especially prominent examples of the relationship between the central regions and relative specialization in producer services (Figure 4.5(A)). At the same time, several peripheral regions are grouped into a category, with 11.9–13.8 per cent of their total employment in producer services: Scotland, south-eastern France and much of Denmark, for example. This pattern must be seen in the context of more rapid employment growth in the EEC's peripheral regions, an increase of 1.5 million workers – ten times greater than the equivalent figure for the central regions. This suggests that regional economies within the EEC are evolving in different ways and service industries, in addition to manufacturing, play a significant part in promoting divergent development.

The regional pattern of service industry employment change reveals significantly higher growth rates in the peripheral regions, particularly in France and Italy (Figure 4.5(B)). In the Italian case this may be the result of the entry of the unemployed into low-skill service tasks, into tourist-related activities, or into public services arising from demand generated by demographic rather than economic need. This does not necessarily represent a real contribution to enhancing the 'basic' (export) structure of industry in the peripheral regions. Too many of these jobs will be in consumer rather than producer services. There

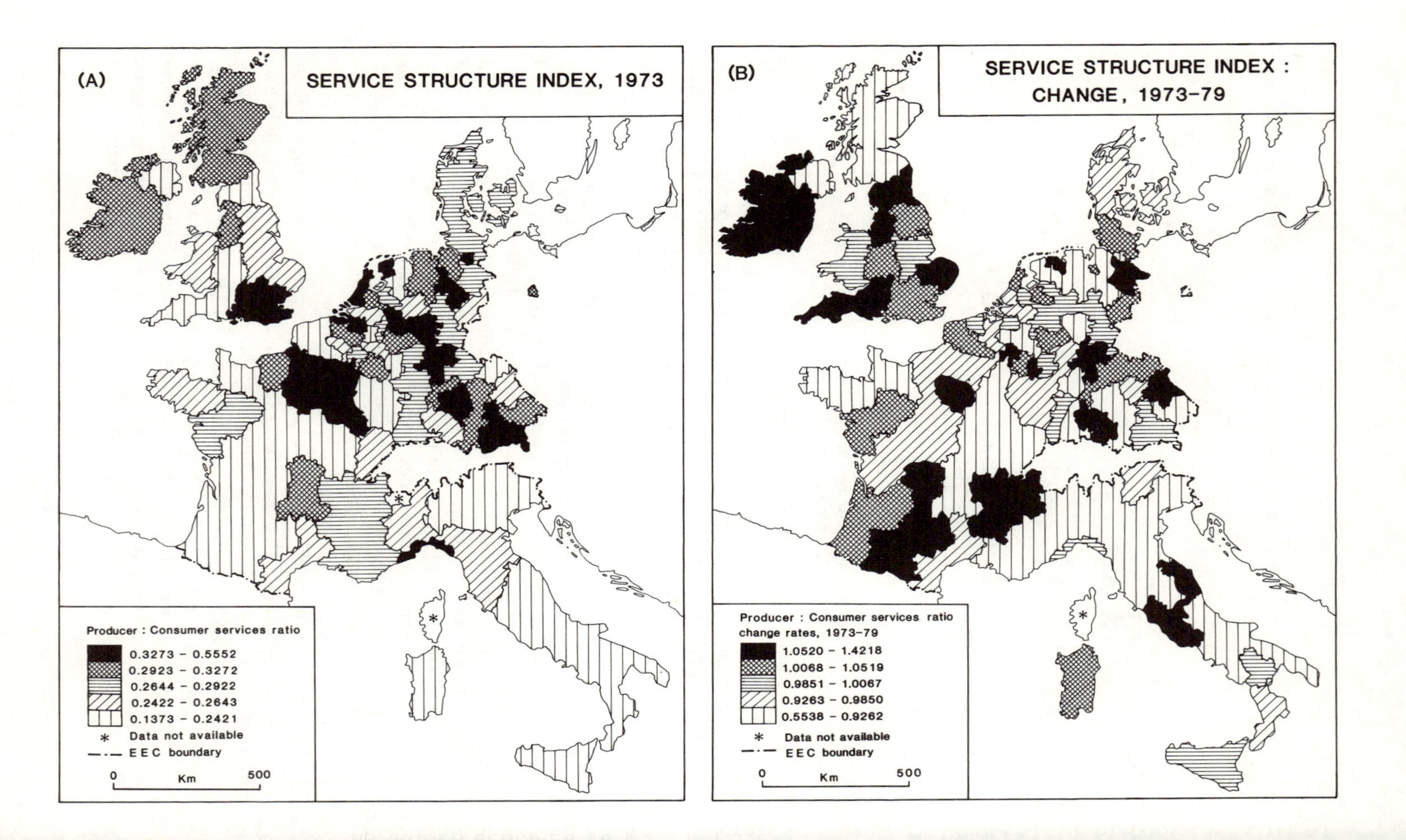
(A)
SERVICE STRUCTURE INDEX, 1973
Producer : Consumer services ratio
0.3273 – 0.5552
0.2923 – 0.3272
0.2644 – 0.2922
0.2422 – 0.2643
0.1373 – 0.2421
* Data not available
EEC boundary
0
Km
500
(B)
SERVICE STRUCTURE INDEX : CHANGE, 1973–79
Producer : Consumer services ratio change rates, 1973–79
1.0520 – 1.4218
1.0068 – 1.0519
0.9851 – 1.0067
0.9263 – 0.9850
0.5538 – 0.9262
* Data not available
EEC boundary
0
Km
500

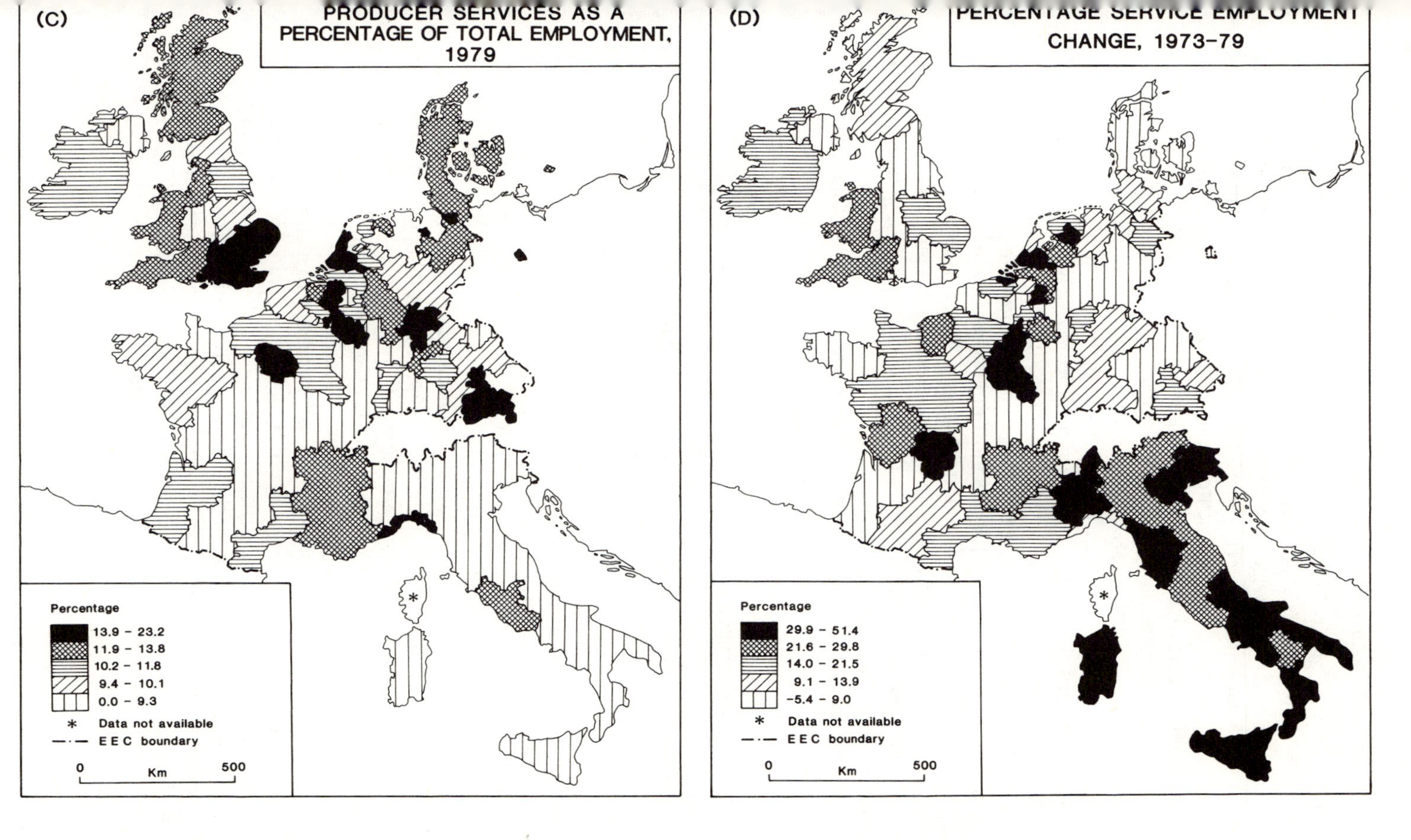

Figure 4.5(A)–(D) Distribution, structure and change of service employment in subregions of the EEC
Source: Keeble *et al.*, 1982, figure 4.7, 97; figure 4.14, 144; figure 4.15, 145; figure 4.16, 146.

are clearly variations in the structure of service industries in the regions of the EEC (Figure 4.5(C) and (D)). The higher the value of the ratio (for producer to consumer service employment in each region), the more likely it is that the service industry structure is favourable to long-term growth (i.e. producer services are well represented and offer the prospect of an economic environment attractive to manufacturing and other services). It is evident that the central regions are moving towards increased relative specialization in producer services (Figure 4.5(D)), while the converse is taking place in the peripheral regions (i.e. increasing relative and absolute specialization in consumer services). The weakness of service industry as a whole in the peripheral regions is also shown by the low values for the indices (Figure 4.5(C)). Keeble *et al.* (1982, 136) confirm the expectation derived from theoretical work that producer services will tend to grow most rapidly in large urban areas and note that the 'more urbanized a region, the more rapidly its service structure is evolving towards a greater relative bias to producer services: the more rural the region, the more rapidly its service structure is evolving towards a greater relative dependence on consumer services'.

Closer analysis of employment change in the subregions using shift–share analysis (see Appendix 3) produces results which are counterintuitive. In other words, although the central regions have a much more favourable sectoral structure of economic activity, they have been 'losing' a large number of jobs through a negative differential shift compared with a positive differential shift in the peripheral regions. It soon becomes apparent, however, that the sectors contributing to structural shift, which has an overall positive value for the central regions, are (broadly) producer services (other services +1.25 million, banking and finance +0.4). These do not figure at all in the structural shifts in the peripheral regions. In relation to differential shifts the central regions perform badly in almost all industry categories (including producer services), while the gains in the periphery are mainly connected with consumer services.

To some extent these results confound all expectations about differentials in the location of service industries at the regional scale and could even be interpreted as representing the development of relative economic weakness of the central regions. However, it seems more likely that, in common with manufacturing industry, both producer and consumer services in the central regions are becoming more efficient in the use of labour and thus able to sustain or increase their productivity in a way which permits them to retain a competitive advantage. Of course, there is no reason to believe that similar improvements are not being achieved by service firms in peripheral regions, so that in order to understand their favourable overall performance it may also be necessary to take account of the contribution of mobile economic activities, especially in France and the UK (see ibid.) where policies designed to encourage a wider distribution of service industries have been used, especially during the 1970s (see Chapter 9).

REGIONAL DISTRIBUTION OF SERVICE INDUSTRIES IN ENGLAND AND WALES

Population-based location quotients for six service industries in the nine economic planning regions in England and Wales, in 1971 and 1981, are shown in Table 4.2. Values range between 1.69 and 0.49 (excluding Greater London) and it is immediately evident that services are far from uniformly distributed in relation to population. For both 1971 and 1981 the majority of the quotients are below 1.00, so that under-representation is the norm in almost all regions, even if only to a marginal degree in certain cases. The south east region, with 34.1 per cent of the population in 1981, is the only location with quotients above 1.00 for all service industries. Although the values of the quotient have all decreased marginally since 1971, in line perhaps with a declining share of national population (– 0.7 per cent change in regional share), the level of overprovision is considerable. If the requirements of the region's population are more than adequately satisfied, it implies that some or all of the service industries represented in Table 4.2 export some of their output to other regions or indeed elsewhere outside of England and Wales.

This possibility is confirmed by the intra-industry variation in the values of the location quotients. For insurance, banking, finance and business services (IBFB) they range between 1.69 in the south east to 0.49 in Wales (1981), between 1.39 and 0.68 for transport and communication (TC) in the south-east and West Midlands respectively (1981), and 0.80 to 1.14 for professional services (PS) in the West Midlands and south-east (1981) respectively. It seems that intra-industry variations in the quotients can be attributed to differences in the extent to which 'immediate' or easy customer contact is important. Hence the inter-regional range of values for the distributive trades, many of which are concerned with direct customer contact, is much narrower than for the insurance and related industries. In the regions which are underprovided with employment in the distributive trades the quotients generally cover a narrower range, between 0.90 and 0.98, and there does seem to be a clearer link with the distribution of population. The much wider range of quotient values for insurance, banking and finance suggests that the association with population share in each region is weaker and it might be more satisfactory to use employment-based location quotients (Table 4.3).

Comparison of the quotients in Tables 4.2 and 4.3 shows that the employment-based values reveal a narrower inter-regional range, although some of the marked differences in over- and underprovision relative to regional employment shares persist. The south-east again emerges with values exceeding 1.00 for all service industries but, without exception, the quotients are lower than those which are population based. One explanation for the 'smoothing' of the difference between regions, especially with reference to the south-east, is commuting from outside the region to some of the rapidly growing employment centres around its outer edges, such as Swindon, Reading, Milton Keynes and south Hampshire. This may also explain the decrease in the values of the employment-based quotients for Greater London into which there is substantial

Table 4.2 Service industry location quotients (population based),[1] economic planning regions, England and Wales, 1971 and 1981

Region	*Year*	*Service industry*[2]						*All services*	*Population share, 1981 (%)*	*Difference, 1971–81*
		TC	*DT*	*IBFB*	*PS*	*MS*	*PAD*			
South-	1971	1.32	1.10	1.63	1.17	1.24	1.28	1.23		
east	1981	1.39	1.17	1.69	1.14	1.14	1.20	1.16	34.1	−0.7
	Change	+	+	+	−	−	−	−		
Greater	1971	1.95	1.03	2.68	1.27	1.52	1.45	1.52		
London	1981	2.12	1.41	2.86	1.28	1.44	1.68	1.63	13.7	−1.59
	Change	+	+	+	+	−	+	+		
East	1971	0.79	0.97	0.72	0.98	1.10	0.99	0.97		
Anglia	1981	0.92	0.97	0.75	0.90	0.95	0.69	0.89	3.8	0.3
	Change	+		+	−	−	−	−		
South-	1971	0.72	0.92	0.67	0.91	0.96	1.25	0.93		
west	1981	0.76	0.94	0.79	0.95	1.09	0.92	0.86	8.8	0.4
	Change	+	+	+	+	+	−	−		
West	1971	0.68	0.90	0.63	0.89	0.82	0.75	0.82		
Midlands	1981	0.68	0.91	0.97	0.84	0.84	0.96	0.85	10.5	0.0
	Change		+	+	−	+	+	+		

East	1971	0.70	0.85	0.54	0.79	0.78	0.69	0.76		
Midlands	1981	0.78	0.89	0.56	0.88	0.80	0.75	0.81	7.8	0.4
	Change	+	+	+	+	−	+	+		
Yorkshire and	1971	0.85	0.97	0.64	0.91	0.86	0.71	0.86		
Humberside	1981	0.83	0.94	0.66	0.91	0.90	0.80	0.87	9.9	−0.1
	Change	−	−	+		−	+	+		
North-	1971	1.05	1.07	0.82	0.96	0.88	0.75	0.95		
west	1981	0.93	0.99	0.81	0.96	0.98	0.90	0.94	13.1	−0.4
	Change	−	−	−		+	+	−		
North	1971	0.85	0.96	0.60	0.94	0.89	1.01	0.91		
	1981	0.79	0.89	0.52	0.87	0.93	0.99	0.86	6.3	−0.1
	Change	−	−	−	−	+	−	−		
Wales	1971	0.83	0.83	0.59	0.91	0.80	0.90	0.83		
	1981	0.75	0.72	0.49	0.94	0.87	1.04	0.83	5.7	0.1
	Change	−	−	−	−	+	+			

Notes: 1 Population based, using data from *Census of Population, 1971* and *Department of Employment Gazette*, February 1983.
2 TC, transport and communication; DT, distributive trades; IBFB, insurance, banking, finance and business services; PS, professional and scientific services; MS, miscellaneous services; and PAD, public administration and defence.

Table 4.3 Service industry location quotients (employment based),[1] economic planning regions, England and Wales, 1971 and 1981

Region	*Year*	*Service industry*[2]						*All services*	*Employment share 1981 (%)*	*Difference, 1971–81*
		TC	*DT*	*IBFB*	*PS*	*MS*	*PAD*			
South-	1971	1.24	1.03	1.53	1.09	1.17	1.20	1.15		
east	1981	1.27	1.07	1.53	1.04	1.03	1.10	1.06	37.4	0.7
	Change	+	+		–	–	–	–		
Greater	1971	1.57	1.05	2.15	1.02	1.27	1.17	1.23		
London	1981	1.58	1.05	2.13	0.95	1.07	1.25	1.22	18.3	0.0
	Change	+		–	–	–	+	–		
East	1971	0.82	1.02	0.75	1.02	1.14	1.03	1.00		
Anglia	1981	0.99	1.05	0.81	0.97	1.03	0.74	0.96	3.3	0.2
	Change	+	+	+	–	–	–	–		
South-	1971	0.84	1.07	0.77	1.05	1.12	1.46	1.08		
west	1981	0.84	1.05	0.88	1.05	1.21	1.01	0.96	8.0	0.8
	Change		–	+		+	–	–		
West	1971	0.66	0.87	0.61	0.86	0.79	0.72	0.79		
Midlands	1981	0.69	0.91	0.79	0.84	0.84	0.96	0.85	10.9	–0.5
	Change	+	+	+	–	+	+	+		

East	1971	0.75	0.92	1.04	0.64	0.63	0.74	0.82		
Midlands	1981	0.80	0.91	0.58	0.89	0.81	0.77	0.83	6.9	0.7
	Change	+	–	–	+	+	+	+		
Yorkshire and	1971	0.88	1.00	0.66	0.94	0.88	0.73	0.89		
Humberside	1981	0.86	0.97	0.68	0.95	0.93	0.83	0.90	9.7	–0.2
	Change	–	–	+	+	+	+	+		
North-	1971	1.04	1.05	0.81	0.95	0.87	0.74	0.94		
west	1981	0.94	1.00	0.82	0.98	0.99	0.92	0.96	13.7	–0.8
	Change	–	–	+	+	+	+	+		
North	1971	0.87	0.98	0.62	0.96	0.91	1.03	0.93		
	1981	0.85	0.96	0.56	0.94	1.00	1.06	0.92	6.3	–0.4
	Change	–	–	–	–	+	+	–		
Wales	1971	0.93	0.92	0.65	1.01	0.89	1.00	0.93		
	1981	0.88	0.84	0.57	1.10	1.01	1.21	0.97	5.0	–0.1
	Change	–	–	–	+	+	+	+		

Notes: 1 Employment based, using data as for Table 4.2.
2 Abbreviations as specified for Table 4.2.

commuting from the rest of the south-east region. However, the effects of this factor should not be exaggerated; it is also likely that differences in levels of unemployment, activity rates for males and females, and the match between job skills available and those demanded also exert some influence. This gives rise to the most consistent underprovision of service industries in the West and East Midlands, especially in relation to the population-based quotients. In Wales and northern England the population-based quotients are also well below average but both regions perform better on employment.

Some service industries have improved their location quotients during the decade: public administration and defence and professional and scientific services in the regions outside the south-east, the south-west and East Anglia; insurance, banking and finance in most regions other than Wales, north and north-west; and miscellaneous services, especially using employment-based quotients, in all regions outside the 'core'. The explanation for some of these changes will be considered in Chapters 7–9; it is sufficient to observe here that the relationship between the theoretical and actual distribution of service industries at the regional scale is, at best, variable.

This conclusion can, however, be criticized on the ground that it has been arrived at by using aggregated statistics. Would the conclusion be different if the data is disaggregated into smaller spatial units or for individual service industry Minimum List Headings (MLHs)? The hypothesis that the location of service industries is a function of the agglomeration of population in urban areas and that the degree of clustering increases disproportionately to the increase in the size of those urban areas is not adequately explored. A simple example of the effect of adopting the former course is also included in Tables 4.2 and 4.3: Greater London has significantly higher values for its location quotients than for the south-east as a whole or by comparison with the other economic planning regions. The 'urbanization' effect clearly applies and the very basic evidence provided in both tables confirms the influence of centrality on the location of all the service industries, especially those with substantial intermediate outputs such as insurance, banking and finance. However, as Greater London's share of population has decreased all the population-based quotients have increased, while almost all the employment-based values have decreased between 1971 and 1981. Employment decentralization has no doubt contributed to this and its effects will be examined in Chapter 9. The location quotient for professional and scientific services has in fact fallen below 1.00 in 1981 (0.95) and the values for distributive trades and miscellaneous services are also much lower than the norm for other services in Greater London.

A number of studies have been concerned with the association between urban areas and the level and diversity of service industry activity (Marquand, 1979; Hughes and Thorne, 1975; Bussiere, 1974; Birg, 1978; Stanback, 1979). Probably the most comprehensive analysis to date has been undertaken by Marquand (1979), who has examined the proportion of service employment in 126 metropolitan economic labour areas (MELA) in Britain, in 1971, to see whether there are any differences between them with reference to the distribution of producer and consumer services, and private and public services. In addition, an attempt is made to show whether there is any evidence for underprovision of services in MELAs or, conversely, some indication of the attraction of some services to locations which offer agglomeration economies. For each industrial group regression coefficients ($y = \alpha x^{(\beta+1)}$) were estimated for service employment in each MELA on total employment and on total population. The significance of London for the national distribution of UK service employment has already been mentioned and Marquand has undertaken calculations which both include and exclude it to see what effect this has on the regression equations and interpretation of the results.

These accord well with the expectation that non-market services will have a looser fit than market services and that consumer services are more closely related to population than producer services (Table 4.4). If there is a significant tendency towards clustering of services $\beta > 0$, then the greater α, the greater the proportion of service employment at all sizes of urban area; the greater β, the steeper the hierarchy; and the greater r^2, the larger the proportion explained by population size (or employment for producer services) and the more that all urban areas conform to the same pattern (for a full explanation see ibid., 1979, paras 4.4–4.7).

Although the differences between the r^2 values for services on population and employment are generally small, the proportion of the variation in

Table 4.4 Regression analysis for British service industry groups in MELAs, 1971

Sector		*Population London*			*Employment London*		
		Included[1]	*Excluded*[1]	*Difference*	*Included*	*Excluded*	*Difference*
Transport and	r^2	0.92	0.90		0.91	0.89	
communication	$\beta>0$	0.1478	0.1343	0.0135	0.1139	0.1054	0.0085
Distributive trades	r^2	0.97	0.96		0.96	0.95	
	$\beta>0$	0.0424	0.0366	0.0058	0.0104	0.0089	0.0015
Insurance, banking,	r^2	0.84	0.81		0.82	0.78	
financial services	$\beta>0$	0.1423	0.1010	0.0413	0.0978	0.0589	0.0389
Professional and	r^2	0.95	0.94		0.94	0.93	
scientific services	$\beta>0$	0.0372	0.029	0.0073	0.0060	0.0031	0.0029
Miscellaneous	r^2	0.92	0.91		0.91	0.89	
services	$\beta>0$	0.0309	0.0122	0.0187	−0.0038	−0.0190	0.0152
Public	r^2	0.78	0.75		0.77	0.73	
administration	$\beta>0$	0.0322	0.0180	0.0142	−0.0028	−0.0134	0.0108

Note: 1 Results of calculation with London included and London excluded. ______ = significant at 1 per cent level; _ _ _ _ _ = significant at 5 per cent level.

Source: Marquand, 1979, extracted from tables 4.1 and 4.2.

distributive trades between MELAs explained by population is higher than that explained by employment. In every case the effect of excluding London is to lower somewhat the proportion of the variation explained (it should be noted that all the r^2 values shown in Table 4.4 are highly significant). For the mixed services such as insurance, banking and finance the r^2 values are correspondingly lower than those for the distributive trades, with the coefficients for population marginally higher. The higher coefficients on population for public administration and for professional and scientific services reflects the presence of health and education services in the latter and the large non-market component in the former. It is notable, however, that the service for which it is assumed that administrative considerations, including in many instances the need to be accessible to consumers, are pre-eminent; that is public administration and defence achieve the poorest level of explanation.

There is clear evidence for a hierarchical distribution of services (Table 4.4) with almost all groups, whether regressed on population or employment, having positive values for β. The significance levels achieved by the values for transport and communication (London included), and by insurance, banking and finance and the distributive trades (whether London is included or not) suggests a strong hierarchical effect, especially for the coefficients based on population rather than employment. In relation to the latter only transport and communication has a value which is significant at the 1 per cent level. The differences between the values of β when London is included are all positive, i.e. the inclusion of London exaggerates the hierarchical distribution in a way which causes the curve to skew upwards. This effect is strongest for those

Table 4.5 The service sector in the British conurbations, 1971–5

Conurbation	*Service employment*[1] *(%)*	*Change 1971–5 (%)*	*SIC 21–3 (%)*	*SIC 24–7 (%)*	*Change, all employment (%)*
Greater London	72.5	0.2	−6.9	5.6	−6.7 (4.0)[2]
Manchester	61.5	6.3	−2.1	11.9	−1.1 (0.7)
Glasgow	61.1	12.8	−2.4	24.8	10.3 (0.5)
Merseyside	59.7	1.0	−11.9	11.4	−3.1 (0.6)
Tyneside	57.5	9.6	0.8	15.5	3.1 (0.4)
Birmingham	52.4	7.3	−2.3	13.4	−3.1 (0.7)
UK	57.9	9.7	1.7	14.6	2.6 (22.2)

Notes: 1 Includes gas, electricity and water (SIC 21).
2 Total employment in 1975 (in millions).

Source: Hubbard and Nutter, 1982, derived from table 1, 210.

services with a mixed or producer function: insurance, banking and finance (difference 0.0413 for population, and 0.0389 for employment), although miscellaneous services (0.0187 and 0.0152 respectively) also incorporate the same effect despite the inclusion of a large number of market-oriented activities.

The distinction between Greater London and the remainder of the south-east is mirrored by other major British cities. Glasgow, Merseyside and Manchester all had a larger proportion than the national average (57.9 per cent) of their employment in services, and Glasgow and Tyneside had a larger-than-average increase in the 'growth' services (Table 4.5). As already noted, however, these are the services tending to reveal the largest locational bias towards particular regions, and the large proportional increase for Tyneside, for example, may only represent a 'catching up' process from a relatively small base of employment in, for instance, professional and scientific services.

The relatively consistent pattern of locational imbalance when the location of the main service industry groups is analysed is more difficult to detect when individual service MLHs are examined. A few examples are given here for illustrative purposes.

Distributive trades

Contrary to expectations, a large number of the MLHs representing the distributive trades have location quotients below 1.00.* Central place theory leads us to expect that employment in retail distribution would be evenly distributed, especially in relation to population; the population-based quotients are indeed mostly above 0.90 but are substantially above 1.00 for Greater London and the south-east. Many more of the employment-based quotients exceed 1.00, and the East Midlands, Yorkshire and Humberside and Wales have deficit values for both types of retail distribution. The quotients for wholesale distribution of lower-value goods like food indicate a closer association with population and employment than those for the wholesaling of petroleum products. The higher value-added of the latter not only permits distribution from a smaller number of locations, including refineries, but also allows a wider range of distribution modes to be used such as pipelines.

The two remaining MLHs in the distributive trades – dealing in coal, oil, builders' materials, grain and agricultural supplies, and in other industrial materials and machinery – also reveal less inter-regional variation. Easy access by clients to the sources of the diverse goods and services provided is clearly an important factor in location. Hence East Anglia (1.61) and the south-west (1.38) have much higher quotients than the south-east (1.03) for dealing in coal, agricultural supplies, and so on, while the West Midlands (1.45) and Yorkshire and Humberside (1.31) have the highest quotients for dealing in other industrial materials and machinery. The type of service and the characteristics of the market seem, therefore, to go some way towards explaining inter-regional variations in service employment. The somewhat surprising imbalance in the pattern of employment in retail distribution possibly reflects differences in the type

*Unless otherwise indicated, the location quotients cited in this section are the population-based values.

of retail services demanded in different areas of the country as well as the non-local demand arising from, for example, the tourist trade in London.

Insurance, banking, finance and business service

There is much less discrepancy between MLHs in insurance, banking, finance and business services with respect to the degree of overconcentration in Greater London and the south-east relative to other regions. The discrepancy is largest for advertising and market research which has a location quotient of 2.27 in the south-east and 0.31 and 0.36 in the south-west and East Midlands respectively. The quotient for Greater London is 4.44, a value only marginally higher than the 4.35 for central offices not allocatable elsewhere (2.24 for the south-east). These are primarily producer services, whereas those which could perhaps be best described as 'mixed', insurance, banking and bill discounting, and other financial institutions, do not reveal such extremes, although inter-regional imbalance remains significant with quotients exceeding 1.00 outside the south-east the exception. Examples include other financial institutions in Yorkshire and Humberside (1.16), property-owning and managing in the south-west (1.04) and other business services in the West Midlands (1.07).

MISCELLANEOUS SERVICES

Miscellaneous services comprise a diverse set of MLHs, ranging from clubs and laundries to motor repairs and other services. The values for the location quotients reflect this diversity and seem to incorporate socio-cultural as well as economic influences on the location of individual services. Hence clubs reveal quotients above 1.00 in most of the regions outside the south-east, East Anglia and the south-west, with a high value for the north (2.21) and Wales (1.59). Public houses also contradict the pattern typical of almost every other MLH and have high quotients for the West Midlands (1.37), the north-west (1.34) and the north (1.14). In common with the services within the distributive trades which are mainly orientated towards final demand, it is the peripheral regions which have low location quotients for dry cleaning and related services or for motor repairing and distribution. Some of these services involve the expenditure of discretionary income and the demand in the more prosperous southern regions may well contribute to the above-average coefficients for East Anglia, the south-west or the East Midlands. Agriculture is also prominent in these areas and represents another important source of demand for the repair and servicing of agricultural machinery. A similar effect can be detected in the case of hotels and other residential establishments, in that the regions with a significant tourist function, such as Wales (1.15) or the south-west (1.99), have positive coefficients, while the West or East Midlands have much lower values (below 0.60).

In common with the analysis using regional location quotients it is possible that aggregated SIC data leads to spurious conclusions about the goodness of fit with MELA population (or employment) and the observed hierarchical effect. Marquand (1979), therefore, goes one step further and applies the same methodology to individual MLHs within each SIC (Table 4.6). The activities

Table 4.6 Regression analysis for activities within British service industry groups in MELAs, 1971

Activity and nature of service		*Population London*			*Employment London*		
		Included[1]	*Excluded*[1]	*Difference*	*Included*	*Excluded*	*Difference*
Consumer							
Retail (820,821)	r^2	0.97	0.96		0.95	0.94	
	$\beta>0$	0.025	0.022	0.003	−0.012	−0.013	0.001
Public services (non-foot-loose) (872,874,906)	r^2	0.96	0.95		0.95	0.94	
	$\beta>0$	0.042	0.041	0.001	0.008	0.011	−0.003
Cinemas, theatres (881)	r^2	0.74	0.69		0.70	0.65	
	$\beta>0$	0.298	0.261	0.037	0.036	0.199	0.037
Motor repairs, garages (874)	r^2	0.92	0.90		0.91	0.89	
	$\beta>0$	0.009	0.012	0.003	−0.020	−0.013	0.007
Mixed							
Business and personal services (860,861,862,863,709)	r^2	0.85	0.82		0.83	0.79	
	$\beta>0$	0.138	0.095	0.043	0.096	0.055	0.041
Professional services (871,873,879)	r^2	0.84	0.81		0.83	0.79	
	$\beta>0$	0.098	0.059	0.039	0.065	0.032	0.033
Insurance (860)	r^2	0.82	0.79		0.80	0.76	
	$\beta>0$	0.278	0.202	0.026	0.179	0.156	0.023
Banking (861)	r^2	0.80	0.76		0.78	0.73	
	$\beta>0$	0.096	0.043	0.053	0.058	0.008	0.050
Producer							
Transport (703,704,705, 706,707)	r^2	0.79	0.76		0.81	0.78	
	$\beta>0$	0.249	0.250	−0.001	0.227	0.238	−0.011
Wholesale and dealers (810,811,812,831,832)	r^2	0.89	0.87		0.91	0.89	
	$\beta>0$	0.116	0.107	0.009	0.099	0.098	0.001
Business services (864,865,866)	r^2	0.68	0.62		0.71	0.65	
	$\beta>0$	0.155	0.088	0.067	0.148	0.092	0.056
Research (876)	r^2	0.21	0.17		0.22	0.18	
	$\beta>0$	0.088	0.039	0.059	0.084	0.052	0.047

Note: 1 Results of calculation with London included and London excluded; _____ = significant at 1 per cent level; _ _ _ _ _ = significant at 5 per cent level.

Source: Marquand, 1979, tables 4.6 and 4.7.

consist of one or more MLHs (see note to Table 4.6) and are further classified according to whether they are consumer, mixed, or intermediate services. As a general rule, the proportion of the variation in an individual activity explained by population or employment decreases from consumer through to producer activities and the exclusion of London also reduces the level of explanation. Business services and research have the lowest r^2 values, while retail or motor repairs and garages have values in excess of 0.90. It appears, therefore, that the conclusions drawn from the aggregated service industry groups are borne out at the level of individual activities. Nevertheless, activities such as hotels (not included in Table 4.6) have much lower r^2 values (0.51–0.57) than the average for consumer activities, and this demonstrates the persistent difficulty of generalizing too much about the location of services. In this instance the distribution of population or of employment are unlikely to be adequate independent variables; the distribution of tourist attractions would probably be more appropriate. The fit to employment is superior in the case of producer services (Table 4.6) but to population for mixed and consumer services.

The slope of the regression line (β in Table 4.6) is increased when London is included except for cinemas and theatres (on employment) and transport (as a producer activity on population and employment). The line differs significantly from unity at the 5 per cent level or better in several cases but especially for such producer activities as wholesalers and dealers, transport and for mixed services such as insurance. The results of the analysis based on aggregated data are, therefore, again confirmed.

As a broad generalization location quotients with a value of less than 1.00, whether employment – or population based, occur most frequently among MLHs represented in the regions outside the core. Such evidence generally reinforces the conclusion that the type of service activity being analysed is significant in relation to its locational pattern and, therefore, the degree of over- and underprovision. In common with the aggregated data, however, it emerges that the population-based location quotients show a wider range and more extreme values, but the range is narrowest for consumer service MLHs and widest for producer service MLHs.

Nevertheless, this more disaggregated approach has complicated the interpretation of the location of services and its relationship to central place theory. Explanations for some of the variations which have been noted deliberately have been kept brief at this stage because there will be further elaboration in subsequent chapters. Suffice to say that, whether related to population or employment, the distribution of service industries does not fit easily under the umbrella of central place theory; it is more relevant to understanding the location of some services than others for reasons already cited. In some instances – many of the miscellaneous services, for example – the spatial distributions seem to arise from regional differences in social values or consumer taste as well as underlying discrepancies in purchasing power and the availability of surplus discretionary expenditure. The latter tends to be expended on non-essential services. The location of public sector services also fluctuates despite expectations that these would be the most uniformly distributed service activity between regions.

EMPIRICAL EVIDENCE FROM ELSEWHERE IN EUROPE AND NORTH AMERICA

It remains to be seen whether the relationships indicated in the detailed analysis undertaken during recent years in Britain are confirmed by the results of related studies undertaken elsewhere.

As part of a comparative study (reported in Marquand, 1978) of the tertiary sector in the member states of the EEC a number of reports were prepared (see, for example, Bannon and Eustace, 1978) using techniques designed to test similar assumptions to those explored using British data. Using 131 urban zones in Belgium, a good fit between employment/population and service industry (subdivided into sectors) was found and the slopes for all are greater than unity. In common with the British results employment provides a superior level of explanation for the location of insurance and financial institutions, transport, and communications and commerce, while the steepness of the slopes from largest to smallest urban zones decreases from producer to consumer services.

The report on service industries in Denmark also conforms with expectations from the British results. A total of thirty-nine cities (included in rather larger trading areas somewhat similar to MELAs) were used in the analysis, which produced high r^2 values for retail trade, followed by financial and business services, public services and, at the bottom of the scale, wholesale distribution. The slope of the regressions is considerably above unity for business services, decreasing through financial to public services. In the French report regression equations were fitted for thirty-three agglomerations and a distinction made between foot-loose and other services. The former includes, for example, banking and insurance, public administration and business services. The relationship between employment in these sectors and total employment is statistically significant with a slope greater than unity. Producer services are shown to be increasingly concentrated in the larger agglomerations but for most of the consumer services the level of concentration is weaker. Bussiere (1974) has also arrived at similar conclusions, although using a different framework of units for the analysis.

Unfortunately not every European country fits the general picture. The principal exception is West Germany, where Birg (1978) has found an inverse relationship between size (population) of urban areas and the proportion of services. Hence the cities at the top of the German hierarchy have rather fewer services than those near to the lower end. Birg suggests that this is a product of the way in which the larger cities have grown but it may also be the product of using inappropriately defined statistical units. The high level of administrative autonomy exercised by the *Länder* (provinces) may also be a factor since no one city or region dominates the space economy, in Germany, in the way evident in France, Belgium, Denmark, or Britain.

Although Ireland has an urban system dominated by a primate city and region, the ranking of steepness of slopes for the regression equations is not easily related to those for other EEC countries. In descending order the slopes are: commerce, professional services, transport and communications, public administration, personal services, banking and recreation and, with a value scarcely greater than zero, 'other' services. The relative steepness of the slopes for producer-type services, typical of other European countries, is absent.

Variations in the distribution of basic and non-basic (or export) services have also been considered by Stanback (1979) for the 259 standard metropolitan statistical areas (SMSAs) in the USA. The coefficient of variation is used to express the amount of variation in the share of employment in a given service activity relative to its average share of total employment in the same activity in all SMSAs. The nearer the value of the coefficient to unity, the more even the distribution of the activity between SMSAs (Table 4.7). Retail trade has a coefficient of 0.12 and is clearly a much more ubiquitous service than government administration (0.58), wholesale trade (0.36), business and repair services

Table 4.7 Coefficients of variation for employment shares among SMSAs: USA, 1970

Service industry	*Coefficients of variation*	
Transportation	0.45	
Railroads and railway express		1.17
Trucking and warehousing		0.41
Other transportation		0.72
Communications	0.33	
Services	—	
Wholesale trade		0.36
Finance, insurance and real estate		0.33
Business and repair services		0.39
Retail trade	0.12	
Food and dairy products stores		0.15
Eating and drinking places		0.22
Other retail trade		0.13
Recreation	0.52	
Lodging places		0.47
Entertainment		1.01
Government	1.04	
Administration		0.58
Armed forces		2.24

Note: The coefficient of variation derived from standard deviation of proportion of employment accounted for by a given service activity divided by the mean of shares in that industry.

Source: Stanback, 1979, table 15, 80.

(0.39), or other transportation (0.72). The latter includes seaports and airports, which are major export services in some SMSAs and not others. Equally the coefficient for lodging places (0.47) indicates that some SMSAs will rely more than others on tourist demand, which may also affect retail trade and cause it to contribute to export as well as local market activity.

SMSAs can also be classified with reference to the industrial composition of employment (see ibid.). Manufacturing centres dominate (30 per cent), but 20 per cent of the SMSAs are characterized by relatively large shares of employment in business services and are, therefore, major central places within the national urban system. A further 13 per cent of the SMSAs can be classified as government centres and 5 per cent as medical/educational centres. Both of these also include an export or producer element and, together with 27 per cent of the SMSAs defined as 'mixed' but in which there was considerable specialization in some business services, it is apparent that the majority of the metropolitan areas in the USA are dominated by service industries. Producer services, assuming business services to be an appropriate surrogate, are concentrated to a significant extent in about one in five of the SMSAs. If the distribution of these centres is analysed by size, the principal concentrations of business services are in the largest SMSAs.

THE DISTRIBUTION OF SERVICES AT THE LOCAL SCALE

Much of empirical evidence used thus far has been presented with reference to broadly defined regions or more rigorously defined metropolitan statistical areas within countries. It is possible, and useful, to focus upon rather smaller areas to see whether the location tendencies identified at the larger scale are replicated (for an early example see Knox, 1969). Two examples are provided: first, some data for urban areas, small towns and rural areas in Kent (Kent County Council, 1975), and secondly, an intra-urban example using data for the metropolitan county of Merseyside (Hubbard and Nutter, 1982).

Service employment in Kent

The structure of employment in the major industry categories for urban areas, small settlements and rural areas in Kent is shown in Figure 4.6. Although the data is derived from a detailed analysis of the workplace movement, information available for small areas in the 1966 Census of Population and the patterns which it demonstrates are still relevant to more recent time periods. There are eighteen urban areas which accommodate 75 per cent of the population of the county but over 80 per cent of all employment. But almost 85 per cent of all service employment in the county is located in the urban areas which comprise two distinct groups: the coastal resorts, where miscellaneous services connected with hotels, recreation and entertainment make a large contribution; and the urban areas, mainly in the north-east of the county, which are administrative centres such as Canterbury or Maidstone, or major subregional centres such as Tunbridge Wells or Ashford. It is also worth noting that Kent

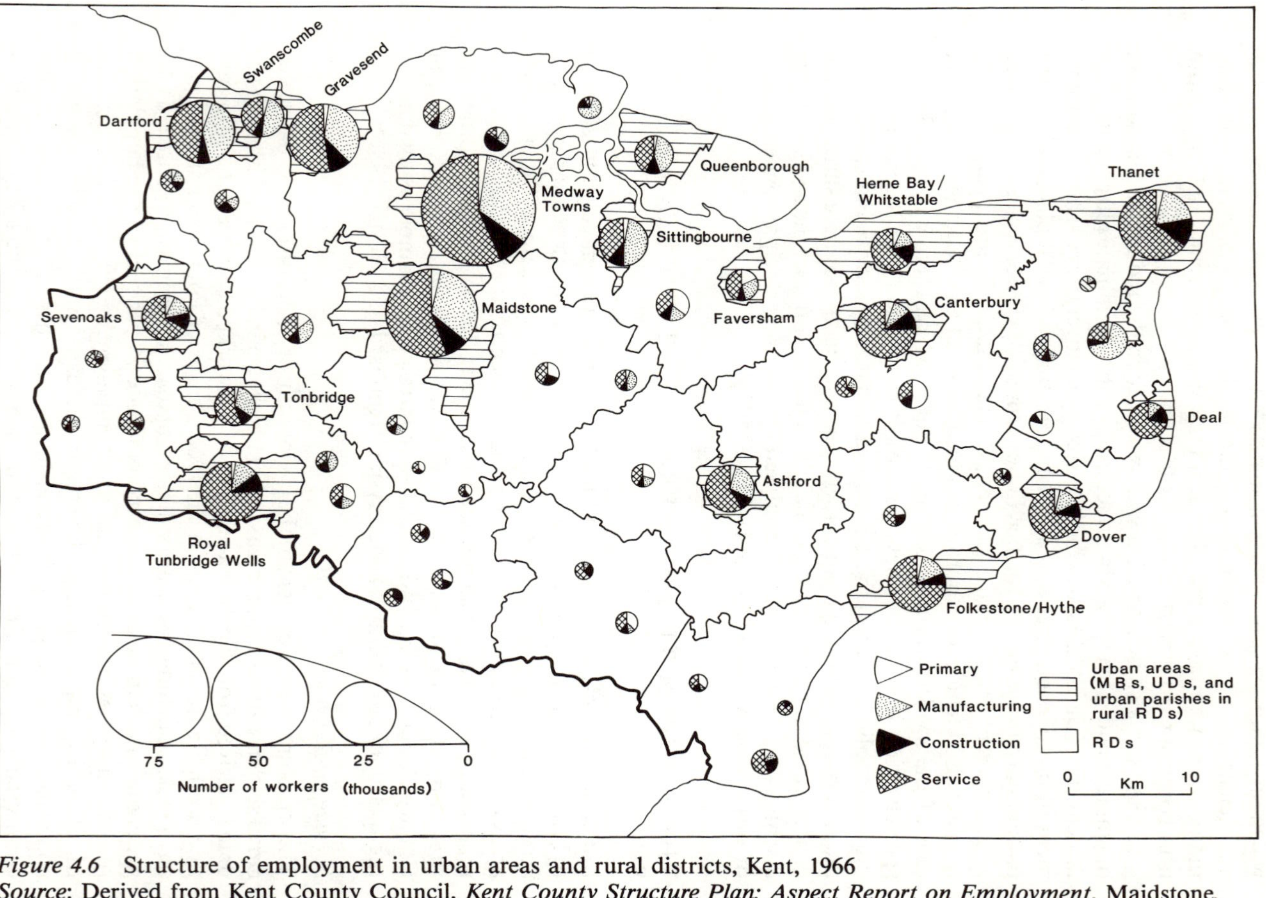

Figure 4.6 Structure of employment in urban areas and rural districts, Kent, 1966
Source: Derived from Kent County Council, *Kent County Structure Plan: Aspect Report on Employment*, Maidstone, County Planning Department, 1979, figure 3, 26.

is contiguous to Greater London and this may have had some effect on the location of non-local services in urban areas within easy reach of the metropolis. Two urban areas dominate the settlement structure of the county, Medway and Maidstone, with the next largest urban areas less than one-half the size. There is a relatively well-defined hierarchy of settlements – with the largest proportion of services in the largest. The small towns, of which there are eighteen, only have 40 per cent of their employment in services, and the rural areas 44 per cent.

The difference between the urban areas and elsewhere can be traced to variations in the propensity of individual services to concentrate in the former. Hence more than 90 per cent of the employment in insurance and banking is located in the urban areas (but this activity only represents 4.1 per cent of all service sector jobs) along with 91 per cent of those in public administration and defence (15.2 per cent of the total). Transport and communications (87 per cent) are also concentrated at a level above the overall share for urban areas of 80.4 per cent. Miscellaneous and professional/scientific services, on the other hand, are only marginally over-represented, and the former is the only subsector found at a higher level than expected in either small towns or rural areas. Evidence from the regional analysis produced by Marquand (1979) would suggest that this is caused by the demand for repair and related services connected with the primary industries which are prominent in the structure of the rural areas (see Figure 4.6). Thus, as at the regional level, there is clear evidence that some service industries are distributed more or less in direct proportion to population (miscellaneous services and distributive trades) and others are closely tied to the distribution of employment.

Analysis has progressed from national to regional and then to county (approximately subregional) level. At each of these scales it quickly emerges that urban areas exert a substantive influence on the location of service industries. The strength of the locational pull which they exert varies according to the type of service activity. As the final stage in this brief overview of the empirical information pertaining to service industry location it seems relevant to consider whether the location of services *within* urban areas is in any way contrary to the behaviour already clearly established.

Service industries on Merseyside

The location quotients (UK based) for the major service industry groups in each of the five districts within the metropolitan county of Merseyside (Figure 4.7) reveal substantial over- and under-representation. The 'centre' of the county is occupied by Liverpool (35 per cent of the total population in 1971), the remaining districts forming an essentially suburban ring. Employment in transport and communications is heavily concentrated in Liverpool and Sefton, which together encompass the Port of Liverpool and its ancillary functions, along with shipping offices located mainly in Liverpool city centre which has the principal concentration of office space on Merseyside. The location quotients in the three remaining boroughs are less than 1.00, although the presence of Birkenhead docks in the Wirral district of the county is reflected in

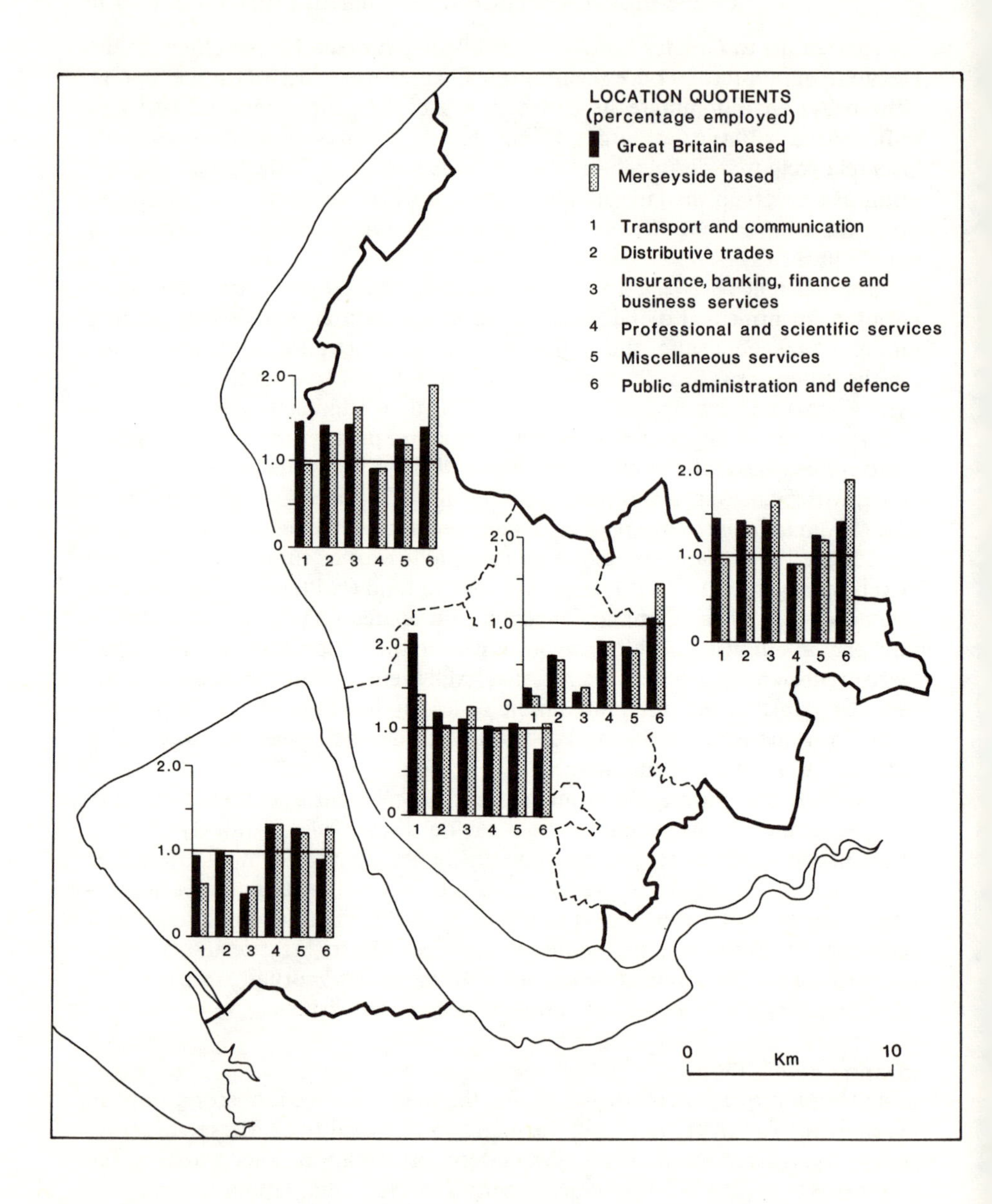

Figure 4.7 Location quotients for the distribution of service industries in the districts of Merseyside Metropolitan County, 1971
Source: Derived from Merseyside County Council, *Merseyside County Structure Plan: Report of Survey, Economic*, Liverpool, County Planning Department, 1979, appendix G.

the value of the quotient. It would be expected that the distributive trades would show the least variability but the quotients range from 1.417 (Sefton) to 0.620 (Knowsley). With its large city centre shopping area, Liverpool has a lower quotient than Sefton. This is a good example of the disadvantage of aggregation; the large quotient for Sefton is almost certainly attributable to the warehouse and wholesale distribution activity (asociated with the Port of Liverpool) in the borough rather than to an excessive concentration of retailing services. On the other hand, the relatively low values for the coefficients in St Helens and Knowsley possibly understate the availability of retail services because, for historical and social reasons, both the volume and diversity of retailing, especially in Knowsley, is poor. Accessibility to retail services within cities, even using this limited evidence, is more variable than central place theory would lead us to expect.

Insurance, banking and finance is concentrated primarily in Liverpool and Sefton; the prominence of the latter is surprising, although the rapid expansion of office space in the southern part of the borough during the early and mid-1970s may have played a part. Professional and scientific services show a much narrower range of quotients (from 0.743 for St Helens to 1.326 for Wirral). The above-average concentration in Wirral is due to the location there of a number of research and development (R&D) facilities. Public administration and defence (mainly the former in Merseyside) is principally located in Sefton, the place chosen for large government investment in office space used by a number of civil service departments, and in Knowsley. The positive quotient for Knowsley arises from a mix of locally orientated central government offices and local authority administrative functions required to service the diverse needs of a borough with some of the most serious unemployment and socio-economic problems in Britain.

Merseyside-based location quotients sharpen the extent of the intra-urban differences in the location of services. Hence St Helens has no location quotients greater than 0.81 for any of the six SICs (Figure 4.7); the quotient is only 0.29 for insurance, banking and finance. Knowsley also has uniformly low values (0.16 for transport and communications) with the sole exception of public administration (1.48). The Wirral and Sefton occupy intermediate positions with at least three SICs repesented by scores exceeding 1.00. The large location quotient for insurance, banking and finance in Sefton is confirmed and, relative to total employment in this category on Merseyside, it is the most over-represented of all activities. As might be expected from its central position in the metropolitan area, Liverpool has location quotients greater than 1.00 for all six SIC categories, with those services which have a substantial mixed or producer component associated with the highest quotients.

SUMMARY

This has been the first of two chapters concerned with some of the theoretical approaches which may be used to account for the location of service industries. The central place model has been outlined and some preliminary comments made on its limitations for modelling the distribution of all types of services. These difficulties will be elaborated further in Chapter 5, which examines some location models which follow directly from central place ideas as well as models which represent alternative approaches to the service industry location problem. The organization of these two chapters should be viewed as a continuum, starting with the relatively abstract approach of basic central place theory and proceeding to models distinguished by a more pragmatic view of the decisions about service industry location in the real world. It is important to note that many of the 'alternatives' to central place theory do not necessarily contradict it; they reflect the diversity of service sector activities and their locational needs or priorities. Indeed such is the magnitude of this diversity that it is unlikely that a single theory could account for the location of all services.

The futility of searching for a single theory soon becomes apparent when the factors affecting the location of services are enumerated. While many need to locate as close to customers as possible, it emerges that services are not in fact uniformly distributed in geographical space. The location factors common to all services are transportation costs, the attributes of demand (especially its spatial characteristics) and economies of scale. These factors go a long way towards creating the spatial distribution of services but are less satisfactory when attempting to explain the predilection of some services to group together to a disproportionate extent at certain locations. It is in this respect that agglomeration economies are an important additional factor in the location equation for service industries. The location of producer services and corporate headquarters in large metropolitan areas is the most distinctive response to this factor which, in turn, means that the same metropolitan areas are attractive locations for consumer services linked to the demand created by producer services, or for public services. The presence of the latter in a large and diverse complex will also be attractive to other corporate headquarters of specialized producer services (see Chapter 5).

Imbalance in the distribution of service industries has been illustrated in the second half of this chapter using employment data. A number of hypotheses derived from central place theory have been examined, and it is shown that consumer services conform more readily to the expected patterns of location than producer services. Irrespective of the scale at which data is analysed (which ranges from subregions of the EEC to individual districts within a UK metropolitan county), an uneven spatial distribution is a consistent and recurrent attribute of producer services.

CHAPTER 5

Service industry location: beyond the central place model

Central place theory, including the later modifications by Berry and Garrison (1958) and Lösch (1954), is heavily dependent upon assumptions about the locational response to demand by consumer and particularly retail services. It is tempting to think, and the empirical material in Chapter 4 may have promoted the belief, that it provides a good theoretical basis for understanding the location and frequency of all service activities. But the theory does not make any explicit reference to producer services or to public sector services which are provided on a not-for-profit basis (Stanback, 1979). Some educational, government and tourist functions may be dictated in their location by environmental considerations, very specific transport requirements, or historical inertia. Perhaps it is not appropriate to think of settlements where such activities are prominent as central places supplying their services to a particular hinterland; rather they are specialized centres with markets spreading very much more widely than central place principles alone would allow. The location pattern of services may accordingly be divided into general service centres and specialized service centres, whose common purpose is to export goods and services and to import goods and services from other areas.

It seems reasonable to assume, however, that the notions of threshold and range which are used to explain variations in the spatial incidence of different types of consumer service will also be relevant for most other types of service activity. All must be able to service the needs of a minimum number of clients; all must be accessible when needed by users; and all must at some time require contact with sources of factor inputs.

In common with consumer services, some producer services are more specialized than others and are, therefore, subject to the same hierarchical influences on locational frequency. Computer consultants, share brokers, investment analysts, or public sector services such as universities and teaching hospitals require expensive investment in plant or equipment to provide their service; extensive market areas are, therefore, essential, for there are relatively few of them, and much of their output is 'exported' well away from their immediate locality. There will be relatively few urban areas able to support these high-order producer services, while lower-order private business and not-for-profit services which are either less specialized or require smaller markets will be established at a greater number of locations incorporating the small number of centres near the top of the settlement hierarchy as well as those lower down. In some cases the density of clients, for example, may be relatively unimportant; the distribution of universities does not reflect the distribution of population or of the eligible school-leavers.

This assumes, of course, that all services which are attempting to enter the market are guided principally by the desire to obtain a central location. The notion of centrality (for an interesting discussion of the wider significance of centrality see Bird, 1977) as a key variable in the location of service industry is not to be underestimated, but there are other variables such as organizational structure or the benefits from agglomeration and urbanization economies which also make an important contribution. Central place theory does not effectively accommodate these.

The central place model also provides an essentially static or cross-sectional framework for understanding location patterns, although Christaller, recognizing this difficulty, does endeavour to allow for the more dynamic circumstances in the real world (see Warnes and Daniels, 1979). Such a limitation is notable in so far as the theory is being applied to the behaviour of economic activities which are part of the most volatile sector in both developed and developing economies. This has been especially true during the last twenty years when the advent of various technological capabilities has modified or provided the capability to change the relationship between suppliers and consumers or between specialist advisers and business clients (see Chapter 10).

On the credit side central place theory provides an extremely useful starting-point for the interpretation of service activity location, especially for consumer services. Because it also introduces the concept of hierarchical distribution of service functions, stepped in the Christaller model but continuous according to Lösch (1954), the theory also has an applied role; for example, in the planning of new shopping centres in relation to an existing pattern of well-established shopping areas or as part of the master plan for a new town. Given the size of the new town, the distribution of population within it and the characteristics of its transport system, central place principles can be used to determine the number of shopping centres required, how they should be spaced, where they should be located, how many functions should be provided for in each centre and how the new town centre should fit into the regional hierarchy of town centres.

GENERAL INTERACTION MODELS

In central place theory it is assumed that market areas will be of equal size for any service firm with a specified threshold and range; consumers will travel from all directions with equal probability. General interaction theory also utilizes this concept of attraction but in a more flexible and less rigorous way than the central place model. The theory is founded on a group of mathematical equations which express the gravitational pull between opposing poles using an analogy with Newton's law of physics. Reilly (1931) recognized that retail location could be improved if the ability of competing centres to influence shopping movements could be estimated. The 'pulling power' of competing centres was expressed as a function of their respective size (measured in demographic terms) and the distance between them: 'two cities attract trade from an intermediate town in the vicinity of the breaking point approximately in direct proportion to the population of the two cities and in inverse proportion to the squares of the distance from these two cities to the intermediate town' (Reilly, 1931, quoted in Davies, 1976, 32). Since this basic formulation, the gravity model, as it is generally known, has been much refined to incorporate the break-points for a network of competing centres, to allow for the effects of variables other than population (such as income or measures of the attraction of a centre), or to allow the distance decay effect explicit in the gravity analogue to have values other than the inverse square. Negative exponential curves have been found especially useful in this regard. Such improvements have greatly extended the value of the general interaction model for application to intra-urban shopping centre location. A typical root formula (Davies, 1976, 33) takes the form:

$$S_{ij} = K_i E_i A_j F(d_{ij})$$

where S_{ij} = expenditure in a centre j by consumers in an area i;
E_i = expenditures available in area i;
A_j = a measure of shopping attractiveness at centre j;
S_j = retail sales generated at centre j;
$F(d_{ij})$ = a measure of travel deterrence from i to j;
K_i = a constant of proportionality which may also be interpreted as a competition term or balancing factor.

Although helpful for the purpose of locating shopping centres, these and similar models are not especially effective at predicting their catchment areas (trade areas) and, by implication, those of individual service functions within the centres. They are also of limited value for the prediction of centre growth potential, and this is also a problem which works its way through to the level of individual services and their locational decisions.

This has given rise to the development of models directed specifically at these two problems: the so-called break-point model and the growth potential model. The former is based on a reformulation of the gravity model and is designed to identify the exact point between competing centres where consumers will choose to travel to one centre rather than the other. Typically it takes the form:

$$D_b = \frac{d_{ab}}{1 + \sqrt{(P_a/P_b)}}$$

where P_a, P_b = size of centres a and b;
D_b = break-point of trade to centre b;
d_{ab} = distance between a and b.

The point which is identified is assumed to apply to all services located in each centre, a clearly unrealistic assumption given that each service has its own range and threshold characteristics. It is also unlikely that the demand surface will be the same throughout the area inside the break-point but there is no indication as to the profile of the distance decay curve. The break-point model is therefore too rigid, a difficulty which has been partially overcome in the probability model suggested by Huff (1963). This model still uses gravity model principles, but assuming that a consumer has several centres to choose from, it specifies the probabilities of choosing each of the centres and this can be mapped to produce probability surfaces for consumers choosing to shop in each centre (see Berry, 1967).

The best-known growth potential model is that produced by Lakshmanan and Hansen (1965) for the prediction of actual sales volumes in a number of major shopping centres in Baltimore (Berry, 1965, also developed a model for Chicago retail centres; see also, Lathrop and Hamburg, 1965; Hill, 1965; Lowry, 1964). Subsequently derivatives of this model have been extensively used to assess the likely impact of adding to existing shopping centres or, more important, the consequences of inserting a completely new centre into an existing system of centres (see, for example, Clarke and Bolwell, 1968; Lewis and Traill, 1968; Batty and Saether, 1972). An early example in Britain is provided by the study undertaken for the proposal to construct a new regional shopping centre at Haydock, a location mid-way between Manchester and Liverpool (University of Manchester, 1966). By incorporating the consequences of introducing a regional centre of different sizes, estimates were produced for the likely impact on 1971 sales volumes (1961=100) at a number of existing and adjacent shopping areas (mainly in local towns as well as Manchester and Liverpool). In making these estimates it was assumed that consumer services, in particular, would locate in the new regional shopping centre, a decision which would undoubtedly be influenced by the estimates generated by the model.

It is not necessary here to consider the detailed operational problems presented by general interaction models. More important is the observation that, in common with the central place models, they provide a framework within which the general location behaviour of services can be understood and predicted. The central place models require more rigorous assumptions than the general interaction models which, for example, do not rely on the need for consumers to travel from a uniform distance and all directions to each central place at each level in the hierarchy. Both approaches work best at the regional or national scale and are more effective at accounting for or predicting the number and location of (service) centres rather than the location of individual service activities, especially consumer services, either between centres or within them. It is also fair to suggest that both types of model, however refined they have been, are essentially static representations of the location patterns created by service and other activities. But service industries are part of a dynamic set of activities which contribute to the constant process of change and adjustment in their own distributions or of activities associated with them. Urban and regional changes emanate from the dynamic characteristics of services and estimations of the number of service functions in a city at some future time, whether using normative or recursive (stepwise) models, hardly approach the nuances of locational choice and behaviour by service industry decision-makers in a world where they must constantly adjust to changing economic, social and political circumstances.

BID-RENT THEORY

Therefore, it is necessary to look elsewhere for theories which attempt to explain the location of individual service activities or their behaviour within urban areas. Probably the most useful insight is provided by Alonso (1960), who proposed a model for the location of activities in which they can offset declining revenue and higher operating costs (including transport costs) by lower site rents at locations increasingly distant from some central point. The amount of rent each activity will be prepared to expend for any site will vary according to the sum it would have to pay to maintain the same profit level, that is would compensate for falling revenue and higher costs (usually referred to as the bid rent). The equilibrium location for an activity occurs where the bid rent matches the rent determined by the market at a site.

Bid-rent functions are different for each type of service activity (or indeed for non-service activities) and some will have much steeper curves than others (Figure 5.1). As a result, some services will be able to obtain more central locations than others because their steeper bid-rent curves will be associated with a willingness to pay more for a central (and thefore more accessible) site than services with 'flatter' bid-rent curves. The rent which each service activity is prepared to pay reflects the utility it will receive from occupying the site; the higher the utility, the higher the rent it will be prepared to pay. Since it is assumed that services are always competing for space, the bidding for sites ensures that the highest and best use occupies each site. In common with most models. Alonso must make some simplifying assumptions the most important being a perfect market for land in which all potential users are fully informed about all the available sites and the orientation of the urban area towards a single, central nucleus. In this hypothetical city accessibility declines uniformly in all

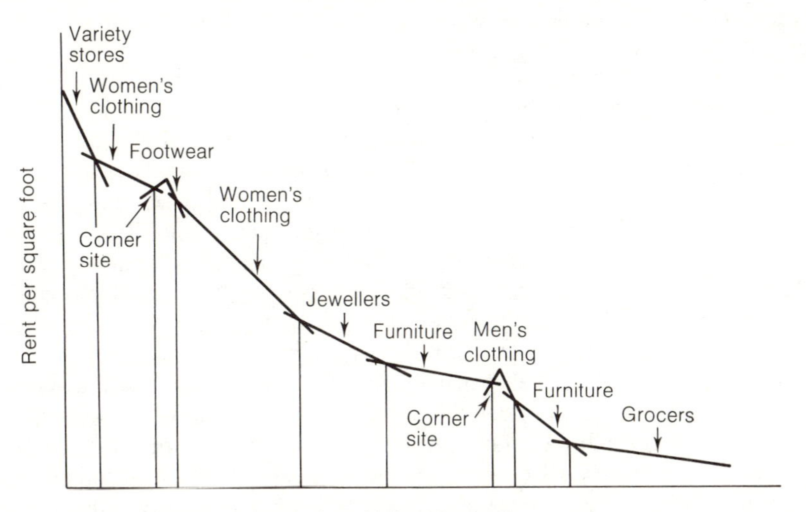

Figure 5.1 Hypothetical rent gradient for retail functions in a large city centre

directions, and the utility of firms and households depends upon accessibility. Alonso does not specify which activities should be associated with different bid-rent curves; it is left to the observer to assess which service activities can be associated with which curves (see, for example, Garner 1967; Scott, 1970). It is also important to realize that accessibility does not take just one form; it can be subdivided into general accessibility (access of the site to city- or region-wide demand), local accessibility (access to a subset of the urban area or region) and special accessibility (the need for access to other services which provide inputs or even to competitors if comparison is a significant factor in consumer choice).

It should be apparent that this model has a more universal application to the location of both consumer and producer service activities. Reference has already been made (Chapter 2) to the large and specialized markets required by many producer activities and, accordingly, they are generally found to occupy the most central sites in urban areas, mainly in the office buildings from which many producer service activities operate. Consumer services as a whole may share the most central locations but reveal a hierarchical distribution along the lines predicated by Alonso (Figure 5.1). Hence department and multiple stores which have high overheads in staff, stock and site rents compete for the locations associated with the highest pedestrian flows (these need not be the most accessible in terms of the work journeys important for sustaining producer services) and, therefore, customer potential. Adjacent to them will be the retailers offering comparison goods such as footwear, fashion and electrical goods, and so on, down the retail hierarchy. Many of these will be located at increasing distances from the peak land-value intersection, including wholesaling and transport services – rail, bus and tram termini, although not the stops or stations to and across the most accessible locations, for example, the underground system in London and the subways in Washington, DC, New York, or Moscow.

In common with some of the other models already considered, there are difficulties when trying to apply the Alonso model to the real-world location patterns of service industries. The model relies on accessibility as one of the key variables distinguishing between the importance of the most central location to each firm or activity. But the desire for general accessibility is not uniform among all firms; some may consider special accessibility, that is the proximity to some of the sources of their factor inputs, more important, while others may attach no value to any kind of accessibility. Variations in the production functions of establishments within the same category of activity may be at least as influential in location decision-making as accessibility.

The Alonso model also relies heavily on the assumption that a free market exists in which there is unhindered bidding for available sites. In practice, the land and property market is far from free; planning regulations, the requirement to have permission or a permit to undertake new or replacement land use or development, locational inertia, subdivision and density controls, the delaying effects created by long leases, or patterns of land ownership all serve to complicate the market for land and ultimately, therefore, the locational opportunities open to service firms. It is also assumed that most service firms, especially in some branches of retailing and in professional services, are motivated in their locational choice by agglomeration economies, while in reality some will be more interested in 'cornering' a specific market area (defined in a spatial sense) and, therefore, concerned to be as 'remote' as possible from their competitors; the suburbanization of some forms of retailing and banking services is a symptom of this. It should also be evident that the centre of a city is, in many cases, no longer the point of maximum accessibility; indeed as a result of negative externalities, such as traffic congestion or overcrowding and delay on public transport services, suburban locations may be relatively more accessible. However, it is interesting to observe that the rent gradient in cities is still very much dominated by the city centre and this also suggests that the location of service activities (or the rents they are prepared to incur in the centre) is governed by factors other than accessibility alone.

EQUILIBRIUM MODELS OF OFFICE LOCATION

Some reference has already been made to the value attached to agglomeration, and in particular the contact opportunities which it provides, by service sector firms. Activities occupying office space are thought to benefit in a major way from agglomeration, and this has encouraged a number of recent attempts to model the equilibrium locations for offices (O'Hara, 1977; Clapp, 1980, 1983; Tauchen and Witte, 1983, 1984). The significance of these models is that they use empirical evidence derived from studies of inter-firm contact patterns and agglomeration economies as features which distinguish the location behaviour of offices from manufacturing activities. But Tauchen and Witte (1984) suggest that agglomeration benefits are assumed to accrue to office firms by the simple expedient of being present in the CBD, for example; no explicit account is taken of the distribution of firms within an agglomeration or of their contact behaviour.

Attempts have, therefore, been made to devise an intra-urban model of office firm location in which spatial distribution and contact patterns are determined simultaneously (Tauchen and Witte, 1983). It is assumed that the contact pattern is determined endogenously by firms making profit-maximizing decisions which will, in turn, have an impact on land values. These requirements (which build upon an initial model developed by O'Hara, 1977) are resolved within a square CBD with a rectilinear road network within which it is assumed that all firms are located and only have contacts with other CBD firms. In addition, the travel costs generated by travel to meetings do not create congestion and the cost of providing office space (the facility-cost curve) is assumed to be increasing over time. In these circumstances the equilibrium allocation of office firm to locations allows each to choose the profit-maximizing number of contacts, each firm has zero economic profit and the office rent at each location is the marginal cost of providing the floorspace. In its abstract form the model provides an exact description of equilibrium but 'does not provide a qualitative measure of the importance of the agglomeration economies in determining the properties of the equilibrium allocation' (Tauchen and Witte, 1983, 1315). This problem is overcome by adopting explicit forms for the functions describing contact benefit and facility cost which also then allows the effects of changes in technology, prices, or public policy to be considered.

The basic model is solved by using data for US cities in the 1–2.5 million range. The resulting curves for one segment (of four) in the square CBD are interesting (Figure 5.2). Since firms are assumed to have contacts throughout the CBD (not just at the centre of it), the contact-expense curve rises as a firm moves further from the centre, partly because of the higher cost of contact

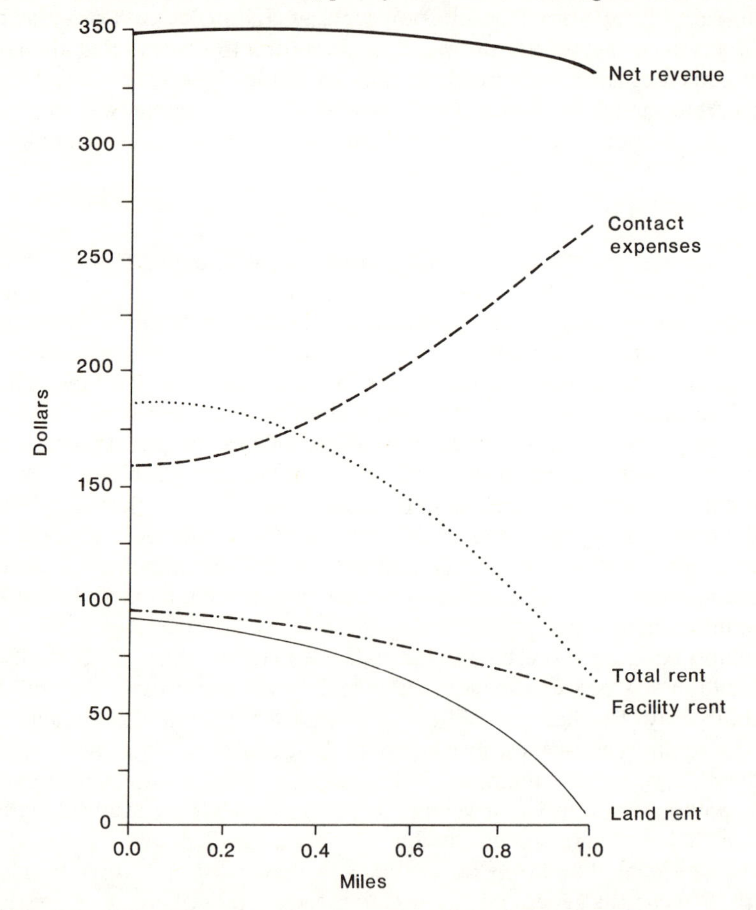

Figure 5.2 Curves for net revenue, contact expenses, total rent and land rent per firm per day generated by the Tauchen–Witte model of office location
Source: Tauchen and Witte, 1983, figure 3, 1319.

with firms at the centre and because of the more limited contact opportunities especially with firms on the opposite side of the CBD. The steady increase in cost disadvantage with distance from the centre of the CBD means that firms will be prepared to pay lower rents as distance increases, so that the resulting curves for total rent, facility rent and land rent are concave rather than convex in the way typical of residential or industrial location models. If the contact-benefit function is allowed to vary by changing the parameters, Tauchen and Witte (ibid.) demonstrate the sensitivity of the equilibrium allocation of firms which may be expressed in an overall increase in the number of firms or by redistribution within the CBD. Thus the medium- and long-range implications of office technology for agglomeration economies should not be underestimated while an increase in contact cost is shown to reduce the relative advantage of locations at the centre of the CBD, quite the opposite of the expected effect. Varying the physical size of the CBD is also instructive: a decrease in size leads to a reduction both in the number of firms and land rents (the latter might have been expected to increase as land in the CBD is made a scarcer resource). This suggests that any attempt to limit the extent of CBDs through land-use controls, for example, has detrimental effects on economies of scale and may discourage patterns of firm density conducive to attractive agglomeration economies and, therefore, the prospect of growth in CBD office activities and employment.

AN INFORMATION DIFFUSION APPROACH

In some quarters central place theory and its derivatives continue to be viewed as excessively rigid. As an alternative it has been suggested that we can better understand the growth and development of a system of cities, and by implication the location of service industries, as an 'accumulation of decisions directly and indirectly affecting the location and size of job-providing activities in the private and public sectors' (Pred, 1977, 19). Such decisions are essentially the product of the way that information is sought, received, interpreted and used by individual entrepreneurs, corporate decision-makers, or central government agencies. Since information and knowledge are the key ingredients for the operation of many services, especially producer activities, their location can best be interpreted within the context of the demand for information, the way it circulates and who exchanges it. The key point about much of the information available to decision-makers (especially specialized information) is that its availability is 'spatially biased'. The diffusion of this information or of innovations between the nodes in a system of cities or between regions then becomes critical for the location of new economic activity, including services, or expansion/diversification of those already in place (Hagerstrand, 1967; Brown, 1968). The majority of decision-making units or individuals are most likely to seek information from existing contacts near to them, or from other sources proximal to those existing sources of information. Thus

> because of the means by which different forms of specialized information circulate through contact networks, the probability of a particular bundle of information being known or acquired varies from place to place at any given time. And, conversely, any actor possessing specialized information at a given location is more likely to have sought or unintentionally obtained it from some contacts or places rather than others. (Pred, 1977, 20)

Specialized information circulates between (and is exclusive to) private and public sources and may also be obtained through visual observation (for a full description see ibid., 20–21). It is used for explicit and implicit locational decisions whether by business services, retail outlets, government departments, or industrialists trying to identify a suitable location for a new or branch plant. Explicit decisions are those made to establish or expand at a particular location; the availability of specialized information may influence the actions of individual entrepreneurs or, if the establishments and related jobs are innovatory, spatial biases in the circulation and availability of specialized information relating to growth-inducing innovations will probably be very important. Implicit locational decisions relate to the purchasing of services or goods whether on a day to day basis or through the award of a contract or subcontract. Many such decisions are routine and invariably involve the utilization of established sources or locations. This reinforces the prospects for growth at one location rather than another because it minimizes the need to identify alternative suppliers and indirectly contributes to spatial bias in information use. There are also non-routine implicit locational decisions which, by definition, require the identification and use of specialized information which has not been needed before. In choosing sources for this information decision-makers will look to places similar to those with which they are already familiar or most like those with which they have already had experience. Therefore, both types of locational decision enhance the prospect for spatial biases in the circulation and availability of specialized information.

The significance of this approach is the possibility that the location patterns exhibited by both users and sources of services are a conscious reflection of the accessibility to, and cost of, specialized information. In a study of the intra-regional diffusion of business services in the New York region Bearse (1978) shows how their geographic spread depends upon accessibility to information and special resources measured by variables such as nearness to the centre, that is New York City, the availability of other intermediate services such as finance and banking, the size of the professional labourforce, and the availability of external linkages to national and regional networks. External linkages are most apparent among establishments employing large numbers of workers (also noted by Pred, 1977, at the inter-regional scale), so that the 'location of large firms is *also* a significant influence on the diffusion of business services' (Bearse, 1978, 577).

Another way to examine this is to consider whether occupational structure varies with city size (Burns and Healy, 1978). There are four reasons why such variations might occur. First, following one of the principal tenets of central place theory, the service industry mix of a settlement changes as successive market thresholds are crossed. We already know that there are differences in occupational structure between service industries (Chapter 3), so it is reasonable to expect variations if a city's service industry mix changes with increases or decreases in city size. Secondly, increasing settlement size is associated with growing specialization of production, even if the output is the same; a large regional hospital in a metropolitan area (which may also have a medical school attached to the local university) will probably employ more administrators and fewer medical personnel to produce the same output as a smaller regional hospital with no teaching role. Thirdly, urbanization economies/diseconomies with increasing settlement size affect the demand for services; public transport may be more prominent because of the effects of congestion on private vehicle trips in larger cities, or office-cleaning services will be much in demand because of the concentration of office space in major cities. Finally, Burns and Healy (ibid.) note that real and money incomes vary with city size and, therefore, demand for services will also vary and lead to the creation of a different occupational mix. They show that service employment in forty-one urban-oriented occupations in 185 standard metropolitan statistical areas (SMSAs) (most of the occupations used are white-collar professional, managerial and related) is not distributed strictly as a linear function of metropolitan size and specialization; occupational mix is clearly shown to occur among metroplitan areas. Indeed for the highest-order occupations, employment grows faster than city size (twenty-four out of the forty-one occupation groups had elasticities greater than 1.0). The lowest elasticities were for the less skilled occupations.

The process of occupational specialization which is a by-product of the influence of economic rather than demographic influences on the location of services may also be encouraged by the effects of agglomeration economies. Some services can be provided at lower cost than their customers can produce them 'in-house' but in order that the latter, which may be corporate headquarters, can take advantage of the skills available they will need to locate near to the source. This offers the prospect of a larger market which may stimulate the development of other specialist services, thus further increasing the attractiveness of the emerging service industry complex for other headquarters. Other consumer services such as restaurants, retail stores, hotels and boutiques, or public services such as transport and medical facilities, will also be stimulated in this way and will reinforce the comparative advantage of the location at which agglomeration is taking place.

This example lends support to Pred's thesis: services requiring skilled or knowledge-intensive labour such as business services, ranging from management consultants, share brokers and advertising agencies to banks, all reveal a propensity to congregate to a disproportionate degree in certain urban, usually metropolitan, areas. Thus the rigid hierarchical organization of interdependencies between urban areas and the number of service functions which they provide, according to central place theory, is replaced by an urban system in which the diffusion of information need not conform to hierarchical principles. As a consequence, there is growing concentration of information-oriented service industries in a limited number of urban areas, but contrary to the tenets of central place theory, these locations need not necessarily all be at the top of the hierarchy. Interdependencies between settlements may exist horizontally as well as vertically. The headquarters of a manufacturing or insurance company may be used to control a widely spread network of plants or branch offices in larger and smaller urban areas as well as those of similar size, provided that the specialized information which they require to be able to function is available.

Some other evidence in support of the information diffusion model of service industry location has been provided by Stanback (1979); SMSAs which had significant business service functions in 1960 experienced the most rapid growth of employment in the same services during the decade 1960–70. Some 115 SMSAs had an above-average share of employment in business services in 1960 and 70 per cent subsequently achieved higher than average shares of job increases, accounted for by business services, during the following decade. The initial agglomerations have thus been reinforced through specialization and diversification, making some SMSAs more attractive for the location of corporate head offices and similar administrative functions. Reinforcement of established business service centres through a process of cumulative growth is also confirmed by the fact that those SMSAs classed as primarily manufacturing centres in 1960 show very limited progress in shifting their economic base in a way which reflects the rapid expansion of business services in general. In the long term this further reduces the ability of such cities to compete for a share of producer service growth, thus increasing the possibility of an ever-widening gap between 'service' and 'non-service' SMSAs.

Concentration of service activities in a relatively small number of metropolitan areas is also evident in Canada (Davis and Hutton, 1981). After making some assumptions about the basic/non-basic role of individual service activities (hotels, motels and other lodging are assumed, for example, to be basic services) for each metropolitan area with a location quotient greater than 1.00, they have calculated the proportion of the employment in that activity which is export orientated. The following equation has been used:

$$E_i = R_i - \frac{N_i}{N} . R$$

where R_i = employment in service activity i in the metropolitan area
N_i = employment in service activity i in the nation;
R = total employment in the metropolitan area;
N = total employment in the nation.

It is assumed that any surplus employment is export orientated, and although this grossly simplifies the situation, it does help to isolate a number of metropolitan areas with a range of service exports. In common with the SMSAs in the USA only a few Canadian metropolitan areas (CMAs) have more than one principal service export. These are Toronto (business services, insurance, real estate and wholesale trade), Vancouver (wholesale trade, transportation and business services), Montreal (transportation, business services and finance), Calgary (business services and wholesale trade) and Winnipeg (transportation and wholesale trade). The relationship between the size of a metropolitan area and number of principal service exports is again evident with most of the smaller cities not included above having one service export, usually provincial government or education or health services (ibid. table 3; see also Polese and Stafford, 1982).

A BEHAVIOURAL PERSPECTIVE

The way in which a service firm resolves the influence of transport, labour and agglomeration with respect to locational decision-making may also be affected by its organizational and operational goals. Profit maximization has long been considered to mould the locational behaviour of firms and the minimum transport cost solution is seen to be particularly attractive. But the profit-maximization principle, and variants of it such as the least-cost location princple, came under fire when two alternatives were mooted: sales maximization and satisficing (Simon, 1959; Cyert and March, 1963; Baumol, 1965).

The profit-maximization model assumes uniform costs, but if these are assumed to vary, then sales maximization may produce a different choice of location. For some of the consumer services such as retailing or public transport market orientation is clearly important and turnover as measured by sales revenue contributes to profits. Market orientation is a significant component of agglomeration economies for service industries and this also gives further support to the sales-maximization theory. In addition, targets for sales which may have been included in the calculations relating to the benefits attached to alternative sites can subsequently be achieved with greater flexibility, by taking into account the sales strategies of competitors or responding to changing market conditions, for example, than would be the case with profits. The future is unpredictable and an optimal location for profit maximization may subsequently be overtaken by events and factors outside the control of the firm; but inertia and cost may prevent it from responding by trying to move to the new least-cost (profit-maximizing) location. A great deal of time and effort may, therefore, be wasted and satisficer behaviour (Simon, 1959) could well offer a more realistic interpretation of location behaviour. Satisficing in this context is 'the idea that decision-makers do the best they can on the basis of such information as they acquire' (Lloyd and Dicken, 1977, 320).

By assuming that firms are willing to adopt satisficing rather than optimal solutions to the location choice problem, Simon (1959) recognizes that the location decision while subject to some objective inputs and assessments, also involves many subjective judgements. These arise, for example, from the environmental attributes of possible sites, from the quality of amenities in the vicinity, the prospect of good labour relations at one location rather than another, or the role of factors personal to the individuals making the locational decision. Objective analyses may indicate several equally adequate locations, thus leaving the final decision to subjective assessments of, for example, the relative environmental attributes of the alternatives. The satisficing model also incorporates the likelihood that most firms or individual entrepreneurs faced with a location decision will only have a limited amount of time to devote to it, or indeed there may be a time limit because a lease may be due to expire, or there may be a desire to minimize the amount of effort and time devoted to the task. This reduces the likelihood of an optimal solution being achieved.

The disadvantage of the satisficing approach, however, is that in contrast to that based on optimization it is clearly not possible to specify the precise location which will be chosen in any given set of circumstances.

A fundamental choice can be made, then, between the application of normative principles to service industry location, as derived from the work of the neo-classical economists, or the more 'flexible' principles embodied in behaviouralist models. Edwards (1983) advocates the use of the latter for several reasons: first, the influence of the environment as perceived by the decision-maker is recognized; and secondly, account is taken of the internal environment of the firm (its efficiency or personnel, for example). Therefore, thirdly, the use of the behavioural model embraces organizational choice of location as distinct from individual location choice. This means that search activity is incorporated since access to and use of information about locational alternatives is crucial to the final location decision. A fourth reason is that there is no attempt to arrive at one optimal solution, merely to identify the most acceptable location from a number of alternatives. In other words, this approach is more closely related to the circumstances influencing the real-world patterns of location; more particularly, it is better suited to deal with the significant influence of inter- and intra-firm communications on the location of service activities (Pye, 1977, 1979; Gad, 1979; Goddard and Morris, 1976).

An existing firm may, therefore, engage in a sequence of actions (Figure 5.3) which may ultimately lead to a location choice that generates greater satisfaction. The decision process model suggested by Edwards (1983) takes these factors into consideration while, at the same time, recognizing the intangible quality of many service sector establishment outputs (as well as inputs) and the diversity of organizational types within the service sector. Pressure on the location of a service firm may come from internal (changes in marketing, organizational structure, or policy) or external sources (cost of inputs such as labour and competition) (see ibid. for details). The location decision process is divided into two stages: the decision to relocate and, subsequently, the choice of new premises. In the case of new firms starting up for the first time the second component of the process (the choice of new premises) can be utilized, thus ensuring its universal value for application to the service sector location problem.

The first stage is brought about by 'a mismatch of the requirements of an establishment and its relationship with the environment in order to fulfil these requirements' (ibid., 1333). It is then necessary for the firm to assess the best way to respond to this mismatch, a process which will be affected by the attitudes and perceptions of the personnel involved, the ways in which they interpret the policies and goals of the organization and how these are translated into space needs, type of premises, or the limitations of the existing location. All of these factors must ultimately be resolved into a set of locational requirements which occupy a central positon in the decision process at the interface between the decision to relocate and the choice of new premises. If the locational requirements dictate a location change, or for a new service firm identification of their first premises, a sequence of search, evaluation, choice

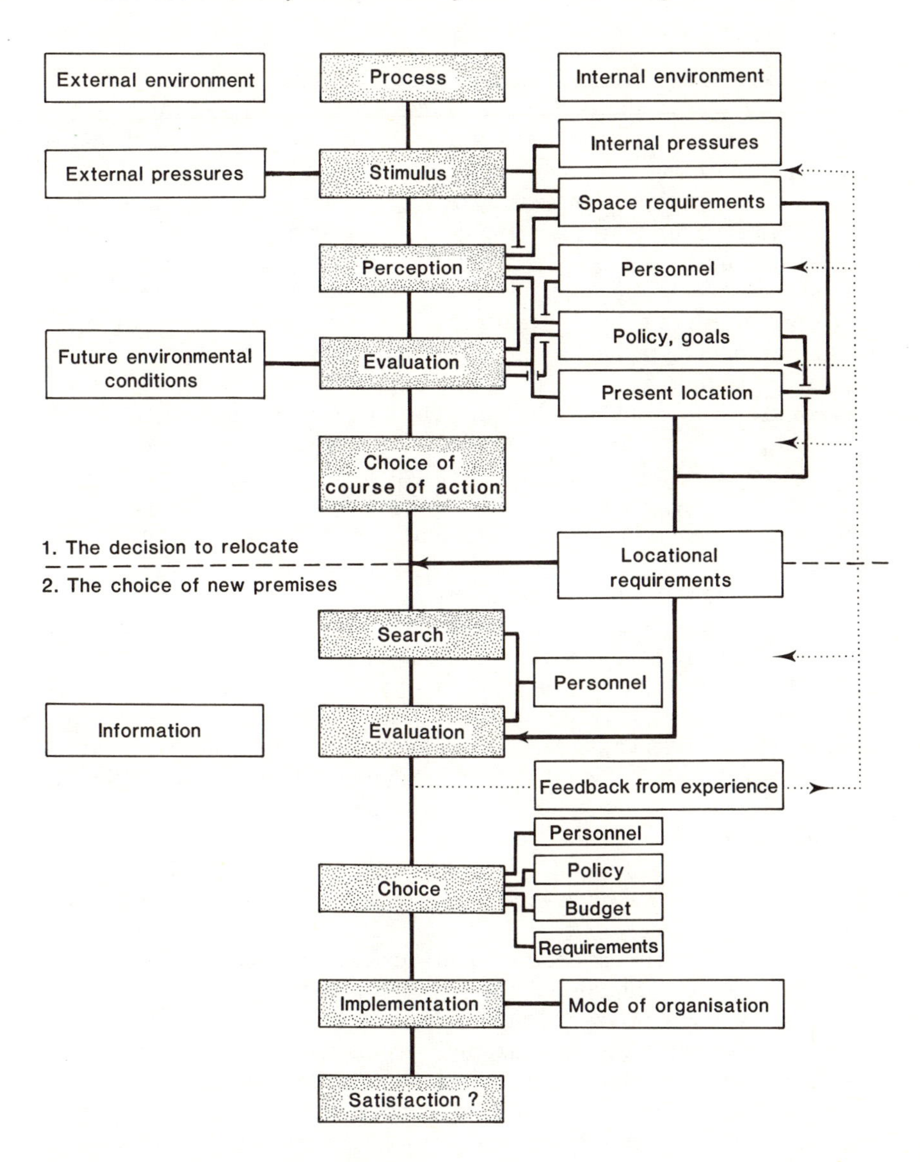

Figure 5.3 A decision process model for office firms
Source: Edwards, 1983, figure 2, 1334.

and implementation begins. The way in which this proceeds and the outcome is dependent once again upon the personnel involved (Figure 5.3), policy constraints, or budgetary considerations. The quality and quantity of the information available about alternative locations and premises (from the external environment) will also influence the course of the location decision-making process.

ORGANIZATIONAL CHANGE

In common with events in the manufacturing sector (see, for example, Hymer, 1972; Taylor and Thrift, 1983) there has been a historical shift from small, single-location service enterprises to large, multi-establishment service organizations. The service sector is, however, still numerically dominated by small enterprises but the location of these is increasingly dependent upon the location of their larger counterparts or competitors with whom many smaller services have strong business links. Private sector producer services most clearly demonstrate this change; banking, for example, was very much a small-scale, local operation during the eighteenth and nineteenth centuries and the growth prospects of individual banks were restricted if they relied purely on increased deposits or lending within their local markets. The range of services which they could provide would also be limited by their size and the deposits which they held and, therefore, it was inevitable that mergers and takeovers would be necessary both to gain access to larger markets and to ensure survival. One of the consequences was that not only would decisions need to be made about which establishments in an enlarged organization should be expanded, closed down, or relocated, but also about how best to promote the enlarged bank's activities, its policies on lending, and so on.

Hence in such circumstances a requirement arises for a set of administrative and specialist services connected, on the one hand, with the control of dependent establishments at several locations (branches, regional or divisional establishments and manufacturing plants), and on the other, with fulfilling a need for independent advice necessary for the effective operation of a larger organization. The administrative functions are usually grouped together at a head office which needs to locate in a way that permits adequate access to the establishments under its control, but also permits the decision-makers located there to have proper access to the information and knowledge which they require before making a wide variety of decisions with both immediate and longer-term implications for their organization. Hence a centrally located administrative department became a prerequisite as, for example, the number of competing banks diminished but their size (number of branches, services provided, assets and turnover) increased. In some cases the number of branches is so large and spatially diverse that an intermediate level of administration, the regional office or headquarters, needs to be introduced. These are located mainly with reference to accessibility to branches under their control but also with an eye on ease of communication with national headquarters. This would tend to favour the larger cities which have the best communications. Domestic banking in Britain is now dominated by just five banks which operate extensive branch networks in England and Wales and have subsidiaries in Scotland with their own branch networks. In 1974 these five clearing banks had some 270,000 employees and assets in excess of £45,000 million.

Service firms dealing with a single business are usually managed with a functional form of organization (Channon, 1978). In dominant business firms a divisional form of organization is used whereby the dominant division itself receives substantial operational involvement and is usually organized on a functional basis. In such firms the central office is large in order to co-ordinate the activities in the major divisions. A third type of firm, the related business firm, also has a large central office but this is not responsible for operational management; rather it ensures adequate co-ordination of interdependent activities between divisions. Finally, unrelated business firms only have small financially orientated central offices, leaving relatively high levels of autonomy at divisional level.

Changes of this kind have led to concentration in the location of certain service industry functions whether in banking, insurance, retailing, or professional services. In the public sector the evolution of large and increasingly complex bureaucratic structures to handle national government affairs has invariably led to the centralized location of a large proportion of the civil servants involved. To some extent, of course, this reflects the close links between the activities of these bureaucrats and their political masters, the elected representatives in the US Congress or in Parliament, for example. But as government departments have become larger, or have been required to provide services at the local scale, intermediate levels of control have also been introduced and these have been geographically more widely dispersed.

More recently there has been a trend among some service industries towards diversificaion (see ibid.). This strategic change in the structure of service organizations was brought about during the late 1960s by increasingly negative government attitudes, in Britain, towards concentration of economic activities, a continuing growth in competitive pressures between service organizations and, in some cases, a need to develop international services in order to be able to compete with, for example, overseas commercial banks which are locating European headquarters or regional offices in London, Paris, or Düsseldorf.

SUMMARY

After elaborating the limitations of the basic central place model in this chapter, some models which utilize concepts incorporated within the theory have been considered. The family of general interaction models requires less rigorous assumptions about consumer travel behaviour, for example, and seem to work most effectively at the national or regional scale. They also work best for predicting the growth or location of service centres rather than the location of individual service activities. The roots of these models also mean that consumer services are covered more effectively than producer services. Bid-rent theory offers an approach which can be applied to individual types of service activity but it also makes assumptions, such as a perfect market for land or total knowledge of all the alternative locations available to decision-makers at one point in time, which are difficult to reconcile with the actual circumstances. All these models – central place, general interaction, or bid rent – rely upon assumptions which overlook, for example, the interference or regulation imposed as a result of planning or zoning controls, locational inertia, land-ownership, tax incentives, or variations in access to (or circulation of) information.

The static and somewhat rigid central place theory and its selective effectiveness as a source of explanation for service industry location has encouraged other approaches. Most important of these is the idea that the location of economic activity is determined by the availability of information, the way it is obtained, interpreted and used by entrepreneurs, corporate decision-makers and central government agencies. Not only is this especially useful in the context of producer service location, but it also helps us to understand why some regions or metropolitan areas attract a disproportionate share of service activities. The studies which have been undertaken show that information diffusion takes place most rapidly within systems which have some common characteristics. Each constituent unit of the system has well-developed linkages to the wider world or to the networks through which knowledge is transmitted. The units in the system reveal a high degree of internal integration, yet are also sufficiently diverse in occupational or social structure to be able to respond to changing demands within the system. Finally, this flexibility confers an openness on the system so innovations have room to enter or to grow. Conversely, each unit in the system is not so specialized (by firm or service industry) that entry of new activities is impeded.

The importance of information and the factors which determine how it is used is demonstrated with a behavioural model of office location decision-making. In this model an attempt is made to incorporate the effects of the firm's environment (both internal and external) on how it arrives at an initial decision about whether or not to relocate, and subsequently the way in which this is followed through to the choice of a new location. In this type of model, as well as those associated with theories of organizational change which are discussed briefly towards the end of this chapter, the largely static, cross-sectional features of the central place model are replaced by more open, dynamic models which are more adequate for aiding the interpretation of the location of individual types of services. They are, however, more difficult to use to explain the location of larger groups of related service industries.

CHAPTER 6

Equity and access: public and private consumer services

INTRODUCTION

Theoretical assumptions about the location of service activities have to some degree been confirmed by the empirical evidence. But it is also apparent that both regional and intra-urban distributions of services, whether examined on an aggregated or a disaggregated basis, reveal substantial deviations from the distributions predicated by central place or information diffusion ideas. Agglomeration, localization, or urbanization economies exert some influence on the location of different kinds of service; the result is spatial imbalance in their distribution. While the reason for this has in itself attracted considerable interest among geographers, it leaves room for many other questions about, for example, the optimum distribution of services by comparison with the patterns which currently exist, the most efficient pattern of service location (this need not be the same thing as the optimum), or the relationship between need and actual access to certain essential services.

The latter is linked with notions of equity (Harvey, 1973). As Pateman (1981) illustrates, however, the concept of equity is far from easy to explain even though it is generally accepted that it essentially derives from the ideas of social justice propounded by Harvey (1973) and others. Viewed in its spatial or geographical context, social justice can be equated with territorial justice which expresses a concern to achieve an allocation of a resource, in this case services, on the basis of need, merit and contribution to the common good (see also McAllister, 1976). It will soon become apparent that the spatial distribution of service activities, either covertly or by design, is variable relative to demand or need. By examining existing patterns and relationships it may be possible to identify spatial inequities, although this does not mean that the best solution should involve spatial adjustments; redistribution through changes in tax systems, for example, may also be necessary (Smith, 1977). Other issues arising from the location of services include ways of effectively matching the supply and demand for services; how to cope with shifts in the origins or structure of demand; how to minimize the conflict between certain services and other users of space in cities and regions; and many other questions which logically present themselves once the basic empirical attributes of service industry location have been charted.

The objective of this chapter, therefore, is to consider rather more closely some of the problems and issues connected with the provision of services, with emphasis on the spatial allocation aspects of the problem even though it is recognized that financial and politial constraints are also relevant to an understanding of the availability/non-availability of services.

THE PUBLIC–PRIVATE SERVICE DICHOTOMY

It is worth reiterating the dichotomy between public and private services (Samuelson, 1954). Private services (and goods) are divided among consumers in such a way that, if one consumer gets more, another will get less. The consumption of many public services does not have this effect, however, consumption by one individual does not reduce the quantity available for another (Peston, 1972). One reason for this is that many public services are provided out of funds collected centrally in the form of national or local taxes. A distinction should be made between 'pure' public services, that is unrestricted availability, and 'impure' public services, whose availability is to some extent governed by the same principles that apply to private services. Everybody may be eligible to benefit from impure public services but there will be spatial variations, for example, in their availability. A further distinction must be made between the provision of services: some are publicly provided, others privately.

This division is not as obvious as it may seem because the construction of a new suburban shopping mall by a property development company can be classed as a public service, while public provision of a new museum incorporates private principles as it may not be equally available to all because of charges for entry or distance from potential users whose tax payments have contributed to the cost of the project. Also implicit in the public–private distinction is that decision-making about the location of private services tends to be dispersed among many individuals who may act as they think fit within the context of national legislation. Public service location decision-making is much more likely to be subject to centralized control whether by central government departments or by local authority administrations. On the other hand, inefficient location choices may be less apparent for public services; inappropriate location choice for the suburban shopping mall is likey to be penalized by unacceptably high costs and possible closure.

IMPORTANCE OF ACCESSIBILITY

By definition, the provision of a service of any kind involves notions of accessibility and making it as easy as possible either for consumers to reach the source of supply or for suppliers to reach their markets. If there are variations in the degree to which a service is available, then the welfare function of the service is not being fully utilized. According to this view, accessibility to services comprises part of the well-being or quality of life possible for individuals or households (Smith, 1977).

But accessibility is a complex variable because of the variety of ways in which it may be defined (see, for example, Whitelegg, 1982; Knox, 1982a) as well as the extent to which it may be changed by human action or decisions. Variation in access to natural resources, for example, is governed by their incidence, so that some countries are better endowed than others through the 'accident' of the earth's geological formations. This leads to inequalities in accessibility to natural resources which can be overcome by the ability of the underendowed to purchase enough to reduce their deficit as well as the willingness of the countries with a surplus to sell. Agricultural trade at the international scale provides many examples of the latter, for example, the fluctuations in the sale of US wheat to the USSR during the 1970s, but it is increasingly evident that inequalities in the access to services, caused by variations in technological developments in computer hardware, bulk pipeline technology, or supplies of natural gas between countries, are also subject to these macroscale influences. Recent attempts by the scientific establishment in the USA to somehow restrict the access of eastern bloc researchers to published research papers containing information which may be used to advance their own scientific knowledge reflects economic, social and politico-spatial variation in the command over such resources and which may, to some extent, be consequent upon differences in wealth, development priorities and accumulated knowledge in contrasting societies.

Much the same is true of the accessibility to private and public services within communities and regions. In circumstances where a service, such as education or health, is supplied by both private and public sectors it is possible for those members of the population with a good command of financial resources to either purchase those services which are not readily available quickly enough, for instance, specialist health care for an urgent medical condition, or a 'better' education, like that often provided by non-state (public in the UK) schools. Such households are also able to select from the wide range of competing services, say, within a segment of retailing, the one which most adequately meets their needs at a particular point in time. If in the future their needs or priorities change, they are also able to switch suppliers accordingly. All this is in sharp contrast to those in the population with much more limited access to their own financial resources; these are far more exposed to the vagaries of service provision either by the private sector which, in its search for profits, may neglect to locate in areas occupied by the less wealthy (whether rural or urban) or by a public sector which is overstretched and unable to provide adequately for the needs of all (Rosenberg, 1983).

Such adequacy may include being located sufficiently close to the less wealthy, so that they require to expend only the minimum amount of their time budget on obtaining the services they need, leaving as much as possible free for other pursuits such as work or searching for work. The nearer a household is to schools, libraries, medical services, shops, or parks, the greater the opportunity for minimizing travel costs and releasing income for expenditure on the consumption of services such as leisure (sports activities, cinemas, clubs, and so on). Indeed Hagerstrand (1970) and Carlstein *et al.* (1978) argue that accessibility to services (as well as to jobs) can only be properly understood with reference to household time budgets since, for most of us, our activities follow a very regular and highly circumscribed 'path' during the typical day or week – a path which incorporates spatial and temporal components. This influences the level of effective accessibility to a service, a concept in which physical proximity does not necessarily equate with freedom to think of making a journey to obtain the service. A good example is provided by a housewife who may not have a job, but has a car readily available (the household's second car in some instances) for undertaking service-oriented trips, although she finds it difficult in practice to find time (gender constraint) to make such trips because of a time budget which involves a regular commitment to getting the family prepared for school or work, taking the children to school and collecting them during the afternoon if not old enough to make their own way, and transporting the children to sports, social and other events, or preparing lunch for a husband who comes home during the midday break (see, for example, Pred and Palm, 1978; Whitelegg, 1982; Knox, 1982a). If the effective accessibility of these individuals is limited, then that of housewives who do not have a car available during the day or who are unable to drive will be even more restricted (Koutsopolos and Schmidt, 1976). From the other side of the fence the ability to use a service will depend upon the hours when it is actually available. Stimson (1980) has illustrated this with reference to provision of general practitioner services in a southern part of Adelaide (Figure 6.1). There are considerable temporal variations in availability and the surgeries only have utility for potential users if they are both convenient in a spatial sense and available at the time required. Early morning and evening opportunities are far fewer than around the middle of the day and some surgeries only open during one of the periods identified.

As Whitelegg (1982, 54) puts it, the 'outcome of this process can be measured in terms of 'success' or 'failure' in achieving a satisfactory level of consumption of those goods and services which any given society, at any given time, recognize as essential for general welfare and well-being'. But such success or failure is not the only criterion upon which accessibility may be assessed, it also relates to specific physical connotations such as the distance separating the consumer and the supplier, the locational distribution of competing or complementary sources of a service as well as to the time required to overcome distance to a service or its limited spatial distribution. Once again, however, the ability to overcome the constraints on availability imposed by the spatial distribution of public or private services depends on a wide range of variables including value of time to individuals and their households, the income of those households, and the amount of income which is discretionary and may be expended on 'non-essential' services like leisure activities or a visit to a restaurant and the resources available to permit a search for the best among several alternatives – or whether the household has its own transport, such as a car, or is totally dependent upon public transport.

Accessibility will also be determined to some extent by the attributes of the transport network, both highway and public transit, in so far as the costs attached to using it (for example, tolls, congestion, directness of route, fares, number of changes necessary between origin and destination and waiting times at transfer or interchange points) may encourage or deter potential users of a service from travelling to obtain or use it. Again some consumers or suppliers

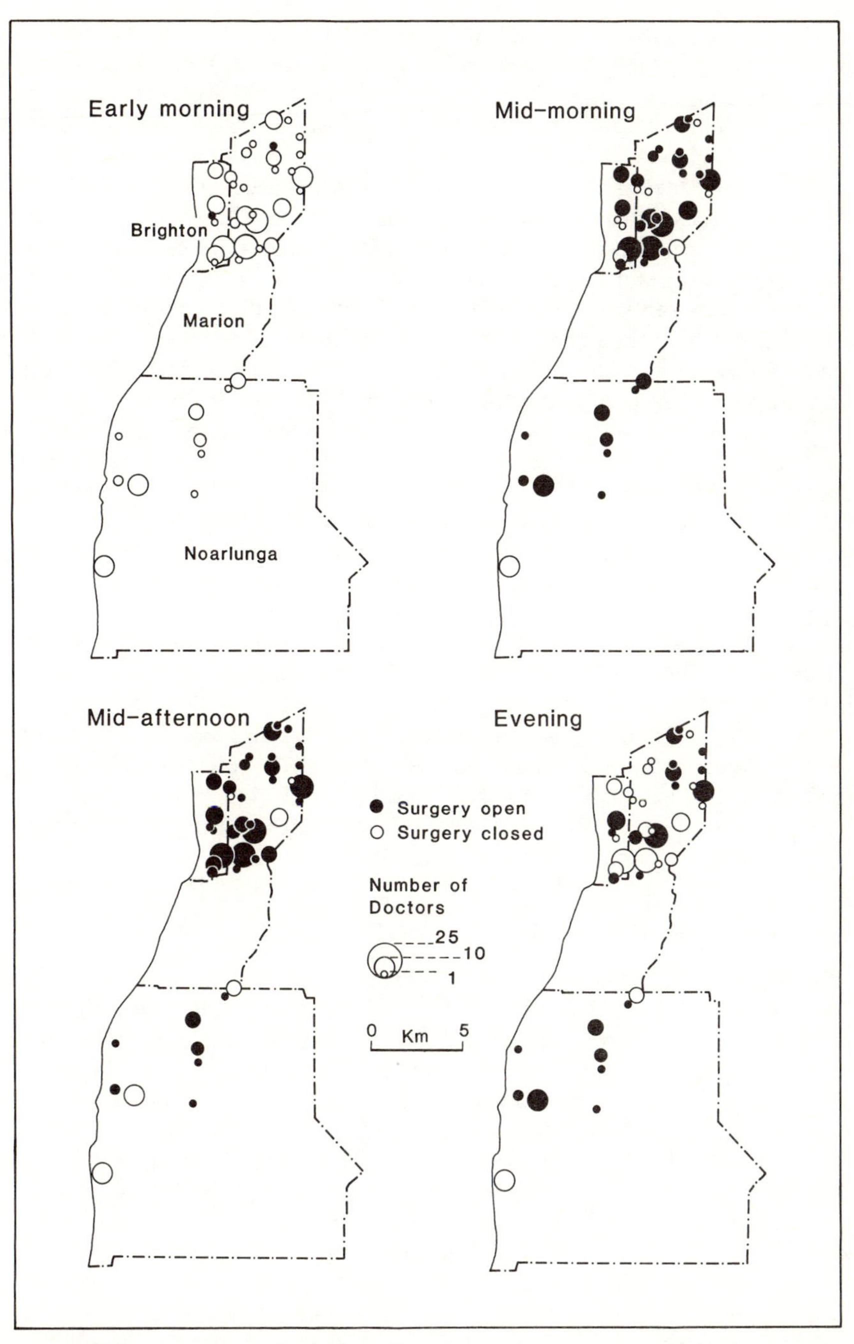

Figure 6.1 The spatio-temporal provision of general practitioner services in the southern suburbs of Adelaide, 1977
Source: Stimson, 1980. figure 5, 894.

will be better placed – in a financial or locational sense – to minimize any possible disadvantages arising from the characteristics of the transport network. Residents of rural areas have particular accessibility problems arising from availability of public transport services (Moseley, 1979; Haynes and Bentham, 1979).

Behavioural factors will also influence individual accessibility to services: the perceived advantages or disadvantages of using one method of travel rather than another; attitudes towards the use of public transport; the ease with which parking can be obtained (perhaps at no cost to the user) at one shopping centre rather than another; and the importance of reputation as regards safety and danger when deciding whether or not to attend a first-division football match at an away ground, or the importance attached to the quality or ambience of a shopping centre, supermarket, or cinema when deciding whether or not to use it. These are just a few of the possible behavioural influences which may ultimately determine the 'real' as opposed to 'absolute' accessibility of competing services.

The measurement of accessibility is clearly complex (for a detailed discussion see Jones, 1981) and its effects on the availability of services to different segments of the population are difficult properly to unravel. In such circumstances it may be wise to take some specific examples which illustrate the inequalities consequent upon varied accessibility to a number of services which have received attention in the literature. A wide range of services has been examined during recent years, including primary medical care (Lankford, 1971; Knox, 1978; Phillips, 1979a; Bohland and Frech, 1982); hospital services (Morrill *et al.* 1970; Whitelegg, 1982; Gould and Leinbach, 1966); recreational facilities (Smith, 1985); fire services (Adrian, 1983; Hogg, 1968); child care facilities (Freestone, 1977); educational services (Herbert, 1976; Kirby, 1979; Walker, 1979); and other public services (Knox, 1982a, 1982b).

THE EXAMPLE OF HEALTH SERVICES

Whether provided by the private or the public sector, health services are usually organized on a hierarchical basis. This makes them amenable to analysis and planning for new facilities using the principles of central place theory (Knox, 1978; Stimson, 1980; Earickson, 1970; Shannon and Dever, 1974). This analogy should not be taken too far, however, since professional codes will often prevent market mechanisms from operating fully (Herbert and Thomas, 1982). At the lower end of the hierarchy are the general practitioner services which are often described as providing primary care. General practitioners (GPs) may operate individually but they are increasingly working in small groups or partnerships (60 per cent of GPs in the UK now comprise groups of two or more), with the result that the services they provide are distributed among a smaller number of locations. It is clearly at this level that user accessibility is most important and where any problems connected with it can give rise to inequalities in health care consumption.

In cities the ideal distance between a doctor's surgery and patient should be no more than 0.75 km (Knox, 1982a) for mothers with pre-school children and for the elderly. Both these groups may not have access to private transport and will, therefore, have to walk or use public transport; the latter may be difficult unless a route passes conveniently near and travel to the doctor's surgery does not entail one or more transfers. It is well documented that as distance increases from surgeries consultation rates decline (Hopkins *et al.*, 1968), partly because of the disutilities of travel and partly because of the reluctance on the part of patients to call on their doctors to make long journeys for home visits. There is clear distance decay apparent in the distribution of households consulting general practitioners in Adelaide (Figure 6.2). About one-third of the trips illustrated were undertaken in association with other purposes such as the collection of children from school or shopping. It is also worth noting that a trip to consult a GP may generate further travel; in the Adelaide example 70 per cent of the consultations generated a prescription and the requirement for a follow-up journey to a drugstore, and 20 per cent led to referral to higher-order medical services (Stimson, 1980). This generalization will be complicated, however, by the greater reluctance of blue-collar workers to consult general practitioners compared with white-collar workers, who are also more interested in preventive medicine (Knox, 1982a).

The actual distribution of general practitioners has received extensive scrutiny in American cities (Shannon and Dever, 1974; Lankford, 1971; Earickson, 1970; De Vise, 1973), Australian cities (Stimson, 1980; Morris, 1976; Cleland *et al.*, 1977) and in British cities (Cartwright, 1967; Knox, 1978; Sumner, 1971; Phillips, 1979b). In the USA practitioners provide their services on a fee-paying basis and their location behaviour, therefore, depends on their assessment of effective demand in different areas; in other words, the doctor will 'attempt to choose a location which will maximize his total 'expected' income' (Lankford, 1971, 244). Hence per capita income, morbidity rates, population size, the range and quality of existing physician services and the likelihood of the population wishing to purchase medical assistance must all be weighed by the physician in his locational decision. Early attempts to explain the distribution of practitioners demonstrated a connection with per capita incomes, and later with population size, but Lankford criticizes both of these findings on the ground that they do not explicitly incorporate the spatial attributes of the distribution of doctors. He, therefore, attempts to explore, for US urban and rural areas combined, the relationship between the gross distribution of physicians, the pattern of availability (physicians per capita) and the possibility of changes through time.

The first thing to emerge is the role of population size rather than income as the main locational influence on the service function of US doctors; it explains some 80 per cent of the variation in the distribution of practitioners. The relationship between the number of practitioners and total population is curvilinear, with a disproportionate increase in the former with each unit increase in population. This is a product of the higher thresholds and ranges of more specialist physicians who will tend to locate in the most accessible and largest population agglomerations. Physicians do not avoid locations in

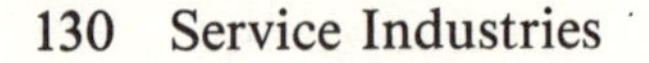

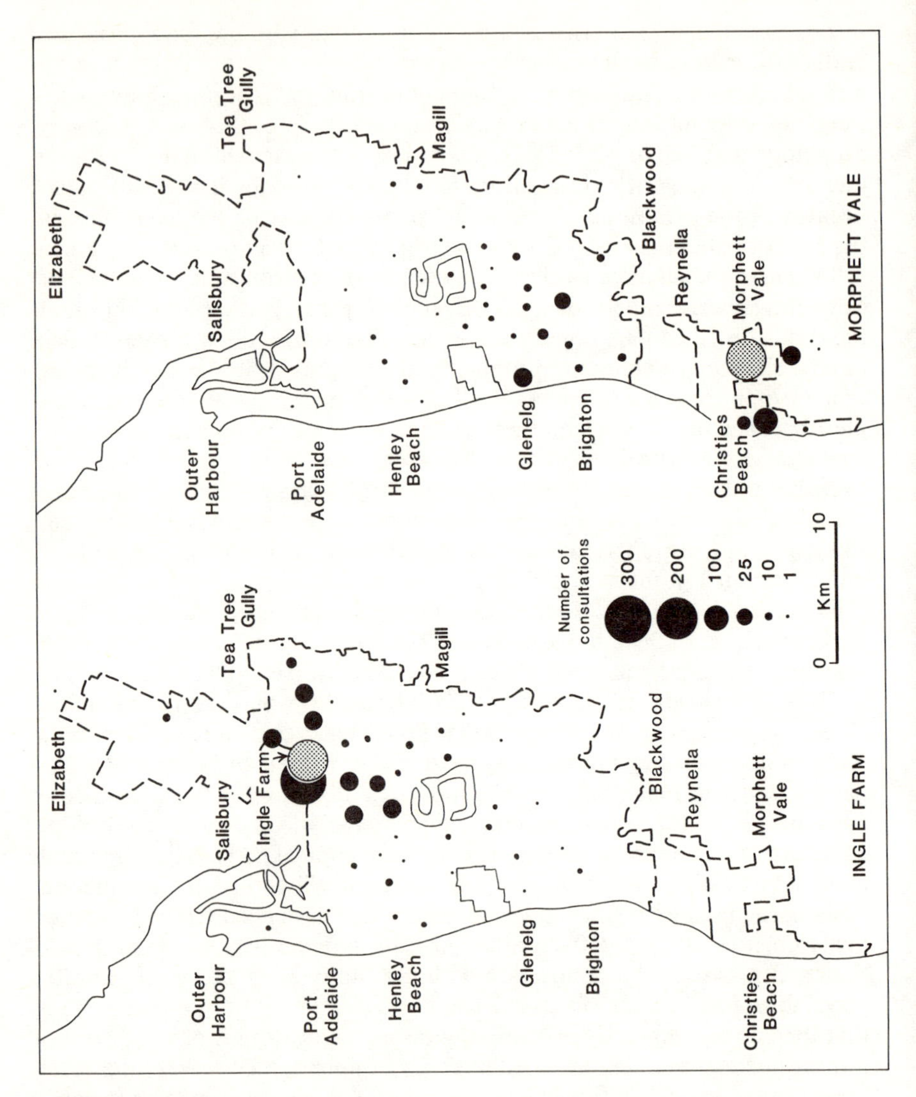

Figure 6.2 Location of general practitioners consulted by households in two suburbs of Adelaide, 1976
Source: Stimson, 1980, figure 7, 899.

or near low income areas and gravitate towards high-income areas (see ibid.); the notion that areas with predominantly young or elderly age structures attract doctors can be dismissed. Regional rather than local population size is the key location variable and the proportion of the population which is non-white seems to have a negligible influence on physician location. These conclusions are derived from the application of a curvilinear model to data for physicians per capita in 1950, 1960 and 1969. It is also possible, however, to test the hypothesis that areas which have experienced rapid population growth will, over time, attract a disproportionate increase in the number of medical practitioners. Lankford considers the results disappointing, in that even allowing for a twenty-year lag effect the relationship is poorly developed. It seems, from his analysis, that all the variables expected to be significant for the location shifts shown by physicians, such as upward movements in age structure or income, the proportion of the population non-white, or increases in urbanization, have relatively little effect.

There is, therefore, a good deal of inertia in the location patterns exhibited by general practitioners as well as in higher-order facilities such as hospitals, and this, in circumstances where population changes through demographic and migrational effects are constantly occurring, is an important contributory factor to inequalities in access. Population outmigration has left some areas overdoctored for a number of reasons: the conservative ethos of the medical profession (Hart, 1971); the possible consequences for career development if general practice is undertaken in less fashionable areas of a city; the relative ease with which the older, larger houses in the inner areas of cities can be converted for use as surgeries and residential accommodation for the doctor and his family; the cumulative investment represented by existing surgeries; and the generally poor provision for purpose-built facilities in newer suburbs and estates, where few houses are suitable for conversion and where planning permission would probably be difficult to obtain.

The distribution of general practitioners' surgeries in Glasgow, in relation to the social ecology of the city, reveals the effects of some of these considerations (Figure 6.3) The rationale for Lankford's (1971) conclusions from his static analysis is clearly illustrated. It is evident that a large number of surgeries are grouped in the older 'very deprived' inner areas, while both private and local authority housing areas constructed around the edges of the city since the last war are relatively underprovided. This is especially the case in local authority estates such as Easterhouse and Drumchapel, which have social and economic problems at levels which easily match those prevalent in the better served inner city areas. As Knox (1982a) indicates, personal mobility is low in such estates, so that round trips of more than four miles to visit a doctor's surgery may not be uncommon, and for families rehoused from inner city tenements but unable to register with (or unable to find) a suitable local surgery, return journeys of eight miles might be necessary. A similar situation exists in Edinburgh where the great majority of surgeries are located in a number of clusters around the CBD in neighbourhoods comprising old but sought-after housing with a diverse social ecology (see ibid.). Again large tracts of outer housing estates, both public and private sector, are poorly served.

In a detailed study of general practitioner location in Adelaide it is also shown by Stimson (1980) that the older inner areas have more GPs than the rapidly growing outer suburbs. Knox (1978) is cautious about making generalizations from the four cities (Glasgow, Edinburgh, Dundee and Aberdeen) which he examined but suggests that, in addition to the discrepancies between inner and outer areas, very few surgeries are located in CBDs, although elsewhere in the

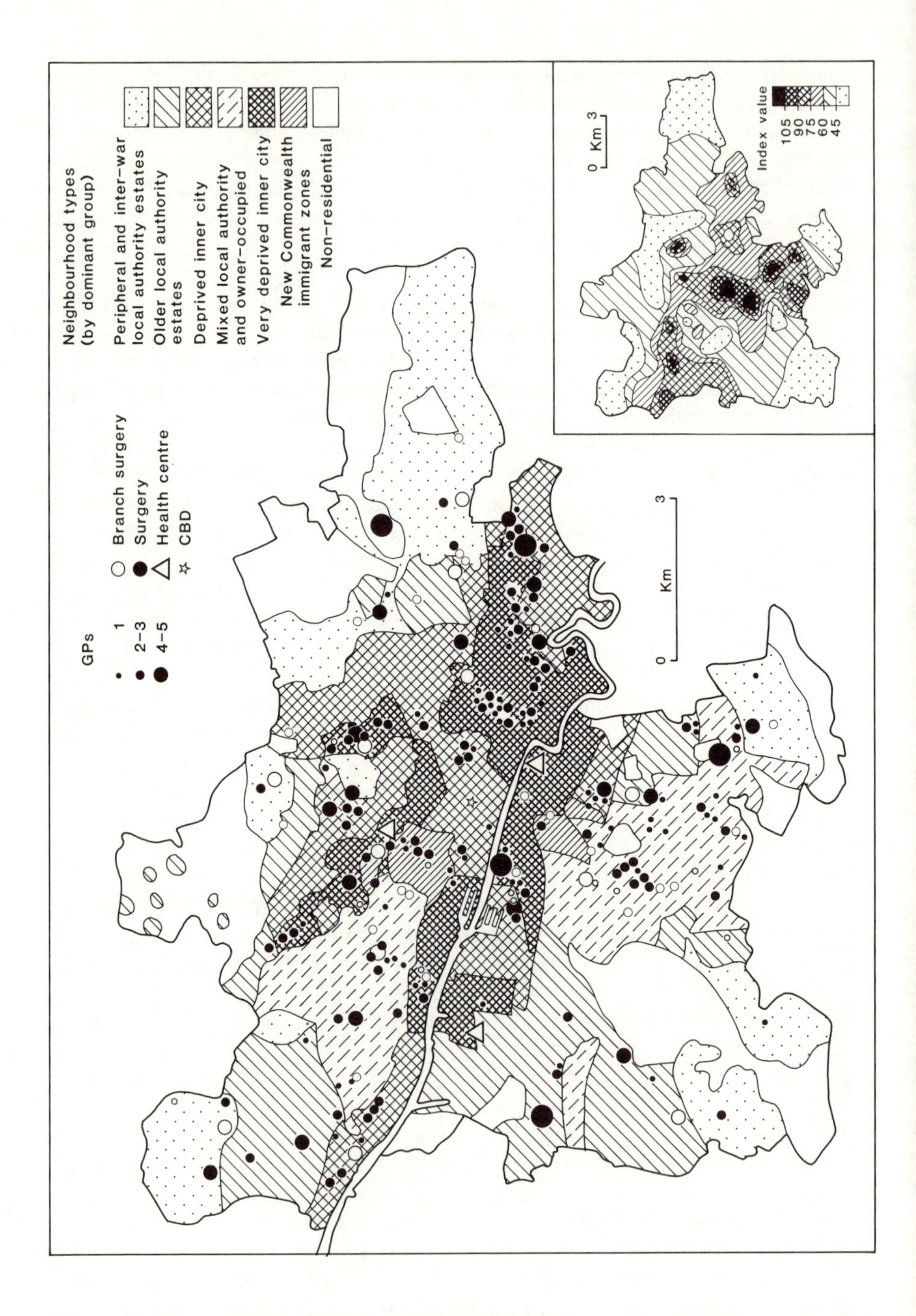

Figure 6.3 The social ecology of general practitioners' surgeries in Glasgow
Source: Knox, 1978, figure 4, 423.

four cities the distribution of surgeries is associated with the incidence of business centres. Also the inverse care law (Hart, 1971; Knox and Pacione, 1980) that the availability of medical care varies inversely with the needs of the population served, is not well supported by the empirical evidence. Indeed the distribution of doctor's surgeries 'tends to discriminate unsystematically between housing classes' (Knox, 1978, 424) in the cities studied, and Cleland *et al.* (1977, 53) concludes that: 'there exist "doctor rich" and "doctor poor" areas in Adelaide. ... This pattern of maldistribution is typical of most cities and presents for Adelaide a pattern not dissimilar to many North American cities.'

The implications of observed spatial distributions of GPs' surgeries for accessibility have so far been inferred but it is also possible to measure it more objectively to produce a surface of relative accessibility to health services (see insert in Figure 6.3), which can then be related to social ecology. The index of accessibility (I) is computed from:

$$I_i = \frac{A_i(t)(\%)}{M_i(\%)} 100$$

where I_i index of accessibility;
$A_i(t)$ = time-based index of accessibility for neighbourhood i;
M_i = market potential of neighbourhood i.

A value for I greater than 100 indicates relative local overprovision and a value less than 100 denotes underprovision of doctors (for details of the derivation of the index see Knox, 1978, 424–6). While the pattern for Glasgow is relatively complex (Figure 6.3), overprovision of primary health care is confined mainly to older residential neighbourhoods with mixed social and tenurial status such as Pollokshields, near the centre of the city, or Scotstoun on the western side of the city. The most generously served area of the city is a local authority housing estate, Croftfoot, while some of the lowest levels of accessibility are found in central, middle-class neighbourhoods such as Pollokshaws.

Access to dental care for schoolchildren

Bradley *et al.* (1978) devised two measures of accessibility for a study of school dental services (SDS) for primary school children in Newcastle upon Tyne. Part of the service is provided by a number of school dentists who visit schools and, if necessary, refer patients to health clinics for further treatment. In these circumstances parents do not have to choose between alternative locations, so that a measure of access is developed in which distance to schools and the hours worked at individual clinics are related and:

$$\text{Access to SDS from school } i = H_j/d_{ij}$$

where d_{ij} = distance from school i to the nearest clinic j;
H_j = total number of weekly hours worked at clinic j.

Schoolchildren may, however, opt to use national health service (NHS) dentists who are able to select their own locations as they think fit; practices in and adjacent to middle-class areas of Newcastle upon Tyne dominate the pattern. In theory, parents can choose to take their children to any of the dentists available, so that all the families with children in the city are potential patients. Hence the measure of accessibility should employ the concept of population potential:

$$\text{NHS dentist potential at school } i = \sum_{=1}^{85} H_j/d_{ij}$$

where d_{ij} = distance from school i to dentist j;
H_j = number of weekly hours worked by the dentists.

The patterns of access for the two types of dental service are illustrated in Figure 6.4, where the values are scaled to produce percentages of the highest computed access index or potentials. Because the school dental service includes a number of mobile caravans, extra access is clearly available in the west and north of the city, while the peak potential for the general dental service are concentrated in one major area, the middle-class neighbourhoods of Jesmond and Gosforth. There is, therefore, a very clear differential access to general dental services according to social class but this is not the case for the school dental service; it is 'class-blind' (see ibid.), that is it does not favour any class in its location practice, but neither does it complement the general service by providing dental facilities where the former is absent. Nevertheless, Bradley *et al.* show that there is a significant correlation between dental health at a school and access to general dental services and that the demand for the school dental service tends to be higher where dental health is lowest (see ibid.). There are also significant correlations with working-class areas, so although its location pattern is not biased, the school dental service does provide a complementary service.

Whether largely subject to the vagaries of the market or influenced by the goals of publicly funded provision, it seems that the provision of primary health care is far from equitable. In theory, hospitals, police, education and fire services are available to all as it is intended that they should contribute to the public good (Samuelson, 1954; Head, 1974). In practice, because spatial access is a major factor in the quality of the service consumed by users, such services cannot (or do not) fulfil the ideals of a public good. In large part, then, the problem stems from accessibility, aided by the differences in attitude towards health care among different socio-economic and ethnic groups in the community. As the American, British and Australian studies have shown, inequalities arise at all scales of analysis and serve to underline the discrepancy between defining goals for the provision of a consumer service such as health care and matching these goals to actual provision according to need.

While GPs can give immediate advice and basic treatment, they clearly are unable to provide more specialist diagnoses, treatment and after-care. Because of the economies of scale and threshold demand requirements, more specialist health services may be found at the next level in the hierarchy. Not as large as hospitals, but more sophisticated than GP operations, the specialist health centres tend to locate adjacent to major hospitals near or within central business districts, or in the vicinity of major suburban shopping malls or hospital

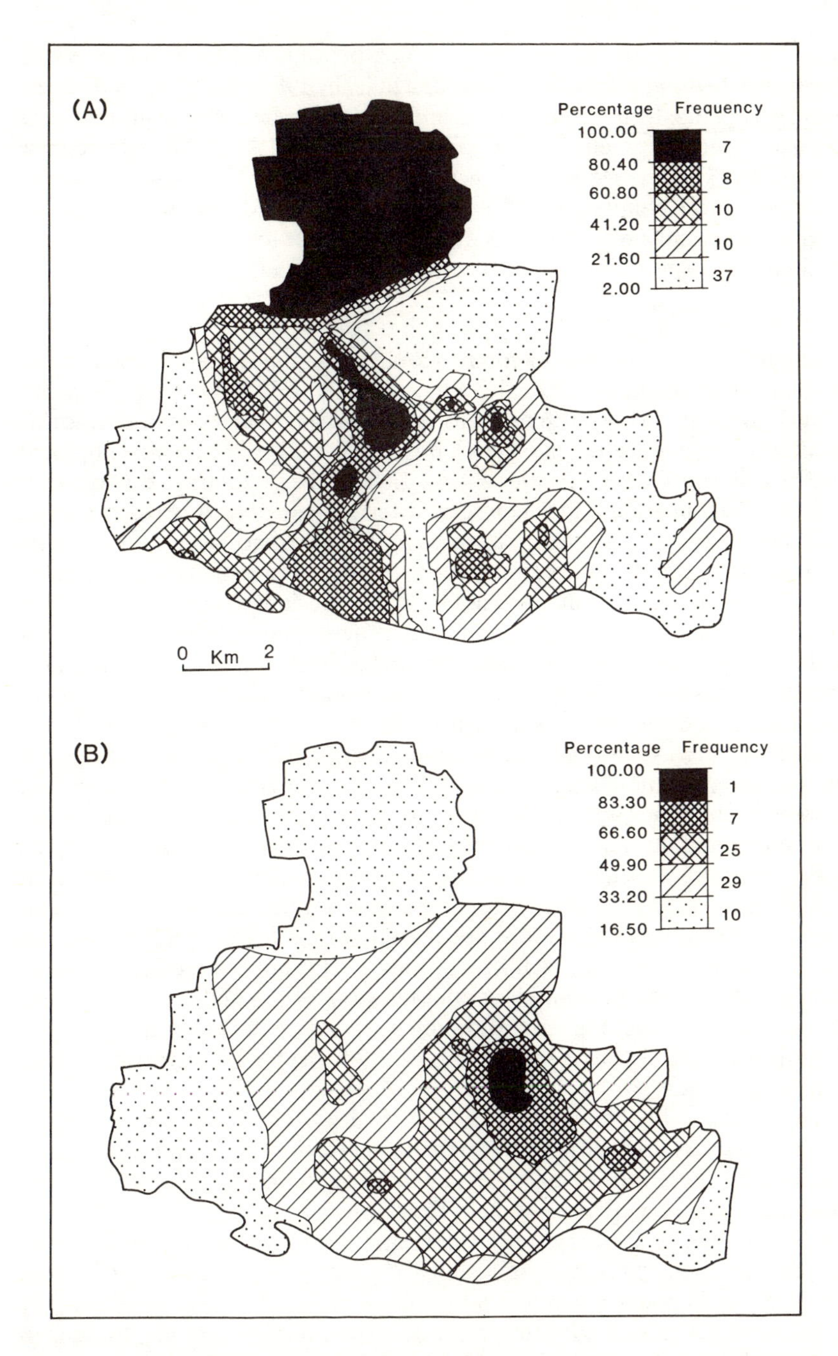

Figure 6.4 Accessibility to school dental services (A) and distance potential to general dental service (B), Newcastle upon Tyne AHA
Source: Bradley *et al.*, 1978, figure 2, 535; figure 4, 537.

complexes. In Britain the increase in the number of health centres accelerated during the 1970s and there were more than 800 by 1980. General practitioners may also operate from health centres and there has been a steady shift towards groups of general practitioners (Department of Health and Social Security, 1980). In 1977 some 40 per cent of general practitioners practised in groups of four or more compared with just 17 per cent in 1961. Accessibility to the metropolitan transport network is one of the notable location factors for such facilities, and specialist medical centres seek to benefit in the same way.

Access to hospital services

At the top of the three-tier hierarchy are the general hospitals which provide sophisticated medical services and employ highly skilled and specialized personnel. Whitelegg (1982, 79) explains that these large-scale facilities must be at a 'location where adequate back-up facilities exist, where such equipment can be efficiently deployed for that fraction of the population experiencing that particular problem'. Inevitably the advantages of proximity to users offered by GPs are to some extent compromised in the case of both specialist medical centres and hospitals; patients may have to travel long distances to obtain the treatment they require. This need not be highly specialized treatment since the growing tendency is for hospital outpatient departments to treat minor cuts and fractures because GPs are not perceived as being able to help adequately and the delays and difficulties which patients may encounter when trying to make an appointment for treatment.

Since research shows (see, for example, Knox, 1982a) that there are social, income and psychological factors affecting the willingness of patients to seek treatment, the added distance to reach hospital services may be a further deterrent to use, and ultimately to the population's well being. In recent years the highly centralized location of hospitals has been changed by the development of regional or subregional hospitals at suburban locations in American, Canadian and Australian cities but there is often a time-lag between population shifts and the establishment of new hospitals. This not only reflects the constantly changing distribution of population within cities and regions, but may also be a response to changing thresholds for particular medical techniques or amended administrative assessments of the most cost-effective ways of providing health services.

It can reasonably be hypothesized that patients will need to travel further to hospitals than to health centres or general practitioners' surgeries because the latter are more widely dispersed than the former. However, whereas patients are free to choose a GP according to their own judgement and acceptance of travel and other costs, this is not, in most cases, possible for hospital use. Over 50 per cent of the physicians in a study of Chicago hospitals (Morrill *et al.*, 1970) referred their patients to just one hospital and this was not necessarily the nearest. Patients might, therefore, be able to minimize travel distance to their GP but not be able to do so for hospital consultations or treatment (see Haynes and Bentham, 1979). Although there are limitations on which hospitals physicians may be affiliated to, Morrill *et al.* (1970) comment on the surpris-

Table 6.1 Relative location of patient homes, physicians' offices visited and hospitals visited: Chicago, 1965

Area of residence of patients	*Visits to physicians' offices (%)*		*Patients seeking care at hospitals (%)*	
	Closest to home	*Next closest to home*	*Closest to home*	*Next closest to home*
Other suburbs	55	17	65	15
Inner suburbs	35	32	40	18
ALL SUBURBS	46	19	53	17
Outer city	18	23	21	25
Inner city	18	25	23	20
CHICAGO CITY	18	23	22	23
METROPOLITAN AREA	33	21	41	19

Source: Morrill *et al.*, 1970, tables 2 and 3, 163–4

ing degree to which patients actually do use their nearest hospital; the mean distance in Chicago was 3.3 miles. A factor contributing to this low figure may be the concern on the part of the family doctor to minimize his own travel distance to the hospital(s) to which he is affiliated, so that patients, who are attempting to achieve the same goal in relation to the location of the GP, will also benefit indirectly (Table 6.1).

The Chicago study also shows how patient perception of distance to hospitals is distorted by factors such as income, religion, or race (see ibid.). Such factors may be especially important in metropolitan areas; in small towns and rural communities with one or, at most, two hospitals within reach there is very limited opportunity to exercise choice. In metropolitan areas like Chicago, New York, or Los Angeles it is possible for hospitals – which operate within a market framework – to be selective according to ethnic and religious factors, or to financial means. Hospitals which serve only patients unable to pay using conventional medical insurance, such as Cook County Hospital in Chicago, are distinct from those which may consciously try to restrict use by black patients or those hospitals run by religious denominations and, therefore, favoured by patients of the same religion. Consequently many trips to hospitals may ignore intervening opportunities: Cook County Hospital is some way removed from the major black communities in inner south Chicago; Mercy Hospital attracted fewer black patients than expected from the proximity of black areas (Figure 6.5); and there are fewer Jewish hospitals than Catholic or Protestant hospitals in Chicago, so that the former make long and apparently irrational trips from Jewish residential areas such as West Ridge (Figure 6.5). Access to the hospital system in American cities is, therefore, highly differentiated and variable and imposes disproportionate accessibility costs on those least able to afford them. Travel savings of as much as 50 per cent might accrue to patients if there were

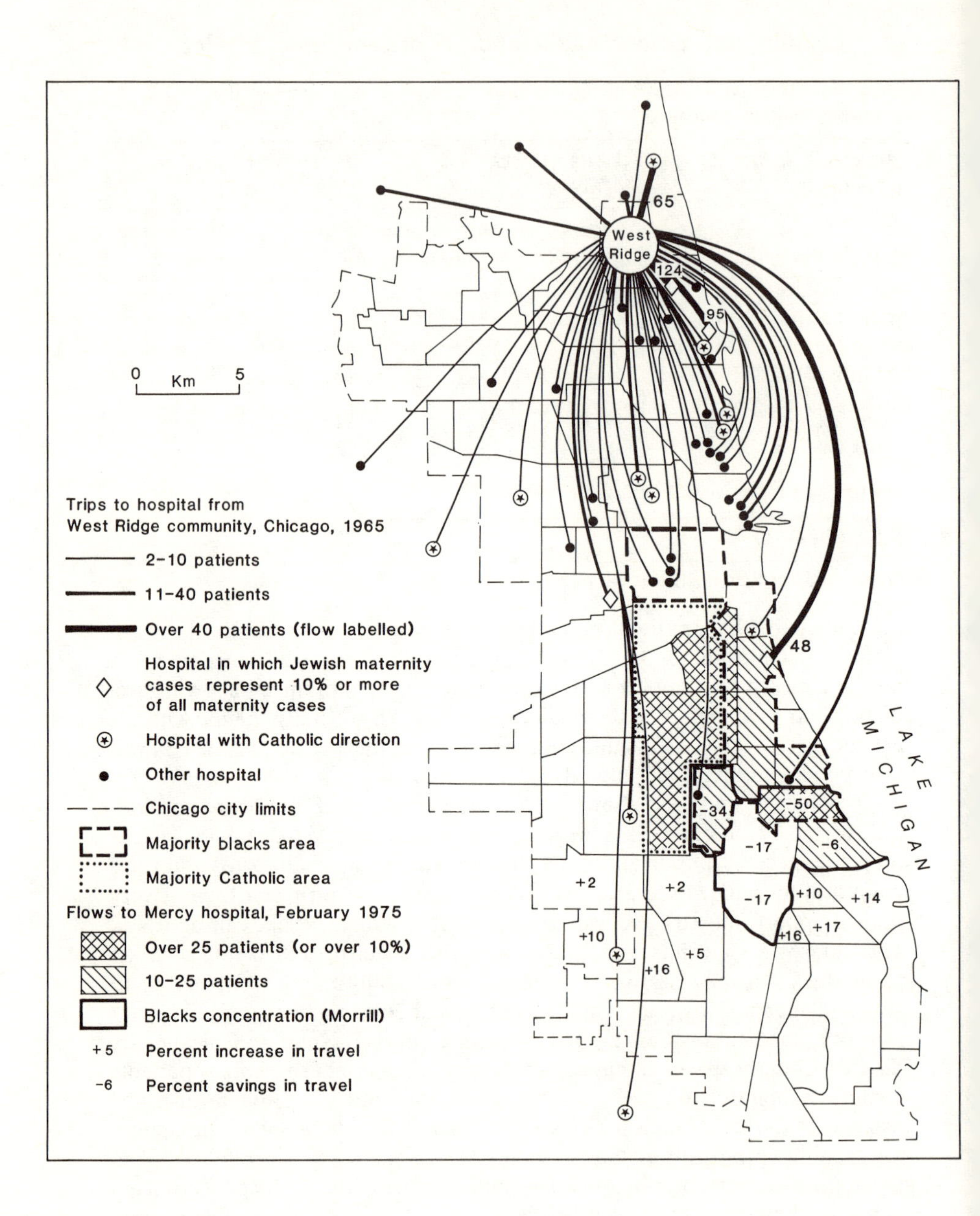

Figure 6.5 Spatial attributes of trips to hospitals in Chicago
Source: Compiled from Morril *et al.*, 1970, figure 3, 168; figure 4, 169; figure 5, 170.

free access to the entire hospital system (see ibid.).

The latter has been the case in the UK, where there have been two reorganizations of the publicly funded health service which, despite growing inroads by private medicine, still provides by far the largest share of health care services. Radical changes to the hierarchical system of publicly provided health provision first introduced during the immediate postwar years (1948) were made in 1974 when the national health service was reorganized into a hierarchy of regions (fourteen), areas (ninety) and districts. It was believed that the new organization would lead to a fairer distribution of health service resources between regions, disadvantaged areas and medical specialities. However, it soon became clear that the new organization was not meeting the administrative objectives and, in 1982, the 205 health districts, which comprised the lowest level in the hierarchy, were replaced by a smaller number of health authorities directly responsible to their regions (thus totally removing the areas previously forming an intermediate tier in the hierarchy), a recent overview is provided in Drury (1983). The new health authorities are responsible for larger areas than the former districts, so that in the case of Blackburn (Whitelegg, 1982) parts of three districts have been combined into one new authority (Figure 6.6). The two largest hospitals are located in the major urban area, Blackburn, surrounded by a number of minor hospitals providing limited health care and referring more difficult cases to the two central hospitals.

It is worth citing the principal objectives given by Blackburn Health District (1979) for the development of health care in its area (Whitelegg, 1982): centralization of children's hospital facilities to provide an improved level of care for both medical and surgical patients; raising consultant medical beds to the norm and centralization of the facilities to bring maternity services up to the current standards; providing surgical support for the centralized children's unit and the medical beds at Queen's Park Hospital by relocation of facilities now at Accrington Victoria Hospital and closing Withnell Hospital; optimizing scarce medical manpower by the restriction of acute specialities to no more than two centres; developing community hospitals at Clitheroe, Longridge (Ribchester), Accrington, Darwen and eventually Blackburn to improve the level of primary medical care and to make the best use of district general hospital facilities; and progressively developing Queen's Park Hospital as the definitive district general hospital for Blackburn Health District.

These objectives have been specified in detail because they embrace a common purpose, that of increased centralization and rationalization of health care provision (excluding GPs, of course, who are responsible not to the NHS, but to the Family Practitioner Committee). This is a theme present in many similar reorganization programmes implemented by other health authorities since the mid-1970s. The effects of the Blackburn reorganization on the spatial distribution of the availability of various kinds of treatment and the emphasis given to the emergence of a major district general hospital almost certainly mean substantial modifications to accessibility for several different categories of user.

Whitelegg has collected information about actual travel patterns of

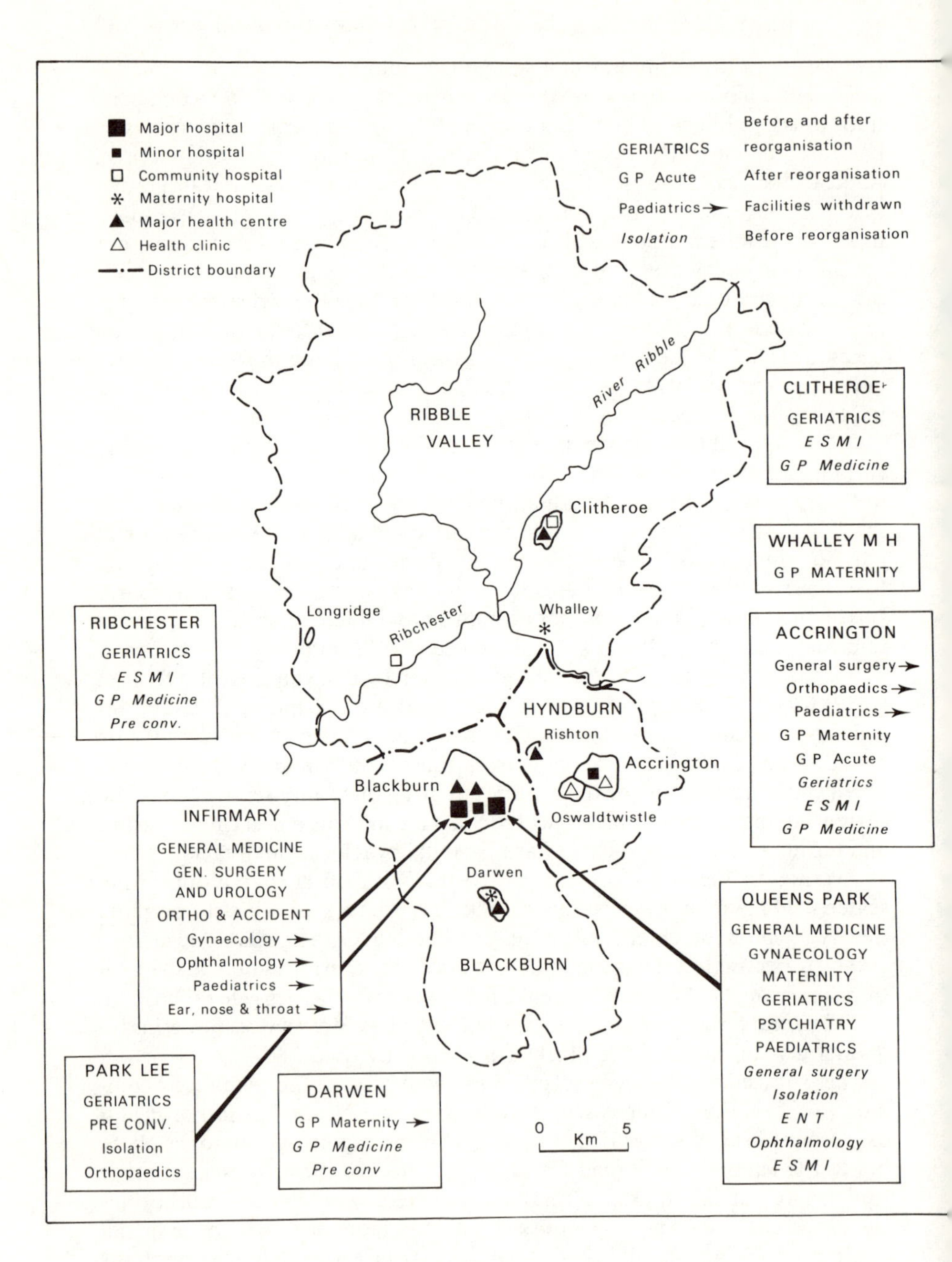

Figure 6.6 The effect of reorganization of health districts in the Blackburn area on health centre and hospital services
Source: Compiled from Whitelegg, 1982, figure 4.2, 82; figure 4.3, 84; figure 4.4, 85.

outpatients, and visitors (to inpatients), to health care facilities in the area (see ibid.). Some of the data has been used to estimate the impact of the reorganization on accessibility to different types of health care (Table 6.2). Visitors will invariably travel from similar origins to those of the patients, they will need to undertake travel which reflects the location of the specialist facilities and will make journeys requiring the use of public and private transport. The new pattern of facilities will generate savings in journey distance (3580 passenger/km) which far exceed the increases (874 passenger/km) but only for general medicine and geriatric facilities. The bulk of the savings accrue to the latter, although visitor frequency is much lower than for general medicine. The requirement to attend hospital for the latter is likely to be more widely spread among the population of the health authority, so that the real travel savings are more limited than the overall decrease implies. The reductions in travel distance shown in Table 6.2 are, therefore, selective and less dramatic than the difference between increases and decreases suggests.

Table 6.2 Effect of district plan for reorganizing Blackburn Health Authority on travel distances to medical services

Speciality	*Average length of stay (days)*	*Visitor norm*	*Round trip distance, difference (km)*	*Travel increase/ decrease (passenger/km)*[1]
General medicine	13	4.28	14	(780)
ESMI/Geriatric	100	0.7	24(16)[2]	(1680)(1120)
General surgery[3]	8	2	16	256
Orthopaedic surgery	10	2	16	320
Paediatrics	4	4.5	16	288
Gynaecology	5	2	1	10
Total (decrease)/increase in travel distance				(3580)/874

Notes: 1 Length of stay × round trip distance × number of visitors.
2 Savings accruing from relocation in Ribchester and in Accrington (value shown in brackets).
3 There are some travel savings for Blackburn residents (approximately 15 per cent of the increase value).

Source: Whitelegg, 1982, table 4.5, 102.

Despite operating under very different regimens, the health care services in the USA and UK are seen to be committed to equality of opportunity in obtaining the help that patients require. Most of those who have written on this subject conclude by assessing the lessons for policy which their results suggest. Lankford (1971), for example, suggests that in order to achieve better access to primary medical care in the USA the number of physicians can be increased or patient–physician mobility improved. There is little point in trying to achieve the former unless some controls or incentives are used to influence the locational behaviour of the additional physicians; Lankford mentions 'nationalization' or the provision of subsidies for physicians prepared to practice in areas which are underprovided but concludes that this would only generate a cumbersome bureaucracy and poor results. In essence, the entrepreneurial ethos should be allowed to prevail with physicians free to locate as they please, leaving the alternative as intervention in the travel costs of patients by, for example, allowing the cost of travel to medical services to be declared as a federal tax credit. Alternatively, free transport to medical centres could be provided, especially from deprived rural and intra-urban areas, but here too the administrative overhead might prove burdensome.

Policies for the distribution of GPs in the UK are largely based on the average list size in medical practice areas many of which cover complete cities which will clearly display great internal variations in population density, socio-economic attributes and epidemiological characteristics. Improvements in access to GP services could, therefore, be achieved by producing a finer mesh of medical practice areas combined with more discriminatory methods for identifying their size relative to the needs of the population within them. Under present regulations it is possible for practice areas to be designated for special finance available to the GPs practising there but few exist and the incentive payments available do not assist with the cost of employing ancillary staff. These and other difficulties are not helped by the fact that policies for the provision of NHS medical services are largely devised and implemented by the system's administrative personnel and by the medical profession. The root of the problem, although explained in an American context, has been well expressed by Torrens (1978, 92), who states that the 'public has usually been placed in a position that was subservient rather than superior to the health care professionals, managers and bureaucrats' and 'have been so intimately involved with the working of the health care system, they have tended to think that the system belonged to them, rather than to the public that gave it origin in the first place'. It is not clear whether the greater involvement of the consumer in decision-making about health facility location, for example, would resolve any of the prevailing inequalities which perhaps require much more radical redistribution of incomes or attempts to enhance personal mobility if any progress is to be made. Whitelegg (1982), for example, provides a wide-ranging overview of the alternatives for a programme of 'positive action' which should include a spatial dimension if they are to succeed.

FIRE SERVICES

While the location of medical services essentially involves balancing accessibility by patients and visitors with thresholds for economic provision of more advanced and expensive equipment and treatment, the location of fire services is much more dependent on their ability to respond effectively to fire calls and related alarms. Therefore, an assessment of equity in this case, may be based on the degree to which areas are outside the zones around fire stations demarcated by a maximum response time. For urban fire services a five-minute response to all areas is often the maximum (Hogg, 1968; Rider, 1979), with lower limits for high-risk areas such as industrial complexes or CBDs. In a study of the distribution of seventy-three fire stations in Sydney (Figure 6.7), in relation to the actual incidence of alarms (10 per cent sample of

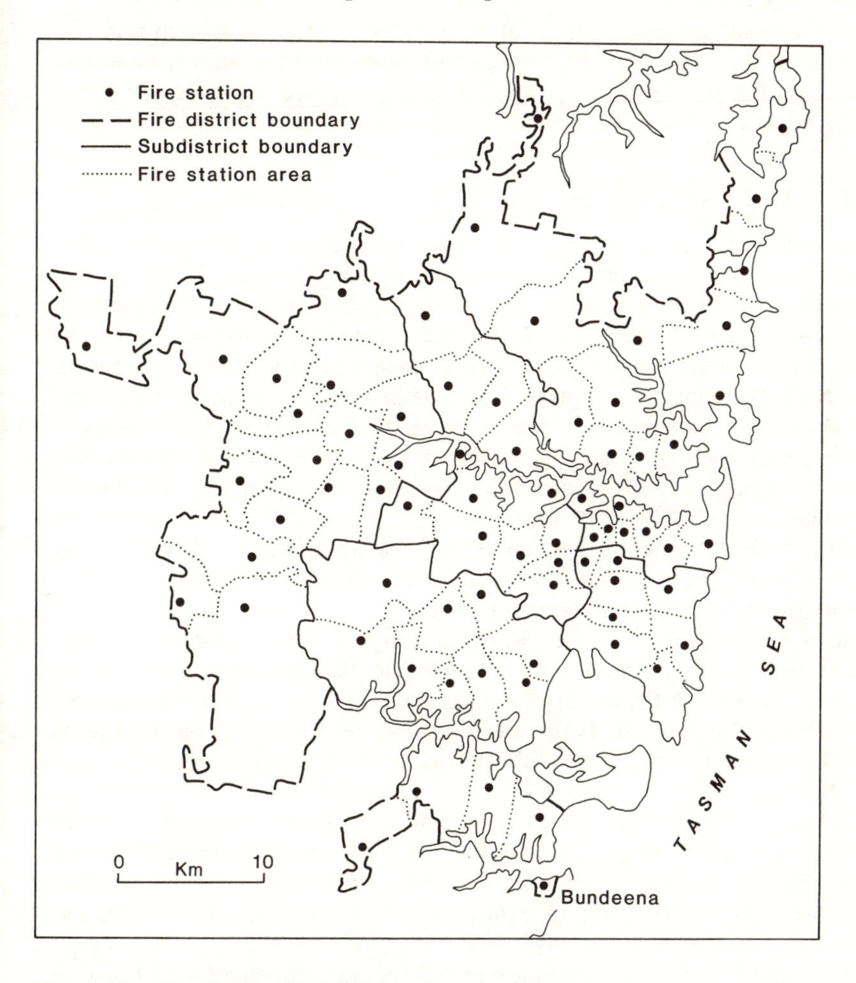

Figure 6.7 The distribution of fire stations in Sydney
Source: Adrian, 1983, figure 4, 1088.

1978 alarms), some thirty-eight of the 253 traffic zones in Sydney had response times from the nearest fire station in excess of five minutes (Adrian, 1983). These included a major industrial belt in the south-west of the city where response times of three minutes were required, while in practice all the demand zones were outside the five-minute limit.

As with other urban or indeed rural services, the pattern of fire station location has been overtaken by the rate of urban development, leaving some areas overprovided and others with 'service poor' fire station coverage. Adrian compares the 1978 distribution of fire stations with two theoretically optimal distributions and finds that the existing pattern most closely approximates a minimax distribution, that is a system of stations which minimized the maximum station-to-zone response time (see ibid.). He concludes that the needs of lower-risk suburban fringe, often residential, areas have been traded off against the needs of the higher-demand inner and often industrial areas of the city. The net result is that many areas are not receiving adequate fire service cover if a five-minute response rate is taken as the maximum.

MODELLING ACCESS TO PUBLIC SERVICES

In the examples cited above, as well as for public libraries, community centres, schools and leisure centres among others, the main problem is how to locate a given number of these facilities in a way which allows the best geographical access to them or (for fire or ambulance services) from them. The units of supply are points and the units of demand are areas, nodes, or sometimes grid squares. These units of supply and demand can be manipulated mathematically and a number of geographers among a large number of contributors from other disciplines, such as operations research and management science, have developed models designed to identify the optimum distribution of service facilities (Abler *et al.* 1971; Scott, 1971; Massam, 1975; Taylor, 1977; Bach, 1980). Most of the early examples of these models are connected with studies of the location of private facilities, such as warehouses or manufacturng plants (Lea, 1973), where it was assumed that demand and the inputs were known in advance and the task was to minimize the cost of transporting fixed amounts of inputs or outputs. It was, therefore, not necessary to take account of the location strategies of other firms. The demand for public services does not necessarily depend on the location of other facilities, and the absence of competition then allows location choice which minimizes transport costs (i.e. attempt to locate optimally).

Hodgart (1978) identifies three forms of the spatial problem for model-builders: first, to assign z facilities freely where it is assumed that no facilities already exist in the area; secondly, to locate x additional facilities while taking any existing facilities into account (sometimes described as the incremental facility problem); and thirdly, given x existing facilities, how they may be reorganized by closing badly located facilities and opening others elsewhere. The first problem has attracted much attention but hardly gives solutions which are easily applied to most real-world situations where some facilities, whether hospitals or public libraries, will already be in place and the location of new facilities will have to take this into account. As the empirical evidence has shown, there are many circumstances where the third spatial problem would be worth solving but implementation is likely to be difficult for political or planning reasons, or because of the unacceptability of writing off investment in perfectly adequate (in a functional sense) existing facilities.

Optimizing models are developed in the context of a number of objectives and constraints. The maximization of demand is a possible objective in the interests of efficient utilization of a public service. This is appropriate for public services for which the demand is elastic such as a leisure facility, but will lead to inequitable provision of those services for which demand is inelastic such as schools or hospitals. Consumers located in peripheral areas are likely to have a poor service. Minimization of travel cost is likely to create a more equitable distribution of services provided that if, for example, one group in the population is known to use a particular service more than others, such as the elderly with reference to health care, then this is given more weight in calculation of the aggregate travel costs. But this may not be the same as minimizing total travel costs because the influence exerted by some groups may be out of proportion to their numbers in the total population. A third possibility is the minimax solution, which attempts to minimize inequality arising from the distribution of service nodes. A location is chosen which reduces the longest journey of any consumer to a minimum; an acceptable solution for those so affected, but these are likely to be a minority. This solution may, therefore, cause the majority to travel further than necessary in the interests of reduced travel for a minority of the population.

Covering models, on the other hand, set out to ensure that a particular service is available to all at a predetermined standard (White and Case, 1974). It is assumed that the value or quality of the service decreases with distance from source and the standard of service can, therefore, be defined in terms of the maximum distance or time appropriate to the service. Emergency fire and medical services fall into this category but other services such as public libraries could also be located in this way. These models are also another way of incorporating equity objectives into the location of public services, although the smaller the response time or distance, the greater the likelihood that a solution will favour the main population centres, thus contributing to an inequitable solution.

All four of these models operate on the assumption that users will travel to the nearest location for a particular service; for reasons evident from the earlier discussion in this chapter it is probably more realistic to asume that demand from any one source will be divided between a number of locations. The proportion consumed will decrease as the distance to succeeding alternatives increases. This is the intervening opportunities approach (Schneider, 1959), based on models of spatial interaction. It is a more complicated approach because, for example, the catchment areas for service facilities of the same kind will overlap. Most often used for modelling the location of services responding to inelastic demand, the intervening opportunities model may also be applied to elastic demand (Abernathy and Hershey, 1971) in which the total number of trips is dependent upon distance or travel time to the nearest location.

Operationalizing these models is often a complex process and not without difficulties which may affect the interpretation of the final outcome (Cooper, 1963; Wilson, 1976; Teitz, 1968; Hodgart, 1978). A simple example is cited by Hodgart (1978) for a model which locates a new facility, taking account of related facilities already in existence (Figure 6.8). The search from a starting position S is constrained by existing locations A, B and C because movement towards location M, which would minimize total travel to all facilities, would encroach on the catchments of existing centres and increase aggregate travel. Therefore, the choice of starting-point is important, and several 'runs' with

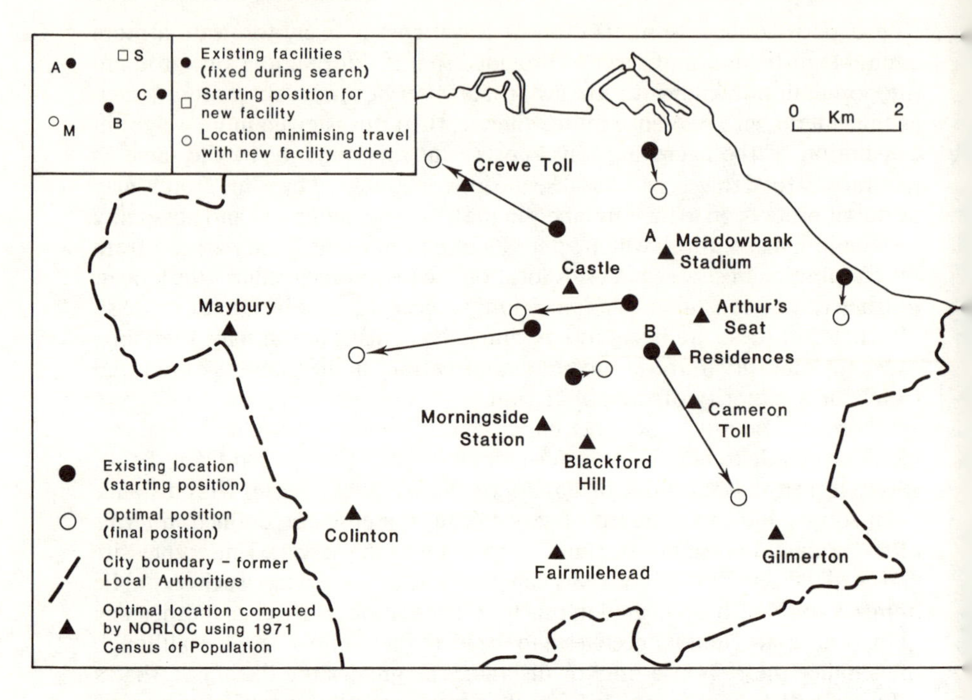

Figure 6.8 Comparison of actual locations of public swimming-pools in Edinburgh, in 1971, with locations which minimize aggregate travel
Source: Compiled from Hodgart, 1978, figure 5, 41; figure 9, 42.

different starting-points will be required in order to identify the optimum location pattern. Such models can certainly show that better solutions exist for providing public services and they can be used to indicate any inequitable characteristics of existing distributions as well as in any locational patterns proposed for the future. In a plannning context the ability to generate different solutions within given objectives and constraints provides an opportunity to identify those which achieve the most acceptable combination of efficiency and equity. As shown in Figure 6.8, there is a considerable discrepancy between the actual location of swimming-pools in Edinburgh and the distribution of locations which would minimize aggregate travel. Such a pattern might also disadvantage the less mobile groups in the population, and Hillman and Whalley (1977) have argued that this is a good reason not to provide large, diverse leisure centres at a small number of locations.

PRIVATE CONSUMER SERVICES – RETAILING

Among service industries as whole, retailing has probably received more attention from geographers than any other single economic activity. Consequently there is a large and ever-expanding literature devoted to retail geography (see, for example, Dawson, 1980, 1983a; Davies, 1976, 1979; Scott, 1970; Simmons,

1964). This can be broadly divided into three categories the first of which comprises studies which examine the distribution and structure of retailing in relation to the spatial and hierarchical principles of central place theory. Since the patterns which reflect central place principles depend on a number of assumptions about the behaviour of consumers, there is, secondly, also a substantive literature on various facets of consumer behaviour and perception in response to individual retail opportunities and to shopping centres as a whole. The third dimension of retail geography is that concerned with retail planning and public policies towards changing locational needs following the organizational and operational changes affecting the industry. The consequences of the location of retail services for questions connected with access and equity have received much more limited attention even though it can be suggested they are just as relevant here as they are for the provision of public services.

Although it has already been suggested that retailers make location decisions which contribute to the public good, in making these decisions they clearly attempt a realistic assessment of the economic viability of operating at one place rather than another. While profit maximization is undoubtedly the underlying principle governing the location of retailing in order to achieve it, retailers must be fully aware of all the possibilities open to them as well as to the consumers upon which, ultimately, they depend. The environment in which private consumer services operate is, in a spatial context, very variable: population is unevenly distributed; accessibility to all locations is not equal; consumer tastes and preferences vary between areas and at different times of the year; and consumer spending power is unevenly distributed and subject to quite rapid changes as the economic prospects of cities or regions are affected by, for example, the opening and closing of manufacturing plants or – usually with more short-term but none the less significant consequences – industrial disputes. In these circumstances profit maximization is an ideal for which retailers may strive to varying degrees, depending on their size and growth strategies. Small retailers may be satisfied with a location which will give them an adequate standard of living rather than maximum profits if only because the risks entailed in making locational adjustments may outweigh the subsequent improvements in profit margins. Pred (1967) draws attention to our very limited knowledge of the behavioural foundations of locational decision-making by entrepreneurs in the tertiary sector, a shortcoming which persists and which may largely be the result of the unwillingness of many retail enterprises to discuss their location strategies in detail (Davies, 1976).

It is, therefore, not at all clear whether retailers are really motivated by spatial variations in demand when making location decisions or, conversely, by their ability to shape demand by choosing locations on their terms. The latter is a real possibility because of the organizational changes taking place in the retail industry; the demise of the small retailer is well under way (Phillips, 1964; Kirby, 1975; Dawson and Kirby, 1979) and has largely been caused by the expansion of supermarkets, superstores, hypermarkets, and multiple groups operating at regional and national level in the durable and non-durable sectors (Parker,

1975; Thorpe, 1977; Gower Economic Publications, 1977; Davies and Bennison, 1978). Many of these groups are operated by such large companies that they are more concerned to make locational choices which take account of the existing distribution of the companies' retail outlets, or the land which they own and the costs of developing different sites. Indeed they may be able to bring about changes in unacceptable land-use zoning or restrictions on the design of retail structures because the communities in which they wish to locate are unwilling to risk losing them to some other location.

Despite this uncertainty about the relative importance of economic and behavioural variables for retail location patterns, there is a substantial literature, derived mainly from marketing geography, on methods of store location (Nelson, 1958; Applebaum, 1965; Kornblau, 1968; Davies, 1976; Kivell and Shaw, 1980). Most of the impetus for this work has come from the USA which has a large and more diverse retail system than most other countries as well as more space for development and a less rigid land-use planning system. The main principles involved in the process of selecting a new location for a single retail outlet have been listed by Davies (1976) as, first, the general position of a potential location within the city; secondly, the kind of business complex that a shop should be located in (alternatives might include conventional shopping centres in the CBD or along major arterials or at suburban subcentres, or space in local or major regional shopping malls); and thirdly, the most desirable site within each of the alternative centres.

Trading opportunities in the city are in a constant state of flux, so that monitoring of the firm's existing location(s) within the city along with identification of possible new locations is a recurring requirement. This involves at least four elements: estimates of existing trade area capacities including socio-economic attributes of the population and their patterns of expenditure; an assessment of market penetration, particularly by competitors who may, for example, have fewer stores in some areas than others and therefore leave opportunities for filling gaps where underprovision is thought to exist; the differential growth prospects in different parts of the city arising from changes in demographic characteristics of the population, redistribution from inner to outer cities, areas scheduled for new residential development, or trends in expenditure on different retail services; and the relative accessibility of the existing location in relation to new highways, changes in public transport schedules, improved parking facilities at stores operated by competitors, or more favourable customer–parking-space ratios permitted by other, adjacent municipalities. Data for these analyses is obtainable from secondary sources such as population censuses, or special censuses such as *County Business Patterns* in the USA or the *Census of Distribution* in the UK. In most cases these will have to be supplemented with more detail about customer behaviour, attitudes towards competitors and preferences, for example, with primary surveys conducted either by in-house location analysts or by external marketing or retail consultants.

Differences in the kind of retail service provided, market segments served and scale of operation from one store, one man ownership to multiple-store, corporate ownership mean that shops may prefer to be located in particular types of business complex. Therefore, they may need to obtain a number of items of information about, for example, the size of retail centres which will have implications for range and threshold; the latter may exclude some centres because it is too low to meet the requirements of certain types of store. Very important will be the functional composition of a shopping centre because spatial association with competitors may be good for sales (especially for retailers providing comparison goods such as shoes or women's fashions). Conversely, a retailer may wish to avoid certain other types of store or to identify locations where competitors are not yet established.

The flow of consumers in a business area will contribute to the sales potential of each shop located within it; therefore, the layout of a centre will need to be considered relative to the needs of a particular form of retailing. Some activities such as those dependent on comparison shopping will prefer a highly clustered layout, department stores will prefer layouts which provide the space which they require and some will prefer a layout which permits them to benefit from proximity to 'magnet' stores such as the major multiples or department stores. The layout of a centre may reflect its age and retailers looking for a new location may avoid older examples which do not incorporate the latest ideas in shopping centre design. Older centres may also be located in areas of the city experiencing urban decay and will not provide the image which a retailer considers acceptable to customers.

It is likely that more than one business centre will provide conditions suitable as a new location; the final choice will require a detailed evaluation of the actual sites at each of the alternatives. Such an exercise can be undertaken by the firm's own location analysts or be contracted out, but the end-result should include details of fixed costs such as rents, insurance charges, service charges for garbage collection, cleaning of shop frontages, maintenance, or property taxes, together with any limitations on the use of the premises through planning, health and other locally enforced by-laws or ordinances, or detailed measurement of the access characteristics of the premises such as distance from car-park exits, escalators, bus stops and the suitability of the store frontage for the display procedures to be employed.

This scenario for identifying a new location for a retail store comprises a logical sequence from macro- to microscale factors. In practice, the location search process will not be divided into discrete stages, but will probably involve elements of each, depending on the speed with which decisions must be made and whether the research involved is conducted by the company's own location analysts – who can call upon accumulated knowledge and tested techniques which suit their particular kind of retail service – or by outside consultants. The ultimate objective of the location analyst is to identify the right size of store for the location, and this involves forecasting retail turnover which should ideally be as high as possible. Davies (1976) provides examples of some of the procedures used, especially by store location analysts in the USA.

There have been relatively few store location studies undertaken in the UK. Kivell and Shaw (1980) attribute this to the limited options open to retailers in circumstances where planning controls are tight enough to allow little locational freedom and land availability. The prospects for retail development on virgin sites are also severely curtailed by land-use conflicts of a higher order than those extant in the USA because of the much smaller land area in the UK. The studies which have been undertaken are of limited depth (Kivell and Shaw, 1980) and rely on inadequate information about, for example, consumer behaviour. Recently, however, there has been growing concern among planners and some retailers about the impact of new retail practices, such as edge-of-town superstores, on established city centre and suburban shopping areas. This has motivated a growing number of impact studies (Bridges, 1976; Lee *et al.*, 1973; Thorpe, 1977) but these are more concerned with *what is* rather than analysing store location choice. Something may be learned about store location from these studies but only indirectly. It is also fair to say that, in common with studies in the USA, the main outcome of store location analysis in the UK has been refinement of methodologies and techniques of analysis rather than identification of any underlying spatial regularities in the location of retail services.

The outline of the stages involved in store location choice is essentially applicable at the intra-urban level. At larger scales commercial viability remains paramount, but when decisions are required about alternatives at the inter-regional level, it will be necessary to take account of factors similar to those relevant at the local level but which also occur at the larger scale. Hence regional patterns of national economic growth will be important (just as local patterns of economic growth will influence intra-urban location choices), along with any differentials in operating costs (arising from construction or leasing cost differences, for example) and presence or absence of policies for new settlements or the expansion of existing ones. At the lower, inter-urban level factors such as city size and the locations chosen by competitors, or those already occupied by branches of the same organization, will be relevant.

The accumulated evidence, taking account of our knowledge of the distribution and socio-economic characteristics of the urban population, suggests that there are grounds for the hypothesis that there are indeed variations in accessibility to urban retail services. It would appear that inner urban areas are provided with a higher density of shops and, by implication, superior accessibility than outer areas where retail service densities are lower. But proximity does not necessarily equate with functional diversity or quality of retail services. It has been noted that there are more lower-order centres in the inner areas and, therefore, the range of retailing will be limited to convenience and low-quality durable goods stores. In order to obtain a wider choice and higher-order retail services inner city residents may have to travel to the CBD or to other higher-order centres elsewhere in the city.

An individual's ability to exert an element of choice in these circumstances will depend upon income and mobility. This has led Davies (1972b) to suggest that the functional and hierarchical characteristics of urban shopping centres are a product of the socio-economic structure of a city. Indeed there are subsystems of shopping centres (Figure 6.9) in which there are differences in the

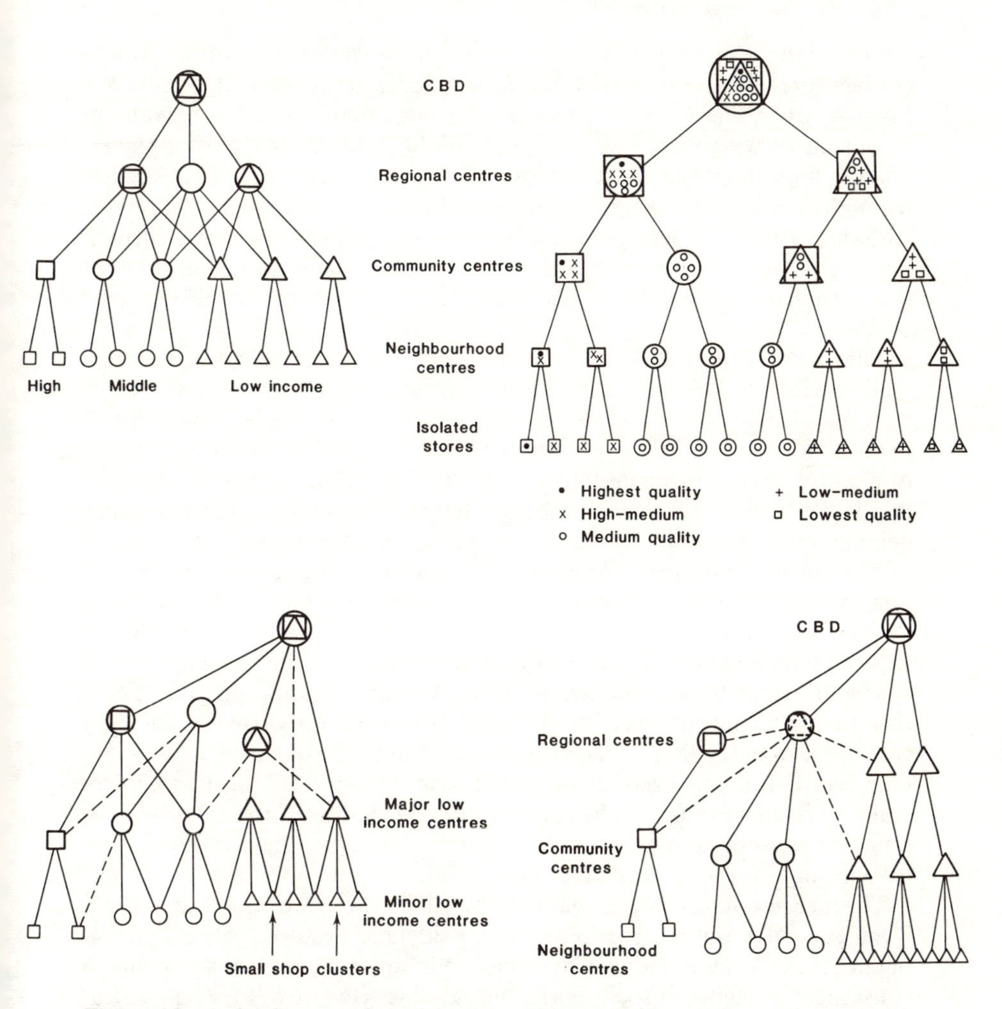

Figure 6.9 A developmental model of shopping centres in relation to income and quality differences
Source: Compiled from Davies, 1972b, figure 5, 69; figure 6, 70.

degree of hierarchical separation. The model is developed from the initial assumption that it is possible to distinguish high-, middle- and low-income neighbourhood centres (the lowest level in the retail hierarchy) which incorporate real and perceived quality differences in the kinds of good provided according to the status of the consumers which patronize them. Since low-income areas are likely to be nearest to the CBD and consumer mobility is lower than elsewhere in the city, the hierarchy of retail centres is truncated (Figure 6.9), so that regional centres are excluded and the large number of low-income neighbourhood centres develop some differentiation in the range of retailing available as noted by Garner (1966), who identified major and minor centres in working-class areas of Chicago. The greater dependence on centres low down in the retail hierarchy also means that a large number of individual corner-stores are able to survive and form a dense network of retail provision in inner urban areas. Such a network does not emerge in middle- and high-income areas where, because of higher mobility and the demand for higher-quality, high-threshold retail services, all four levels in the shopping centre hierarchy remain intact with one notable difference: regional centres compete with the CBD by attracting consumers from inner as well as outer residential areas. This may cause displacement of intermediate-level community centres (Figure 6.9), so that a full structural hierarchy is only present for high-income areas.

The model suggested by Davies, therefore, represents the provision of urban retail services in the form of a composite structure with a number of subsystems with their origins in the variable ecology of cities. This has led Potter (1980) to posit the view that these subsystems will incorporate differences in the quality of retail services in different types of centre. If such quality differences exist, they will influence the extent of shop catchment areas since Schiller (1972) has demonstrated the ability of such shops to attract a disproportionate number of high-income consumers. Potter (1980) identifies three main elements in retail quality: a commodity component which is made up of variations in price or quality between establishments which are offering the same type of good; a service component which comprises entrepreneurial skill or politeness of service; and a structural component which arises from the appearance of shops, their juxtaposition within centres, or standard of window displays. All of these contribute to the cognition of consumers and their subsequent choice of shopping centre. It seems likely that they will also influence the location choices made by retailers who will also have preconceived ideas of their status in the minds of consumers or may wish to change their image or market penetration by selecting locations in suitably fashionable shopping centres.

The measurement of shop quality as well as of shopping centres as a whole is fraught with problems (see Potter, 1979, 1980) but a detailed analysis of retailing in Stockport does indicate that the quality of retail services accessible to consumers is related to socio-economic variables (Figure 6.10). The association between retail area quality level and type of socio-economic area is highly significant ($X^2 = 30.02$, $p > 0.001$). Most of the high quality shopping centres are in the north-west and south/south-east of Stockport, although low-quality centres also occur well outside the inner areas along arterial transport routes. Shopping quality is not just a product of location relative to socio-economic areas, but also incorporates variations arising from age of development, relationship with major highways and morphology of the shopping centre or ribbon.

It is, of course, possible to measure in detail whether there are variations in access to retail services differentiated according to quality or position in the retail hierarchy. With specific reference to grocery stores which meet a ubiquitous type of consumer demand, Bowlby (1979) has attempted to identify groups with differential access. A sample of grocery shops in Oxford (grocery shops which offered fresh meat as well as tinned and packaged groceries, and

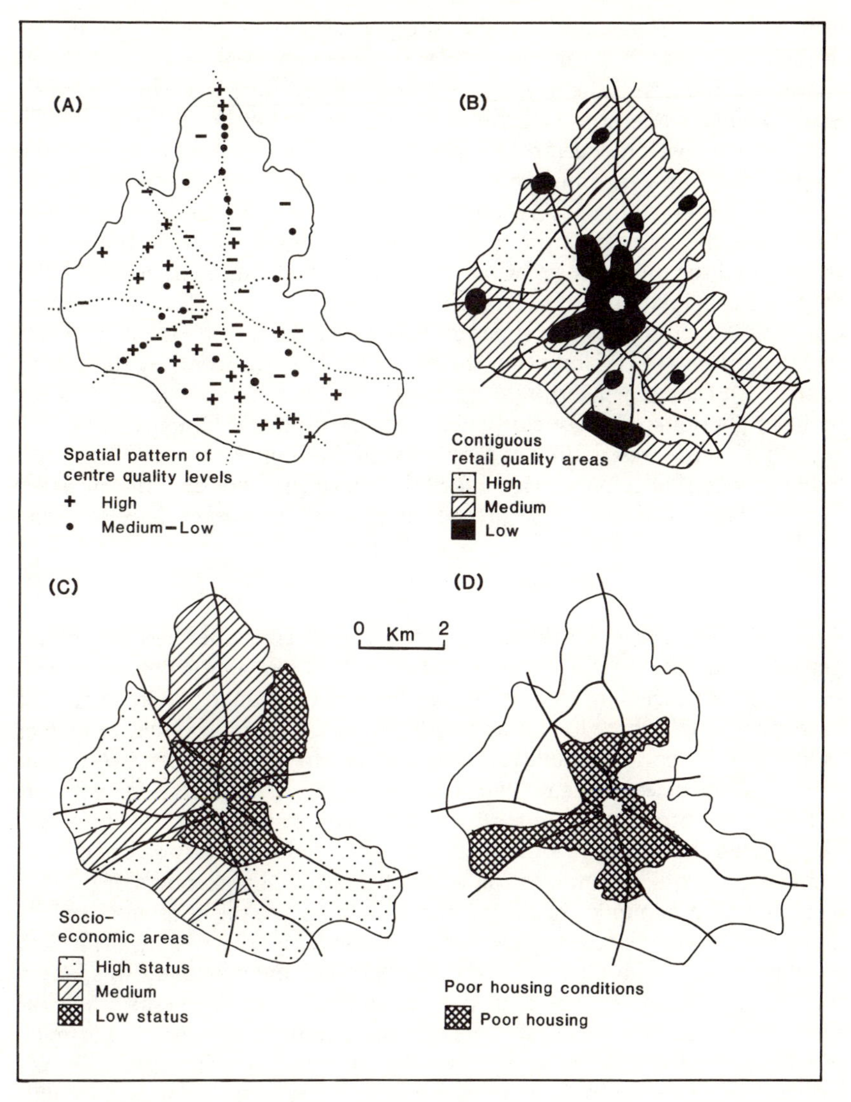

Figure 6.10(A)–(D) Variations in retail quality in the context of urban social patterns
Source: Potter, 1980, figure 6, 223.

which had at least two checkouts, were classified as supermarkets) and a random sample of households was used. Some 98 per cent of households were located within 0.5 miles of a small grocer. The median straight-line distance was 0.12 miles. These figures compare well with those produced by Guy (1977) in a study in Reading (see also Thomas, 1974).

It would seem, then, that access to local grocery stores is not a problem in Oxford, but Bowlby (1979) finds that not all shoppers find access to shops easy or simple. For many consumers, especially the young or high-income households, these shops are simply places to obtain urgent or 'oversight' purchases (i.e. items which should have been obtained during a trip to a supermarket). High-status households tend to use, as postulated by the Davies (1972b) model, the higher-order centres for grocery purchases and these usually involve longer trips. Low-income groups are significantly more dependent upon neighbourhood facilities. They are, therefore, of most use to the elderly, who also value the social role of small stores and considered that convenience outweighed the higher costs of purchases compared with supermarkets. But the purchasing patterns of the elderly, combined with the effects of competition from larger grocers and supermarkets, make the future for small grocers rather perilous. In the long run, therefore, the level of access by the elderly to small grocers may deteriorate at a time when their numbers are increasing in the total population. For convenience goods in general Thomas (1974) suggests that over-representation of families with children of pre-school age leads to high levels of allegiance to local facilities. In general, however, the influence of age structure on local shop dependence levels has not been widely examined.

OTHER SERVICES

Retail services have a very direct dependence upon consumer needs and must make location decisions accordingly. There is, however, another group of consumer services which must also take account of consumer requirements and have increasingly come to compete with retailers. These are activities such as estate agents (realtors), building societies and, to a lesser extent, insurance offices and agencies. As Egan (1983) has illustrated, estate agents dealing mainly in residential properties require to ensure adequate general accessibility for their clients (both buyers and sellers) and this leads to a very steep bid-rent curve and an equilibrium location within the CBD. The steepness of the bid-rent curve is compounded by the fact that accessibility to the information which they provide is part of the service. Hence during recent years estate agents and building societies have become increasingly prominent occupants of space in the principal shopping streets of downtown areas (see also Fernie and Carrick, 1982). They are purchasing general accessibility at the expense of retail services and are able to do so because of their much more limited floorspace requirements relative to the earning potential of central as opposed to marginal locations. There are, of course, negative externalities for the users of shopping areas in terms of a more dispersed pattern of retail services, greater travel distances within the shopping centres and the appearance of 'sterile' street frontages which reduce the ambience of a shopping area. Similar processes are taking place in shopping centres lower down the hierarchy where local rather than general accessibility is the main requirement, along with agglomeration and external economies, which is proving sufficient to squeeze out retail services to slightly more marginal locations.

SUMMARY

While there are certain principles which guide the location of consumer services, whether provided by the market or non-market sectors, this does not mean that all are uniformly accessible to users. In this chapter it has been shown that an equitable distribution of public services such as health care, education, or fire services is easier to specify in principle than it is to achieve in practice. A large number of variables relating to both the consumers and the suppliers of these services ensure a mismatch between need and availability. It is suggested in the last section of this chapter that variable access to private consumer services also occurs because of the expressed locational preferences by activities such as retailing for markets which are stable or expanding, and which provide an acceptable rate of return on investment and turnover.

Accessibility, which can be defined in a number of ways, is the key to measuring the welfare function of a consumer service. The extent to which it acts as an asset or a constraint is partially dependent on individual financial resources; some groups in the population can discriminate more between alternatives or overcome deficiencies of availability by spending more on travel, using a car, or purchasing education or health care from private sources rather than depending upon the facilities provided by the state. Those with more limited financial resources or otherwise disadvantaged, for example, because of age or ill-health, are much more dependent upon satisfactory access to the nearest available services. But proximity may not equate with quality or convenience because of the limitations imposed by family time budgets or the inadequacies of the public transport service. The family car may not be available for use at times which synchronize with those at which the nearest dentist or general practitioner is available for consultation. The spatio-temporal availability of general practitioner services in Adelaide has been used as an example and demonstrates that while there may be an adequate number of practitoners at a suitable number of locations, not all are accessible at all times of the day; some parts of the city have a more comprehensive availability than others. In the USA, where a large proportion of health care is obtained on a fee-paying basis, population size rather than income in a region or city has been shown to explain most of the variation in the distribution of medical practitioners. Highly specialized medical practitioners will locate in the cities which provide access to an adequate threshold population rather than those with the largest proportion of higher-income households.

One feature to emerge in many of the studies of health service location has been the lag between demand and supply. There is considerable inertia in the location patterns of public services which leads, for example, to overprovision in inner cities which are losing population and underprovision in suburban areas which have experienced rapid growth as a result of voluntary outmigration from inner cities or expansion of peripheral housing estates constructed by local authorities. Access to primary health care is, therefore, often far from equitable because spatial access is a major determinant of the quality of the service. Attitudes and the value attached to health care or education further confound attempts to match provision with need.

As the section on hospital services has shown, consumer access considerations have to be balanced against economies of scale. Decisions to rationalize and/or to centralize hospital services invariably involves withdrawal of facilities at some locations and the introduction of facilities not previously available at others. The result is to change accessibility for patients and visitors who may have to travel further in some instances but shorter distances in others. The best solution is one that maximizes patient needs, on the one hand, and creates savings or improvements in efficiency for the health authority or group of hospitals, on the other.

This kind of problem is amenable to solutions using optimizing and covering models because it can be assumed that when locating public sector consumer services they are not generally in competition with other facilities. Elegant alternative solutions can be devised when using such models but converting the patterns which they generate into actual provision is difficult because some facilities will already be in place, so that inertia is again present. Only the location of new facilities can be adequately guided by these models.

The chapter is concluded with a discussion of access to private consumer services, mainly retailing. This has been included in the belief that notions of equity and access are just as relevant in this context. The quality of life for consumers is affected to a degree by access to all types of service which they may need to use. Uncertainties about the relative importance of economic and behavioural variables for retail location patterns, structural changes in the retail industry and variations in the socio-economic characteristics of market areas all lead to variations in accessibility to retail services of any given type. Inner urban areas offer better access but, as for public services, proximity does not necessarily equate with quality or diversity. Income and mobility again determine individual households' ability to exert choice, and this causes differences to develop in shopping centre quality and distribution.

CHAPTER 7

Inter-sectoral dependence, centralization and producer services

INTRODUCTION

Although it has been suggested that it is difficult to make a hard-and-fast division of services between producer and consumer types, it is useful to continue an examination of location in the context of this dichotomy. Therfore, the focus of this chapter will be those services mainly involved in providing their output to other firms as inputs for further production, including wholesaling; a large part of insurance, banking and finance; a diverse group of professional services such as accounting; various types of consulting such as management, marketing, or engineering research and advice; advertising; legal services and trade and other professional organizations; and research and development. These services mainly provide white-collar employment but activities with a large proportion of blue-collar labour such as office cleaning, security services, or food and drink vending machine services are also included among the producer services (Wood, 1983).

REASONS FOR THE EXPANSION OF PRODUCER SERVICES

The ability of organizations to create their output by employing the appropriate specialist staff in-house has decreased as occupational skills and knowledge, especially in information technology-related activities, have become highly specific and therefore expensive to retain. It is now becoming more cost-effective, even in large enterprises, to buy in producer services if only because those providing them are, through their continuing and diverse experience with the specialism, able to provide the most up-to-date and comprehensive service. It is also likely, however, that the expansion of producer service employment has been encouraged by the tendency for business enterprises to get larger, thus making the management task more complex in an administrative sense and, where appropriate, for production.

Business enterprises can grow through internal expansion whereby existing markets are developed further, new markets in other geographical areas are developed, or new product lines introduced (Lloyd and Dicken, 1977). There will be limits to such activity, however, which are set by the response of competitors, the likelihood of market saturation, or vulnerability to the vagaries of demand for the product. A more effective way to grow, and a method which has increased in frequency during the last two decades, is to acquire other enterprises or to merge with them. Although revealing cyclical characteristics which reflect the condition of the national economy, the number of mergers and acquisitions in the USA increased from 3365 in 1955–9 to more than 8000 between 1965 and 1970 and in excess of 5000 during the recession of 1970–4 (Stanback, 1979). In wholesale and retail trade the number of acquisitions and mergers increased from approximately 100 in 1955 to 400 in 1968 and in other services from approximately 60 to more than 600 (Lloyd and Dicken, 1977). Horizontal and vertical integration is also an important part of merger and acquisition strategy but more recently there has been increasing acquisition of firms which are very dissimilar, in product or service, to the parent corporation. Such diversification greatly expands the range of contacts with different suppliers, different methods of production, or marketing practices, for example, and presents the managers of such conglomerates with very complex integration and other problems which require prompt and effective solutions if the ultimate goal, corporate profitability (an important motivation for merger and acquisition), is not to be adversely affected.

As companies become more complex they will modify their internal structure to ensure that chains of commmand, the transfer of information between departments or plants, or decision-making is effectively and efficiently undertaken. The principal stages in the development of organizational structures, starting with a one-person organization through to a third stage represented by an organization with a divisional structure, are shown in Figure 7.1. The most important aspect from the standpoint of the growth of producer services is the growing separation of functions as organizations become more complex. This can take the form of a greater intra-organizational division of labour such as the subdivision of an accounting department into credit control, invoicing, debt collection and corporate taxation; or the creation of a department for corporate client accounts, another for private accounts and a third handling accounting work connected with overseas divisions and clients. Such functional fragmentation may be accompanied by spatial fragmentation, arising from the geographical distribution of establishments acquired as a result of a merger; branch offices; or plants responsible for producing different product lines or serving particular markets. Functional and spatial separation in the organization of economic activities has, therefore, provided a source of demand for producer services whether supplied from within the organization or consumed from external sources.

Headquarters and the demand for producer services

While the individual entrepreneur is able to oversee all his activities from the same point at which he is producing the good or service, the stage II and stage III organization (see Figure 7.1) requires a place, suitably located, from which overall control can be exerted. A headquarters should preferably be in a central position in relation to the units to be co-ordinated and directed and it is not necessary, therefore, for it to be located in the same place as, for example, the principal point of production or the largest concentration of an organization's insurance agents or branches. In order to appreciate the implications for

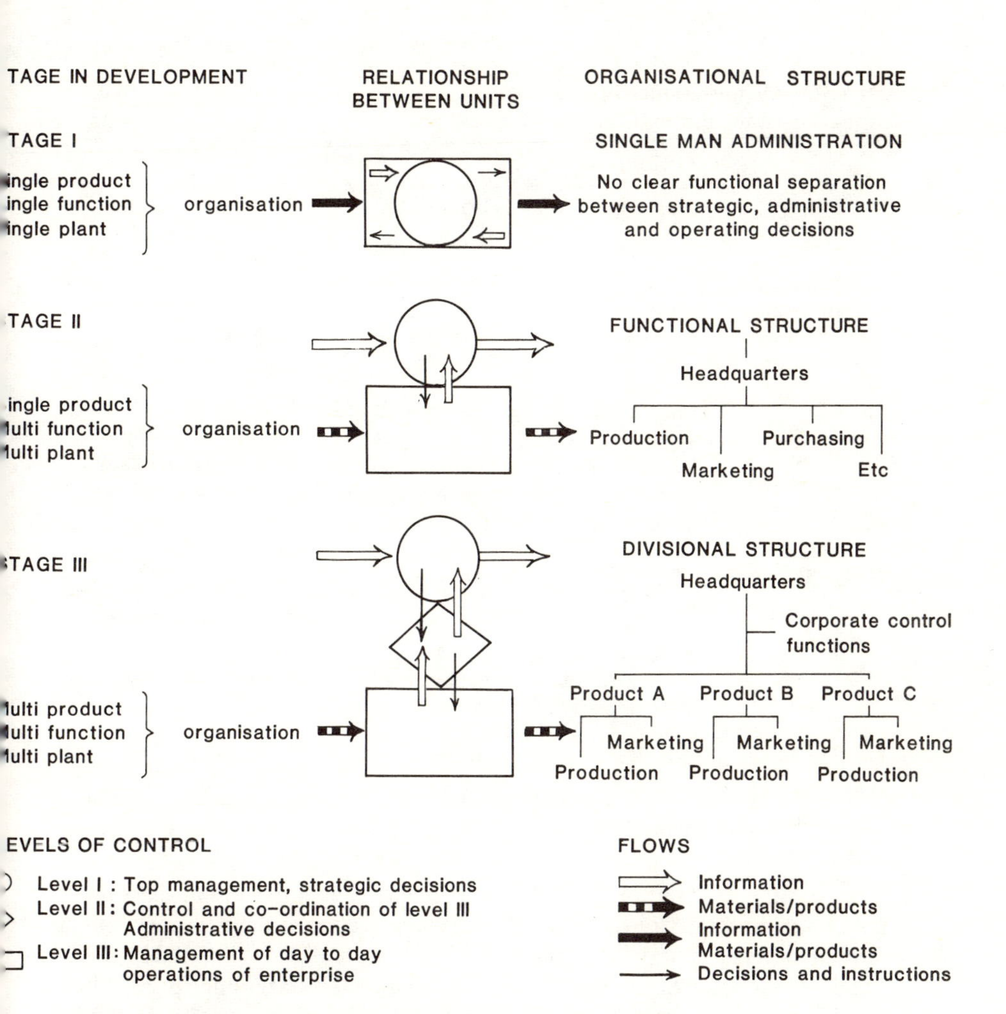

Figure 7.1 Stages in the development of organizational structures
Source: Lloyd and Dicken, 1977, figure 11, 367.

producer services it is useful to briefly consider the role of the headquarters establishment in the activities of organizations. First, they are primarily centres for administrative functions which can be divided into those largely concerned with routine day to day operations such as accounting, provision of supporting office services to divisional and subsidiary establishments, and supplying routine information to other functional units within the headquarters; those concerned with the executive control and direction of the activities taking place at other locations; and those functions which are largely concerned with decisions and planning relating to the development of the organization through product improvements, expansion, mergers and acquisitions or the design and introduction of new services or products. It follows that the headquarters of an organization, irrespective of its size and diversity, generates a

Table 7.1 Organizational structure, contact patterns and locational characteristics of multi-locational, multi-functional corporations

	Organizational Unit		
	Head office	*Group head office*	*Manufacturing plants/branch offices*
Organization–environment interaction	Orientation processes	Planning processes	Programmed processes
Contact networks	Face-to-face, prearranged, government, research, finance, business	Telecommunications regular meetings, familiar individuals	Telecommunications, routine meetings, frequent contact, confined
Functions	Decision-making, planning, negotiations, search, product development	Control, direction of production, data processing, services to head offices, management	Routine office work, processing of materials, handling of goods, construction
Position in hierarchy	National metropolis	Regional centres	Local centres

Source: Stephens and Holly, 1981, table 1, 287.

large volume of internal and external contacts. The latter are at least as (if not more) important for the survival of the organization as the efficiency sought through devising channels and mechanisms for more effective internal linkages.

The relationship between the organizational units of large enterprises, their primary functions, contact networks and locational attributes has been usefully summarized by Stephens and Holly (1981). Headquarters establishments are primarily involved with orientating the firm within its business environment (Table 7.1). Divisional or group headquarters will have some orientation tasks but their main concern is to give control and direction to the production of the good or service with reference to the strategic and other decisions made at the main headquarters. Finally, the manufacturing plants or branch establishments of service firms are largely involved with programmed processes such as the processing of materials or various routine office tasks. The consequences for the contact networks of these different organizational units are also shown in Table 7.1. As a broad generalization there is an inverse relationship between the volume of programmed processes and the incidence of face-to-face contacts and prearranged meetings.

Information availability is, therefore, central to the role of headquarters establishments (Goddard, 1975; Törnqvist, 1973). The information used can be in response to the requirements of programmed situations (actions and procedures familiar to those responsible for routine activities within the organization) or unprogrammed situations in which the organization must respond to, or anticipate, negotiate, or plan, according to its strategic goals, the changing demands of its clients and markets, together with changes in the spatial pattern of demand or supply of inputs (including labour), or new tax regulations or controls on product safety. These are just a few examples of circumstances where the organization cannot use predesigned and tested procedures to meet such external and, perhaps, unanticipated influences on its performance or competitiveness.

The quality of information is closely related to the level of choice among information sources and accessibility to potential contacts (Code *et al.*, 1981). For linkages in which face-to-face interaction is important, such as orientation contacts, the choices available can be measured by using aggregate indices of accessibility derived from an information potential model (see ibid.) which takes the form

$$V_{ij} = \sum_{k=1}^{n} \sum_{l=1}^{m} w_{ji} \,.\, P_{kl} + (d_{ik}^{b} + K)$$

where V_{ij} = the potential interaction of one unit of activity j in areal unit i;
w_{ji} = the weight given to interaction j with activity i;
P_{kl} = the number of employees in activity l in areal unit k;
d_{ik} = the straight-line distance from areal unit i to areal unit k;
m = the number of different office activity types examined;
n = the number of areal units in the study area;
K = distance constant;
b = distance exponent.

This model has been applied to data for metropolitan Toronto in which the weighting of interaction linkages is a function of contact frequencies in downtown (using data collected by Gad, 1979). Potential contacts are assumed to be a function of office type and number of employees, and potential interaction is a function of the employee population, by office activity, by district (thirty-four in total) and weighted by frequency of contact with related and other office types. Even when the distance exponent is varied from 1.0 to 3.0 (most sensitive to distance), downtown clearly emerges as the nucleus of business and related information in metropolitan Toronto (Figure 7.2). One suburban area, Don Mills, is also prominent but is located close to the core area. But information potential varies between functions, so that with a distance exponent of 3.0, over 76 per cent of the aggregate information potential for investment dealers is in central Toronto, 43 per cent for advertising agencies, 51 per cent for public relations consultants and 40 per cent for integrated oil companies.

The range of information which has actual or latent value for the operations of any organization is very wide and it is, therefore, important that those parts of the organization responsible for processing that information should

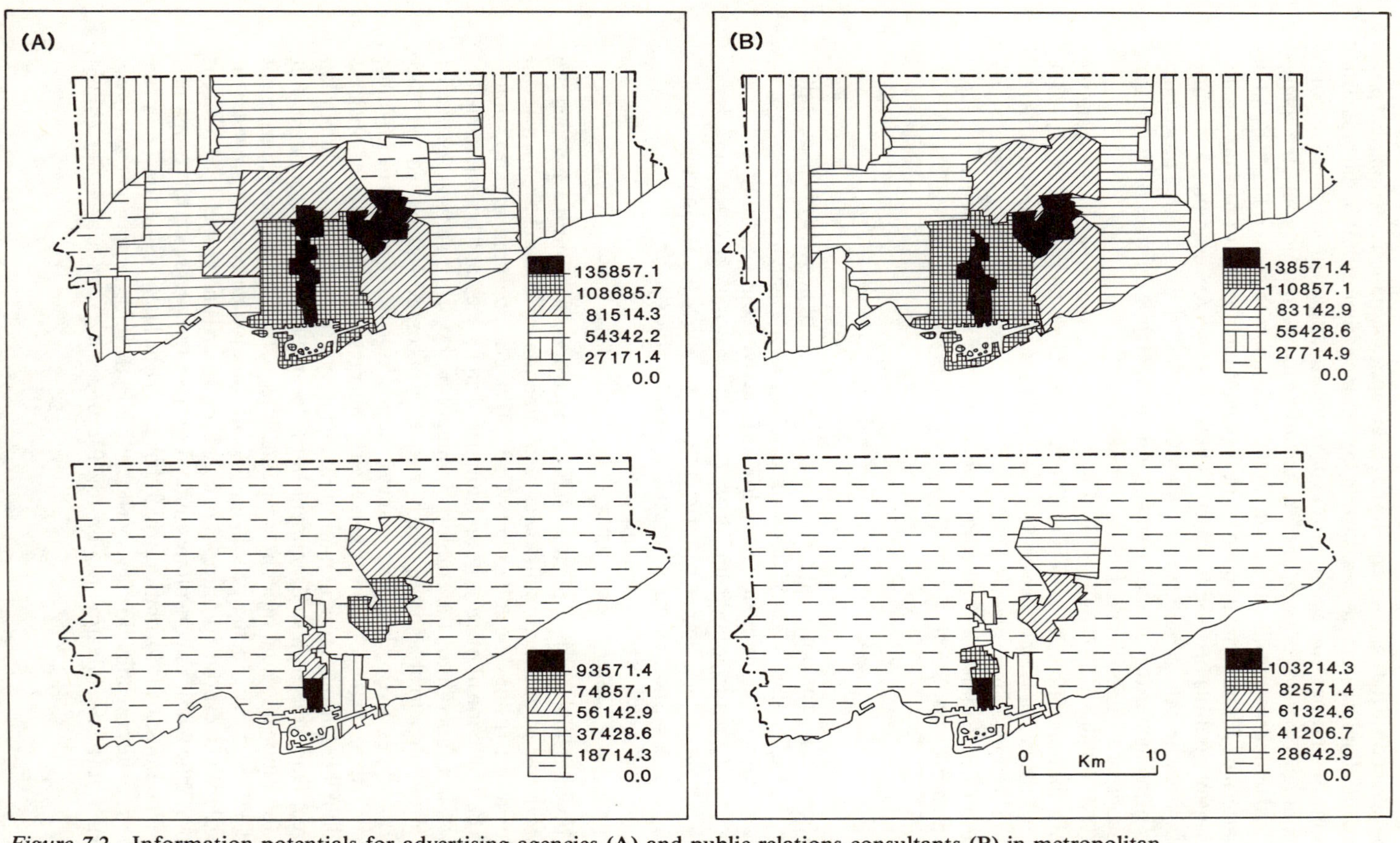

Figure 7.2 Information potentials for advertising agencies (A) and public relations consultants (B) in metropolitan Toronto, with distance exponents of 1.00 and 3.00
Source: Code *et al*., 1981, figure 10, 40; figure 13, 43.

be able to receive and assimilate it, to negotiate over it or exchange it, as quickly and reliably as possible. Location, then, becomes a significant consideration since some places will, all other things being equal, be better placed to gather information than others. Of course, certain kinds of information such as Stock Exchange prices, premiums for insurance policies, statistics of all kinds, contracts for the supply of a good or service, or advice on how to resolve a specified computer software problem can be relayed between spatially remote locations using telecommunications devices such as telex, facsimile, computer links, or the postal services.

But a great deal of information exchange depends upon direct personal contacts which will bolster confidence in the conduct of negotiations and agreement over transactions or allow the exchange of information which might not otherwise be transmitted. Such face-to-face contacts can be achieved by travelling to meetings but this imposes resource costs which may be of unacceptable magnitude (Civil Service Department, 1973) because the individuals involved are likely to be receiving some of the highest levels of remuneration in the organization; time spent travelling is unlikely to be 'productive' (Pye, 1979) and proximity to contacts then becomes a priority and an important influence on location choice. As well as allowing clients to reach them easily, organizations may also want to utilize a range of specialized producer services such as advertising, executive search and selection, or financial planning and advice (see Marshall, 1979).

In all these circumstances management needs to depend more than ever on the specialist inputs from producer service firms. A good example is the growth of financial institutions which hold a growing stake in corporate activities because of their ability to provide debt finance at the scale required to maintain the growth of multi-plant, multi-product enterprises. It is also likely that the growing application of government controls and legislation to all facets of business activity from employment conditions to export procedures has increased the demand for accounting, legal and actuarial services. The growth of manufacturing and service companies has also involved expansion into territories outside their home markets where legal, financial, marketing and other requirements will probably be different, thus creating demand for internally provided advice in these fields or the use of local producer services in each of the countries involved.

One other issue in respect of the demand for producer services is the extent to which complementarity with the manufacturing sector is a major stimulus. Undoubtedly there is a close association between them because, as Stanback (1979) suggests, the very process of industrialization and the associated economic growth would not have been possible without the 'enabling' role played by transportation, by wholesaling services, or by the sophisticated services provided by financial institutions, especially the rise of the banking system. Looked at from this perspective the relationship is not one of manufacturing first, producer services (or indeed services in general) second, but one of mutual interdependence where one set of activities could not properly function without the other. The role of manufacturing as a catalyst for subsequent supply of producer services has probably been declining in more recent times because of the ability of the latter to redirect its output to the individual by way of tax advice, investment counselling, diversification of insurance schemes for savings, loans and school fees. This is reducing the dependence of business and professional services on corporate and similar clients.

THE LOCATIONAL CONCENTRATION OF PRODUCER SERVICES

Changes in the structure of organizations, together with increasing specialization of producer services, have promoted a symbiotic relationship between headquarters establishments (some of which may themselves be producer services) and the growing range and number of business, financial and professional service firms. The symbiosis is manifest in a search for locations which are rich in information, permit high contact intensities and are very accessible to regional, national, or even international telecommunications and transportation networks. Consequently the search for agglomeration economies leads to a high degree of spatial concentration in the location of producer services and this will now be illustrated with a number of examples from studies undertaken in Europe and North America.

A basic indication of spatial concentration of producer services is the distribution of the headquarters of a number of non-industrial company headquarters in the USA in 1981 (Figure 7.3). The fifty largest diversified service companies were included in the annual *Fortune* listings for the first time in 1983 and are particularly significant because they primarily sell skill rather than stocks, money, insurance, or transportation. The range of services which these companies provide is wide and includes broadcasting, stock market and business forecasting, industrial complex designs and the distribution of raw materials to industry. Almost one-half (twenty-one) of the top fifty companies in this group have headquarters in New York, primarily Manhattan; only Beverly Hills (which could be classified as Los Angeles SMSA) and Houston, with four headquarters each, rival the former's dominance. New York has a smaller share of the top fifty diversified financial company headquarters (fourteen) which are spread more widely between the locations shown in Figure 7.3.

In common with financial companies in the UK (Daniels, 1983a) insurance, and to some extent commercial banking, headquarters are more dispersed than agglomeration theories would lead us to expect. Apart from the increasing opportunities to control many facets of organizational activity using modern information technology and telecommunications, these companies will often keep their headquarters at those locations associated with their historical origins. They will still have major, even if subordinate, offices in the major agglomerations such as New York, Chicago, Dallas and Los Angeles. The headquarters of retail companies, utilities and transportation (including air transport, railroad companies and shipping companies) also have a distribution which is more dispersed nationwide and dominated rather less by New York and other cities in the north-east. Nevertheless, if the four maps in Figure 7.3 are laid over each other, it would be shown that the location of US service industry headquarters, at least at the upper end of the size range, is dominated by nine metropolitan areas: New York, Chicago, Los Angeles, Dallas–Fort Worth, Houston, Seattle, San Francisco, St Louis and Minneapolis–St Paul.

Similar features are displayed by the distribution of the headquarters offices of the top 500 manufacturing corporations in the USA (Armstrong, 1972, 1979; Semple, 1973; Pred, 1974; Burns,1977; Borchert, 1978). In 1975 almost 21 per cent were located in the New York SMSA and a further 9 and 4 per cent respectively in the next largest SMSAs, Chicago and Los Angeles. St Louis, San Francisco, Houston and Minneapolis–St Paul are all included in a group with approximately twelve headquarters each, although it is notable that major manufacturing centres in the north-east such as Cleveland and Detroit are prominent in the list of industrial headquarters but less so in the list for service industries. This provides a pointer to temporal trends in the location behaviour of both consumer and business services (considered in Chapter 8).

The distribution of a much larger number of central administrative offices and auxiliaries (24,780) confirms the characteristics revealed by the top 500 companies, but probably because of the wider range of establishment sizes represented, the dominance of the very large metropolitan areas is reduced with New York SMSA accounting for 8 per cent of the total, with Chicago 5.5 per cent and Los Angeles (the third largest concentration) almost 5 per cent (Armstrong, 1979). The preference of headquarters establishments for large metropolitan areas is also clearly demonstrated by data for the UK (Evans, 1973; Goddard and Smith, 1978) and for Western Europe (Dunning and Norman, 1979) and Japan (Abe, 1984). According to Phillips (1982), 47 per cent of the head offices of manufacturing and construction firms in Canada are located in Ontario (mainly Toronto and its satellite cities) and Quebec (29 per cent) (see also Semple, 1977). The other provinces fare rather better for the head offices of sales and service sector firms, with 46 per cent in Ontario but only 19 per cent in Quebec. British Columbia and the Prairie states are the principal beneficiaries, with 17 and 13 per cent of the head offices of sales and service firms respectively.

PRODUCER SERVICES, THE CITY SYSTEM AND CORPORATE COMPLEXES

The expansion of producer services is, therefore, contributing to extensive restructuring of the system of cities (Stanback and Noyelle, 1980, 1982). The most obvious manifestation of this is the appearance of a small number of dominant nodal centres in which the servicing of corporate headquarters and divisional headquarters by a wide variety of high-quality and specialized businesses, professional and financial services is the *raison d'être* rather than manufacturing and related activities. This is not only causing major metropolitan areas such as London, Paris, New York, or Toronto to change their functional structure, but is causing cities which were formerly unimportant as nodal centres for producer services to emerge as significant participants.

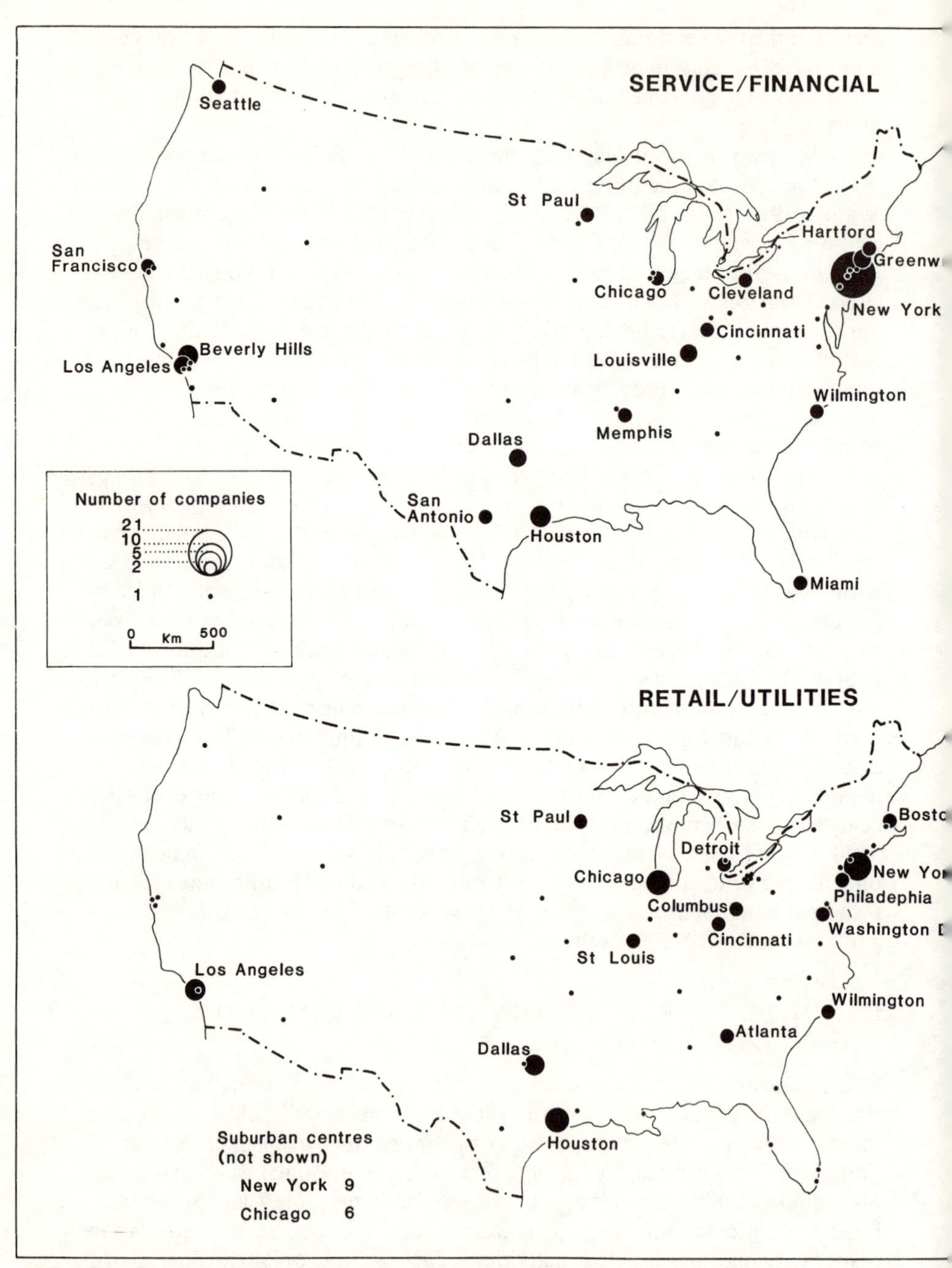

Figure 7.3 Location of the headquarters of the top fifty service/financial, commercial banking/life insurance, retail activities and transportation companies in the USA, 1982
Source: Compiled from data in *Fortune*, July 1983.

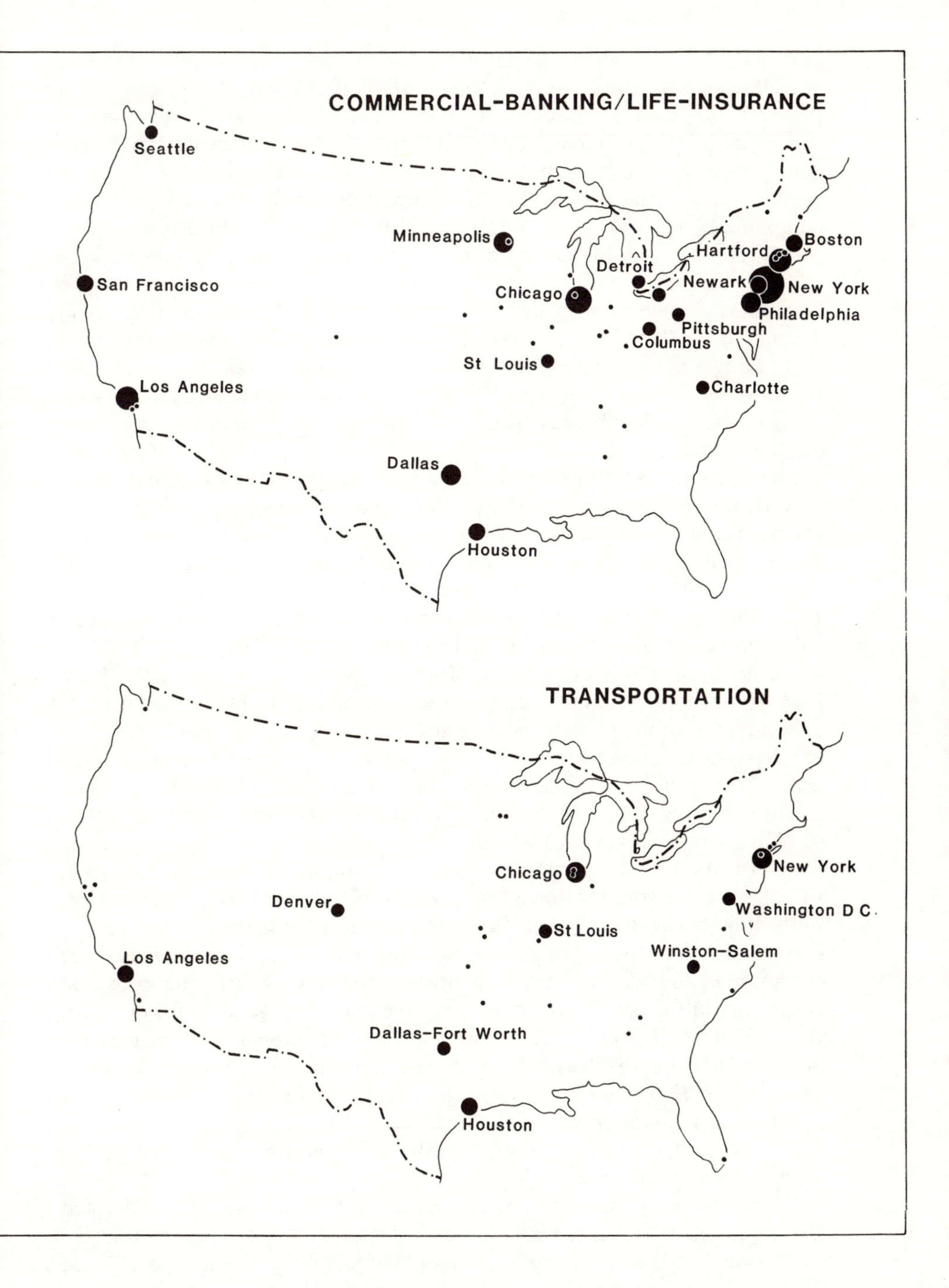
COMMERCIAL-BANKING/LIFE-INSURANCE
Seattle
Minneapolis
Detroit
Hartford
Boston
San Francisco
Chicago
Newark
New York
Philadelphia
Pittsburgh
Columbus
St Louis
Los Angeles
Charlotte
Dallas
Houston
TRANSPORTATION
Chicago
New York
Denver
St Louis
Washington D C
Los Angeles
Winston-Salem
Dallas-Fort Worth
Houston

One of the best examples of this is the rise of the so-called sun-belt cities such as Dallas, Houston and Atlanta during the last fifteen years. Lower-order centres such as Akron, Rochester and Charlotte have also gradually emerged as significant locations for corporate headquarters and their associated or dependent services. Cohen (1979) has described these nodal cities in the USA as complexes of corporate activity in which producer services are the dominant component (see also Harper, 1982; Corey, 1982). Hence Stanback and Noyelle (1980) estimate that 46 per cent of the employment in the largest complexes of corporate activities was located in just twenty SMSAs (these were also the largest SMSAs when ranked by population). However, these same SMSAs only accounted for 30 per cent of total US employment. There are also indications that just as certain metropolitan areas specialized in certain manufacturing functions, corporate complexes will also become specialized in the kind of subsidiaries and divisional head offices which they attract or in the corporate headquarters located there.

The acceptance of the notion of a corporate complex depends upon evidence for interdependence between the activities within the complex. There have been several studies by geographers of the linkages between offices many of which will be producer services; in the central areas of cities such as London (Goddard, 1973), Toronto (Gad, 1979), Edinburgh (Fernie, 1977), Dublin (Bannon, 1973), Melbourne (Edgington, 1982b) or in Wellington, New Zealand (Davey, 1972); and at regional or subregional levels in Sweden (Thorngren, 1973; Törnqvist, 1970), the UK (Goddard and Morris, 1976) and Canada (Polese, 1981). Face-to-face meetings, telephone contacts, the degree of prearrangement of meetings, the number of participants and their status; travel times or duration of telephone calls; whether contacts represented orientation, planned, or programmed activities; and the frequency of meetings or calls have all been monitored using diaries completed by a number of employees in a wide range of office establishments.

With respect to the growth of corporate complexes, the most important finding of these studies, which broadly corroborate each other, is that many of the linkages are a product of functional connections between, for example, banking and finance, or business services and publishing. Some services, such as banking and insurance, also have diverse connections with other proximal activities while others have very poorly developed linkages (see, for example, Gad, 1979; Edgington, 1982b). Note that these functional and 'diverse connectivity' linkages involve producer services in particular and, although Alexander (1979, 25) rightly urges a cautious interpretation of 'spatial proximity as a *necessary condition* for the maintenance of these links', their role as agents in the development of corporate complexes is certainly supported by the available empirical evidence.

Corporate complexes need not necessarily be confined to narrowly defined geographic areas such as central business districts. This has been demonstrated by Browne (1983) in a study of the relationship between high technology and business services in New England. In common with high-technology manufacturing, computer and management services are producer activities which depend heavily on the availability of professional and technical manpower. Access to the expertise in the major universities is also important. Although competition generates high labour costs, computer and management services can be expected to share the same locations with high-technology manufacturing. California and Massachusetts account for a disproportionate share of all US employment in computer and data processing services, in management consulting and high technology employment but the correlation is weak and, although positive, is not statistically significant (Browne, 1983). The principal reason for this situation is the incidence of both services at a high level at a number of locations where high-technology industries are under-represented. The best example is the District of Columbia, Maryland and Virginia which together account for 11 per cent of US employment in computer and data processing and 9 per cent of the employment in management consulting. By comparison, this area only accommodates 3 per cent of US high-technology employment. A similar but less exaggerated concentration of these two producer services is found within New England, where Massachusetts is the one state with better representation than expected.

There is clearly a substantial demand for computer and management services from federal and related agencies located in Washington, DC, and the adjacent states. Perhaps this reflects a strong market orientation in their location which is also revealed by a high incidence in those metropolitan areas such as New York, Chicago, Boston, Los Angeles and San Francisco which have a large number of financial and corporate headquarters. It remains uncertain, however, whether market orientation is more important than the availability of a large pool of professional and technical workers associated with such corporate complexes. In addition, it may well be that the growing penetration of computers into less knowledgeable sectors of the market has increased the need for face-to-face contacts, thus supplementing the established requirement for assistance with software problems, instruction in the installation, use and problem-solving connected with computer operations, or the technical assistance usually required on an ongoing basis when time-sharing computer services are used.

The centralization of many producer services is also encouraged by the fact that, in general, many are small establishments for which labour costs account for 40–60 per cent of revenues. Labour and the value of its time is a major part of the service provided, so that there is much less scope for using economies of scale in the way demonstrated by many manufacturing industries. Hence a management consulting firm can only add to its client list by recruiting further consultants with a resulting increase in overheads more or less equivalent to the increase in revenue. This leads Browne to suggest that the larger the proportion of 'unsupported' or 'unaugmented' time involved in giving a service (i.e. no standard solutions or equipment to assist with providing the service), the more likely it is that producer service establishments will be small (see ibid.). The average size of US management consulting establishments is only twelve employees, while in computing, where some economies of scale are possible (in software services, for example) the average size is between twenty and thirty employees. This compares with average sizes for office and computing machinery establishments of 250, and for electrical and electronic equipment, 150.

Accounting and advertising complexes

There are other examples of inter-industry relationships between producer services and other industries which are manifest in a spatial context. Stanback and Noyelle (1980) have collated some interesting data for the location of accounting and advertising services in the USA. Accounting is a 'top-heavy' producer service with just eight large firms employing some 20 per cent of the total employed by the industry but handling a major share of the auditing and related services required by the corporations listed in the American and New York Stock Exchanges. Much of the work available from federal, state and local government agencies also gets channelled towards the very largest accounting firms and Stanback and Noyelle note that such is the volume of work generated by local government that accountng firms are able to justify the provision of a local office to handle the work involved. Such field offices are operated by all eight principal accounting firms in over thirty of the largest SMSAs which also have national or regional headquarters of manufacturing corporations (and can, therefore, be considered to form corporate complexes).

For example, in Cleveland, Ohio, the 'Big Eight' serve a total of thirty-seven corporate and local government clients, including fifteen firms from the *Fortune* 500 such as TRW, Standard Oil of Ohio and Midland Ross; transportation, retail and utilities firms; banks such as the Cleveland Trust or the National City Bank of Cleveland; and twelve local government agencies including Cleveland City Council, Cuyohoga County Juvenile Court and the Cleveland Regional Income Tax Agency. Between forty and fifty employees may be located in the larger field offices, especially in the major SMSAs such as Los Angeles or New York. At this level the locational strategies of all the very large accounting firms are broadly similar but in medium-sized SMSAs there is a divergence between those firms which attach much higher thresholds to justify the provision of field offices and those prepared to be more flexible. The former prefer to centralize the provision of these accounting services, dispatching accountants to clients for short periods as required. If this trend develops, it will help to reinforce the emergence of 'prime' corporate complexes at the expense of reduced diversity and quality of the producer services available in smaller metropolitan areas.

Advertising services are far less prepared to set up field offices. This probably reflects different operating practices; accountants require first-hand access to client records and files, while advertising firms depend much more on contact with, and knowledge of, the services (ideas, designs, and so on) provided by competitors and, at the same time, requiring access to the media (newspapers, magazine publishers and television companies) which transmit advertising material. Most of the latter, especially those operating at the national scale, are headquartered in New York which is also, therefore, the predominant location for advertising firms; it had ninety-six of the top 200 firms in 1976. These firms captured 90 per cent of the client revenues, leaving the large number

Table 7.2 Location of advertising accounts placed by *Fortune* 650 establishments in selected SMSAs, 1976

SMSA	*Total headquarters, etc, of Fortune 650*[2]	*Number of accounts disclosed*	*Advertisements*				
			Produced in-house	*Placed with local top 200*	*Placed with local branch of top 200*	*Small local firms*	*Placed with out-of-town offices*[1]
Atlanta	110	29	6	8	2	2	11
Bridgeport	182	81	8	—	—	7	65
Buffalo	63	18	—	—	—	10	8
Charlotte	43	11	—	—	—	3	8
Denver	137	20	2	—	8	4	6
Nashville	66	13	2	—	—	5	6
Phoenix	50	23	1	—	5	3	14

Notes: 1 Offices of top 200 advertising firms and some others.
2 Headquarters, divisions and subsidiaries of Fortune 650.
Source: Stanback and Noyelle, 1980, extracted from table 7, 35.

of much smaller local advertising firms to service minor local accounts from supermarkets or car dealers (see ibid.).

The attraction of the New York corporate complex is clearly conveyed in Table 7.2, where the relationship between the number of headquarters, subsidiaries and divisions of the *Fortune* 650* and the placement of advertising accounts is summarized for seven SMSAs. Small local advertising firms are most in evidence among the accounts used by establishments in the smaller complexes such as Charlotte or Buffalo; in the larger complexes advertising is either placed in some other SMSA with one of the top 200 advertising firms or with the local offices of these firms. Only in Phoenix and Denver does it seem that the *Fortune* 650 establishments provide sufficient business to merit the setting up of branch offices by some of the top fifty advertising firms. Most of the out-of-town advertising work is commissioned in New York and Chicago but there are also links with firms in Seattle, Louisville, Winston–Salem and Cleveland (all firms in the top 200).

Research and development activities

Research and development (R&D) is an intermediate service which is often provided from within large national or multinational corporate organizations

*The 500 largest manufacturing corporations, plus the fifty largest retail companies, the fifty largest utilities and the fifty largest transportation companies.

or by specialist R&D firms (Creamer, 1976). The R&D function is crucial for the survival of many companies and is used as a means of defence from competitors; to develop innovations which offer the prospect of new markets; to copy or improve products already offered by others but at lower cost in order to remain competitive; or to make existing products safer and more reliable (Freeman, 1974; Malecki, 1980). These examples provide an indication of the kinds of labour input required, mainly highly trained professional, scientific, or technical workers with a flair for innovation, execution and implementation of ideas using existing knowledge and new techniques. Research and development, therefore, tends to occupy a rather special position within large corporations (which are the most likely to engage in in-house R&D activity) since, the major problem confronting the management of R&D is creating a suitable non-corporate environment and retaining the services of the scientific staff. Locations must be identified which meet this criterion while at the same time fitting in with the most appropriate strategy relative to the organization of the corporation and the type of product.

The locational choices available to corporate R&D are effectively centralized or decentralized facilities (Steele, 1975). A centralized facility, in multi-establishment, multi-product organizations, has the advantage of being removed from any influences exerted by the short-term requirements of production units or existing technology, so that medium- and long-term developments can be the focus of attention. Economies of scale may also accrue to the organization and large multi-disciplinary research teams may be more efficient than smaller groups at several different research complexes. Decentralized R&D involves facilities attached, for example, to each of the divisions within a corporation which may be geographically dispersed but may equally imply grouping together within a limited geographical area. This strategy provides a much closer interface between R&D activities and the production and marketing requirements of any new product and is probably best used with reference to product-related research. This pattern of location may, however, be at the expense of attracting the best research staff who might expect a more stimulating and rewarding research environment in the more forward-looking large and centralized R&D laboratories. Proximity to corporate headquarters may also be relevant because of the significance of R&D for the evolution of corporate strategy and orientation; any pull which this exerts on locational choice may apply to both centralized and decentralized alternatives. Indeed Howells (1984, 23) after reviewing evidence relating to R&D activity within a firm concluded that 'it would appear that the internal structure of the company and, in particular, good communication links between R and D and other corporate activities are a major, if not dominant, influence in the location of research units'. In the specific instance of the UK's pharmaceutical industry this conclusion was corroborated with internal rather than external factors a potent influence on the location of the industry's R&D activities.

In view of these considerations it is perhaps inevitable that R&D gravitates towards corporate complexes in and around metropolitan areas (Malecki, 1979, 1980; Howells, 1984) and, at the regional level, is located in only a small number of major regions in which large urban areas are prominent (Northern Region Strategy Team, 1977; Thwaites, 1978; Organization for Economic Co-operation and Development, 1982). Hence, in the UK, Buswell and Lewis (1970) and Hall (1970) have illustrated the major share of R&D activity in the south-east region and London, or in the Isle de France region and Paris (Brocard, 1972) in France. The most recent estimates for the UK (Howells, 1984) show that the south-east has five times more employment in research services than the next highest region; indeed three contiguous regions (East Anglia, the south-west and south-east) had over two-thirds of national R&D employment in 1971. The position of this part of the country has strengthened over time but with a relative increase in dispersal of R&D activity into the less urbanized areas. There are also gross disparities in the distribution of R&D in Canada (Phillips, 1982). In 1973 over 50 per cent of total Canadian R&D expenditures were made in Ontario, 25 per cent in Quebec and less than 15 per cent in British Columbia. The R&D contacts between private companies and Canada's National Research Council show even stronger concentration on Ontario and Quebec.

Some detailed empirical analysis of the location of R&D in the USA has been undertaken by Malecki (1980) who identifies, using discriminant function analysis, four types of R&D locations. The variables used in the analysis of fifty-eight US metropolitan areas which had five or more R&D laboratories in 1977 are: proportion of firms with headquarters and R&D facility in the same urban area, the proportion of each urban areas labourforce engaged in manufacturing, the number of scientists and engineers per million population employed by the federal government and the volume of R&D performed by local universities per million population. It was expected that three types of R&D location would emerge from the analysis: innovation centres, production locations and headquarters locations. Most of the headquarters-oriented R&D complexes (eight) are in the long-established north-eastern manufacturing belt with the exception of Seattle, Wichita and Minneapolis–St Paul. These are less numerous than concentrations of R&D in areas dominated by manufacturing (twenty-six) which have a much lower proportion (23 per cent) of headquarters-related activity. With some notable exceptions such as Atlanta, Los Angeles and Dallas most of these are again in the north-east in the states of New York, Pennsylvania, Ohio, Michigan and Illinois. The third group are described as innovation centres (eighteen) which are more widely dispersed but centred on urban areas with major university research activities and limited manufacturing, such as Denver-Boulder, Boston, San Francisco and Columbus. Finally, Malecki identifies four urban areas (Washington, Lincoln, Austin and Huntsville) where federal government R&D or very high levels of university research are significant location variables for corporate R&D. In contrast to headquarters and manufacturing R&D complexes, most of the innovation centres are in the south and west of the country where much of the recent expansion of high-technology industries has been taking place (southern California and Texas).

It also seems that there is substantial locational inertia and subsequent reinforcement of R&D complexes through the process of federal expenditure on

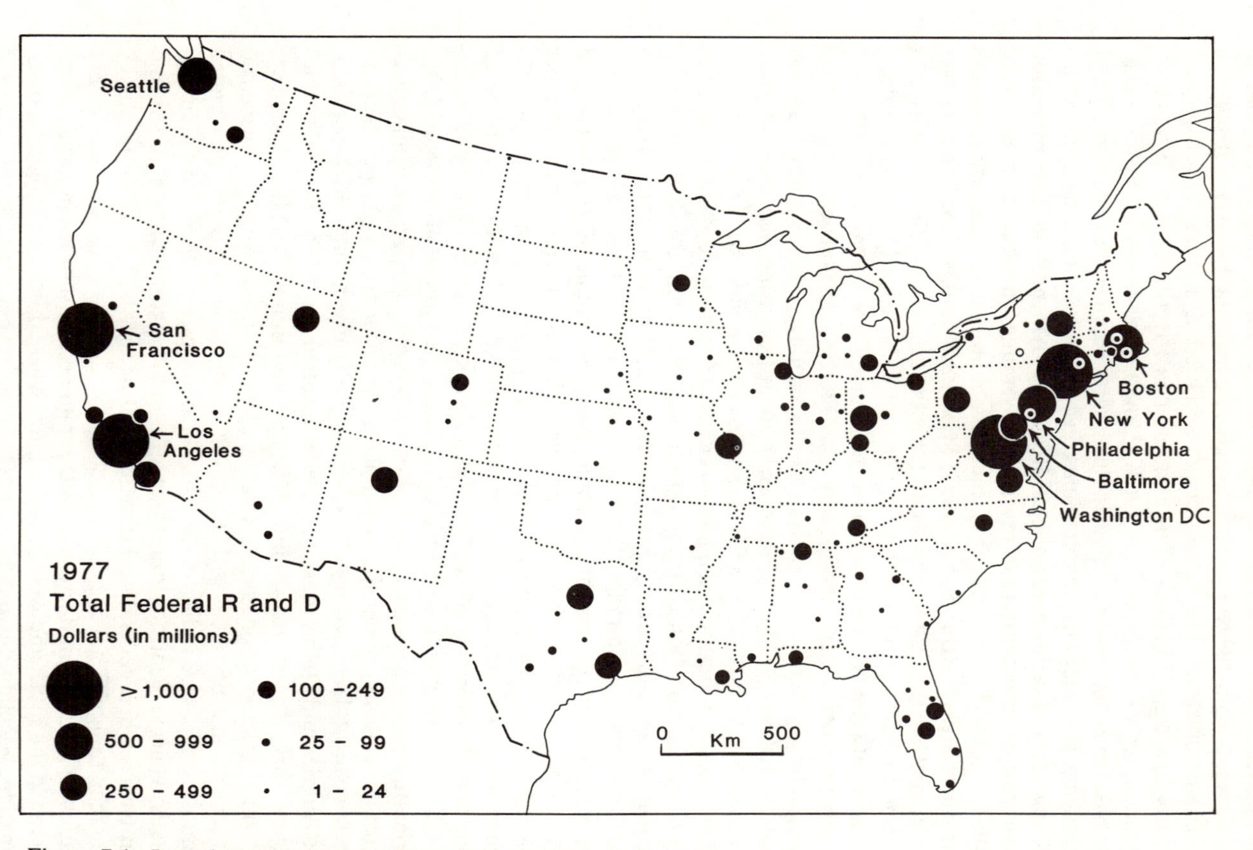

Figure 7.4 Location of federal R&D expenditure in the USA, 1977
Source: Malecki, 1982, figure 3, 30.

this activity (Shapero *et al.* 1969; Malecki, 1982). Such expenditure is directed both at private sector organizations and at the federal government's own R&D facilities. In the mid-1960s just five complexes accounted for approximately 50 per cent of all federal R&D expenditure: the San Francisco area, southern California, eastern Massachusetts around Boston, northern New Jersey and Washington, DC (Shapero *et al.*, 1969). During the interim to 1977 little has changed (Malecki, 1982) with Seattle, Dallas, Philadelphia and Salt Lake City the only notable additions. Some 70 per cent of federal R&D expenditure in 1977 was focused on ten metropolitan areas with Los Angeles receiving almost 20 per cent of the overall total (an increase from approximately 19 per cent in 1963) (Figure 7.4). The result is that a small number of cities and regions exert substantive control on economic opportunities elsewhere in the USA, in the same way that corporate headquarters complexes can be characterized as control points (see, for example, Cohen, 1979; Goodwin, 1965) because of the way in which the firms located within them engage in extensive subcontracting to other, smaller enterprises including producer services within the complex, or to places outside it (see Malecki, 1982, for some examples of inter-urban flows of subcontract linkages; and Beyers *et al.*, 1985, for inter-regional service exports from the Central Puget Sound Region).

The idea that locational inertia exists in R&D expenditures is supported by the inability of the rapidly growing sunbelt cities, such as Houston, to attract a growing share of these expenditures (Cohen, 1977). Almost 94 per cent of total R&D spending by the top 500 corporations in 1975 was attributable to firms headquartered outside the sunbelt. Although this excludes federal R&D expenditures, it suggests that although much of the new industry in the sunbelt is modern, high-technology activity its R&D activity is far less (partly because of federal support) than firms in the north-east which, Cohen suggests, have found it necessary to invest heavily in R&D to develop new products to fend off foreign competitors. Since many are long-established, research-oriented corporations, their commitment to private investment in this direction is considerable and helps to maintain the regional disparities in R&D expenditure despite the changing centre of gravity of US population distribution, urban development and economic activity (Clark, 1985).

CORPORATE COMPLEXES AND REGIONAL DEVELOPMENT

The evidence for the presence of corporate complexes in which producer services have a prominent role and the information about the spatial pattern of individual services such as advertising does, of course, have implications for the comparative performance of regions as well as metropolitan areas. An example is provided by a study of the demand for business services from the eastern townships of Quebec, a small region about 150 km north-west of Montreal (Polese, 1982). Data was obtained from more than 400 private sector establishments in mining and manufacturing, retailing, wholesaling, construction and business services on the use made of twenty-four business services ranging from manpower training to computer services and engineering consultants. Services of this type may be procured from within the same organization (intra-firm transfer), from other firms on the open market, or through provision in situ; that is within the establishments providing information for the survey. It is useful to make a distinction here between organization-oriented services, which are more likely to be internalized, and market-oriented services, which organizations are more likely to buy in. Prominent among the former are legal services, publicity, insurance and accounting, and among the latter, transportation, real estate, advertising, manpower training, architectural services and management consultants. As always with services, however, some business services do not fall easily into either of these groups since some, such as computing or heavy goods transport, can be produced both within the organization and consumed from other firms in the market simultaneously. Nevertheless, Polese finds a close, but not perfect, association between inter-firm demand and market-oriented services (68 per cent of total inter-firm demand) and intra-firm demand which is mainly (61 per cent) for organization-oriented services. Overall almost 57 per cent of total demand is for market-oriented services.

In common with the findings of a study in UK (Daniels, 1984) inter-firm demand for business services contains a significant 'durable' element, including transport, janitorial services, various office equipment services and engineering services (Polese, 1982; see also Jeanneret *et al.*, 1984). Such services are costly to provide or to purchase over long distances but 'non-durable' services are not as prone to this problem and can, therefore, be provided on an intra-firm basis (by headquarters located elsewhere, for example) or by inter-firm contacts from outside the local area or region. Hence one-third of the inter-regional flow of expenditure on business services is made up of imports from Montreal and only 44 per cent on services from within the region. The proportion imported from Montreal is much higher for the non-durable services such as marketing studies (68 per cent), technical studies (87 per cent), legal services (61 per cent), or insurance (58 per cent). The intra-firm market almost totally escapes the eastern townships region which only receives 6.5 per cent of intra-firm service expenditures. Some 70 per cent of all these intra-firm expenditures involve imports from Montreal or, secondarily, Toronto (see Britton, 1974; Burrows and Town, 1971; Jeanneret *et al.*, 1984). Since headquarters establishments or other establishments with control functions tend to purchase from firms in the corporate complex in which they are located, and then pass on some of these purchases to their dependent establishments (Daniels, 1984), there is likely to be an increase in inter-regional differences in producer service availability. An expansion in the volume of inter-regional, intra-firm transactions will reinforce this trend.

Even if continued divergence between the growth of central and peripheral regions or cities seems to be a predominant by-product of producer service growth and consumption, cultural and linguistic factors may prevent such developments reaching their logical conclusion (Polese, 1981, 1983). The inexorable decline in the status of Montreal as one of Canada's premier financial and business cities has occurred during the last decade; Toronto has attracted a growing share of manufacturing headquarters and international corporate and financial functions. For example, in 1971 Montreal's labourforce in financial services could be expressed as 0.72 that of Toronto (1.00) and 0.71 for

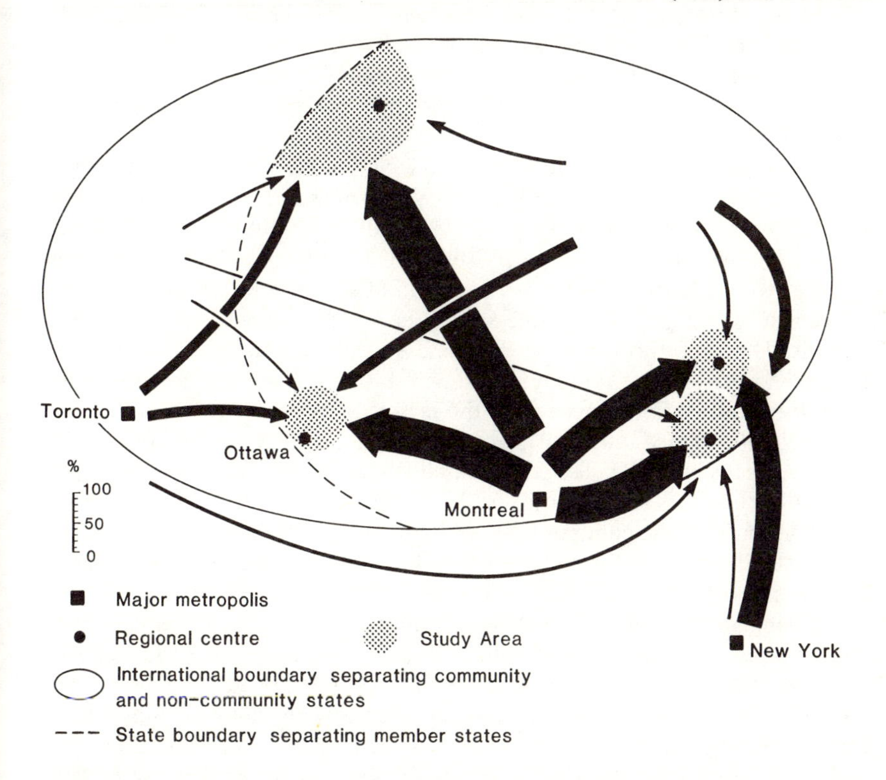

Figure 7.5 Business service flows between subregions of Quebec, 1980
Source: Polese, 1981, figure 3, 497.

business services. By 1981 these figures had decreased to 0.63 and 0.61 respectively (Polese, 1983). But Montreal's position could be worse but for the influence of cultural and linguistic barriers on the locations from which business services are purchased (Figure 7.5). Data for the flow of business services within four subregions of Quebec show how Montreal greatly overshadows Toronto as the principal supplier, even to those subregions closer to the latter. The legal, linguistic and 'cartel-like behaviour' (Polese, 1981, 496) of some professional groups such as accountants contributes to Toronto's relative weakness even though it is still the nation's dominant corporate complex. The only flows of business services which significantly transcend this cultural selectivity are intra-firm transactions.

Significance of ownership and control

The characteristics of ownership and control of the establishments using producer services then becomes critical in terms of the balance between imports of such services and indigenous consumption. Branch plant economies, the type which are prominent in Britain's peripheral regions, are more likely to lead to the import of producer services via headquarters (Burrows and Town, 1971; Crum and Gudgin, 1977); a process which is exacerbated by the growing number of corporate takeovers amd mergers which usually involve the centralization of service supply linkages at the location of the acquiring organization (Leigh and North, 1978; Smith, 1979; Stanback and Noyelle, 1980; Douglas, 1981).

The linkages between business services, for example, which are a dominant group within the producer service category, and manufacturing industries are more complex than is often assumed (Marshall, 1980, 1981, 1982). The conventional view is that producer services appear in response to demand from manufacturing industry but it is possible that differences between areas or regions in the demand for business services depends on the supply in those regions. Hence if the supply of business services is enhanced, there is an improved possibility that manufacturing industry will meet its requirements from local rather than external (to the region) sources.

Using data from manufacturing establishments in three city regions: Birmingham, Leeds and Manchester, Marshall (1982) found that just over 40 per cent of the business service needs of the plants were obtained from specialist suppliers outside the companies of which they formed a part. The services purchased in this way (found to be equivalent to just 1.2 per cent of the turnover of the firms studied) were principally insurance, banking, accounting, advertising and general office services. Almost 80 per cent of the services purchased were obtained from suppliers in the same planning region as the manufacturing plants in the sample, with London cited as the major alternative but mainly for more specialized management consultancy or stockbroking advice. The demand for externally provided business services also varies with size of establishment (defined by numbers employed) and their organizational characteristics: small establishments, those largely producing in small batches or single units, single-site enterprises, or establishments with independent status are all more likely to obtain a higher proportion of their business service needs from outside their own company but with greater dependence on local sources rather than those outside the region in which they are located.

The range and intensity of demand for business services bought in by companies varies widely between the north-west, Yorkshire and Humberside and West Midlands regions even when factors such as organizational structure or size of manufacturing establishment are taken into account. It may well be, therefore, that the business services actually available in each region influences the characteristics of manufacturing industry demand for those services. Unfortunately Marshall did not match the services available in each city region with those actually used, so it is only possible to guess at the significance of the business service profile for manufacturing plant demand. It does seem, however, that irrespective of the quality of the supply, plants with headquarters in London tend to internalize their business services more than the branch plants of companies with headquarters in the same region. The higher the level of decision-taking at a plant, the more use its personnel tend to make of local services; subsidiaries make more use of local services than branches; and headquarters (or single-plant companies) more than branches or subsidiaries. It is likely that knowledge of local suppliers accounts for this rather than other intrinsic differences between the branch plants of local and non-local companies. Consequently the demand for business services, such as insurance, in the north-west region is less than in the West Midlands because of the preponderance of plants (in the sample), with London headquarters in the former and local headquarters in the latter.

It has already been demonstrated that the headquarters, whether of manufacturing or service sector enterprises, show an increasing predilection for concentration in corporate complexes and it seems that the linkages between business services and manufacturing reinforce such complexes. Evidence brought together by Marquand (1978) for member states of the EEC shows that this is not confined to Britain or North America; larger companies in France, the Netherlands, Denmark and Ireland all reported an increasing concentration of decison-makers in professional and managerial occupations, with many of these high-level activities located in the capital cities. Since many specialized producer services depend upon close links with such high-level activities, they are also very centralized relative to the rest of the national economy. In France, Britain and Denmark specialized producer services are, therefore, reluctant to select locations more than two hours from the major national corporate complex in the capital cities; equally manufacturing firms are not prepared to place branches more than two hours' travelling time from sources of specialized services or from headquarters. Imbalances in the regional distribution of producer services inevitably follow from these locational constraints and contribute to a deterioration in the ability of underprovided regions to cope with the problems caused by the decline of traditional manufacturing activities.

INTERNATIONALIZATION OF BUSINESS SERVICES

Until recently the location behaviour of producer service establishments has largely been viewed within the context of individual national and regional economic systems. However, the substantial improvements in international telecommunications, including the use of satellites, during the last decade has allowed producer services to adopt a more international approach to the markets which they can consider accessible and which may use the expertise which they provide. Multinational organizations are usually equated with manufacturing industry but recent studies (Buckley and Casson, 1976; Dunning and Norman, 1979, 1983; Heenan, 1977; Hood and Young, 1979; Cowell, 1983) all point to an increasing involvement by service industries. Dunning and Norman (1983) suggest that multinational enterprises exist to take advantage of certain imperfections in the market which will lead them to make direct investments in foreign countries if, first, they think that they possess a competitive advantage over the firms of other nationalities; secondly, that any advantages they have are best capitalized upon by providing the good or service from within the enterprise rather than by selling or licensing others to provide or use it; and thirdly, that this 'internalization' advantage is best enhanced by providing the service at the point of consumption rather than by production at 'home' and export to some other country (see Cowell, 1983).

The advantages to business services of undertaking foreign-based production are summarized in Table 7.3. Market factors clearly have an important role with respect both to the availability of potential clients and access to skilled labour. By penetrating overseas markets, business services are protecting themselves from the ability of indigenous competitors to outperform them (in cases where they might consider licensing an indigenous firm to provide their service) or are maintaining quality control in those circumstances where there are no indigenous enterprises able to provide the particular service. The search for locations for multinational business service offices, therefore, reflects the influence of market factors and agglomeration in major metropolitan areas.

Almost 90 per cent of the offices (169) of US-based business services in Belgium, in 1977, were located in Brussels; 93 per cent of those in France (134) in Paris; and 95 per cent of those in Britain (261) in London. Banking, management consultancy and executive search and advertising are represented in all three countries and these are the kinds of business service which are highly specialized, involve a good deal of competition for client accounts (for example, in advertising) and are, therefore, trying to use their ownership advantages to obtain a market share. Some of the firms in the sample have offices in all three European countries. An additional distinctive feature, but not surprising in light of the evidence already cited, is that within two of the metropolitan areas, Paris and London, international business service offices are primarily located in the central area, although there is some evidence for those firms with larger offices seeking suburban locations (mainly engineering design, management consultancy and advertising agencies). Coldwell Banker (part of the US Sears–Roebuck group), probably the world's largest property services group, has recently (1985) set up a representative office in central London from which to conduct its European business. Its stockbroking operation, Dean Witter Reynolds, also now has offices in the City and these are rapidly expanding.

A survey of the UK branches of fifteen foreign-based business services with operations in Western Europe revealed a good deal of similarity in the reasons given for adopting international operations (Dunning and Norman, 1979). All offered well-defined 'trademarked' services backed by the experience gained in large US markets and the financial resources of the US parent company. As the European market has expanded a need has arisen for branch offices which reduce the amount of costly transatlantic travel and provide better opportunities for face-to-face contact with clients and a more reliable barometer of market trends and needs. These branch offices serve mainly local markets and are not, therefore, internationally foot-loose (hence the presence of several branch offices of the same US-based enterprise in different European countries), nor are they nationally foot-loose and tend to gravitate towards the centre

Table 7.3 The advantages of international production for business services

Type of foreign production	*Ownership advantages*	*Country advantages*	*Internalization advantages*	*Types of activity which favour MNEs*
RESOURCE BASED	Capital, access to markets, managerial skills, international reputation	Resources (e.g. skilled labour, information)	Secure supply of skilled labour, protection and exploitation of specialist information, quality control	Engineering design, insurance and reinsurance, management consultancy, investment banking
IMPORT-SUBSTITUTING SERVICES	Capital, specialist knowledge, international reputation, access to markets	Labour and other costs, size of local market, need to protect market	Protection and exploitation of knowledge and business contacts, buyer uncertainty, high information costs, quality control	Reinsurance, executive search and accountancy, management and engineering consultancy, branch banking
TRADE AND DISTRIBUTION	General merchandising knowledge, access to suppliers and market outlets	Size of market, access to customers, suppliers and commodity, exchanges, liberal attitudes towards trade	Secure market share, exploitation of business and market contacts	Import and export merchanting

Source: Dunning and Norman, 1983, table 1, 679.

of major cities or to places near to international airports such as the west London area around Heathrow. Initially these offices will develop links with other firms from the USA with French or Belgium operations, for example, but indigenous firms are the ultimate objective, and the opportunities for client diversification will be an important factor in location choice. While markets are clearly important, the final choice of location within Europe may be governed by cultural and political considerations (Heenan, 1977); hence the attraction of London, rather than Brussels, for many US organizations.

Banking has been a prominent leader in the internationalization of services. The number of overseas banks represented directly or indirectly in London has increased by a factor of five between 1969 and 1983 (Figure 7.6). Direct

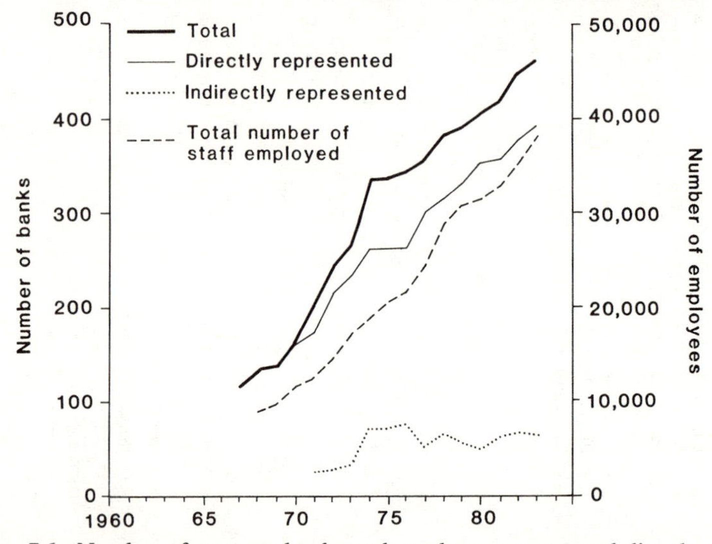

Figure 7.6 Number of overseas banks and employees represented directly and indirectly in London, 1965–83
Source: Compiled from data in the *Banker*, November 1983.

representation, which invariably involves a larger number of employees, has also expanded at the expense of indirect representation. This suggests that foreign banks want to be as 'visible' as possible in overseas banking centres like London and this is best achieved through direct representation. The scale of representation ranges from an average of 165 employees per branch for North American banks (Table 7.4) to less than twenty employees per establishment in the branches of Latin American banks. Almost one-half of the foreign banks in London are headquartered in the USA or Canada (Figure 7.7) and over one-half of the remainder are from Europe. The strength of international banking in London, New York, Tokyo, or Singapore, for example, is backed by comprehensive and diverse insurance services. London is the world's leading centre for insurance; 20 per cent of the business in the world market is placed there (Cowell, 1983), and of the 800 companies authorized to operate, 170 are overseas companies. The majority of the large British insurance companies

Table 7.4 Branches, subsidiaries and representative offices of foreign banks in London, 1983

Area	Establishments		Employees		Employees per establishment
	no.	*%*	*no.*	*%*	
Europe	135	33.8	8,522	28.4	63.1
USA and Canada	85	21.3	14,034	46.8	165.1
Africa and Middle East	51	12.8	1,186	4.0	23.3
Japan and the Far East	44	11.0	1,847	6.2	42.0
South-east Asia and Australasia	38	9.5	3,969	13.2	104.4
Latin America	25	6.3	438	1.5	17.5
Elsewhere[1]	21	5.3	193	0.6	9.2
Total	399	100.0	29,996	100.0	75.0

Note: 1 Mainly Eastern bloc states and Caribbean states such as Bermuda, the Bahamas and Grand Cayman.
Source: Estimates compiled from data listed in the *Banker*, November 1983, table 1, 133.

also regard themselves as international and receive more premiums from overseas than from home business.

Not all foreign banks gravitate towards the corporate complex in London (Figure 7.8). Many have branches in other British cities such as Manchester, Birmingham, Leeds and Glasgow. South-east Asian banks are especially prominent outside London in areas where they can provide a service to large ethnic groups in, for example, Bradford, Birmingham, Leicester and Huddersfield. Fewer North American banks have branches outside the south-east and their branches are principally located in provincial centres with long-established, if much smaller-scale, financial functions: Edinburgh, Manchester, Birmingham and Cardiff.

SUMMARY

Producer service activities have grown and diversified rapidly during the last twenty years. In contrast to consumer services the location of producer services is dominated by centralization in a small number of major employment centres which offer the range of agglomeration and urbanization economies which seem prerequisites for the effective provision of the specialized outputs from such services. Several reasons have been given for the increased demand for producer services, including the need for all kinds of economic activity to utilize the opportunities offered by new technology or the effects of changes in the structure and scale of organizations on the demand which they generate

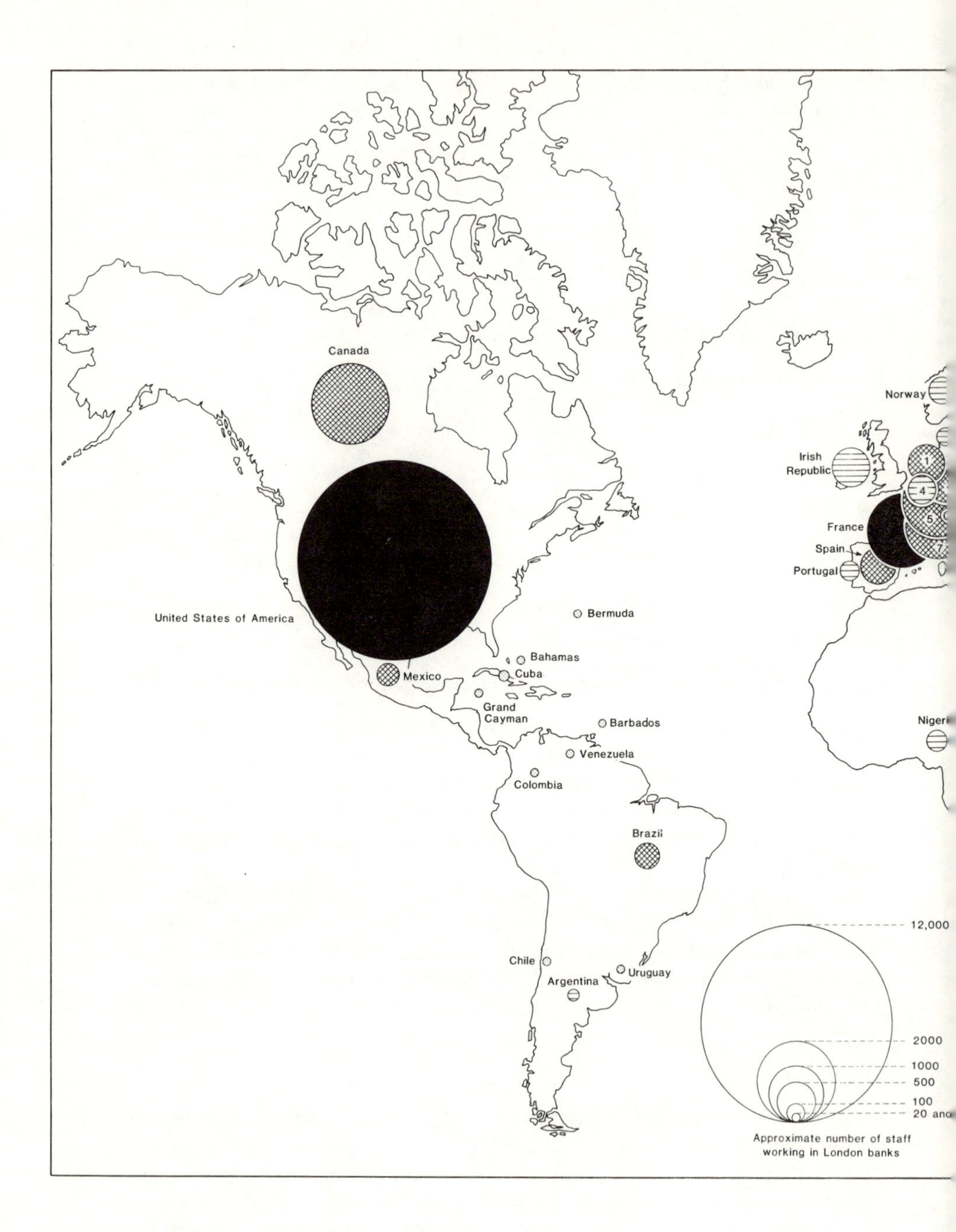

Figure 7.7 Parent country of foreign bank branches and subsidiaries in London, 1983
Source: As Figure 7.6.

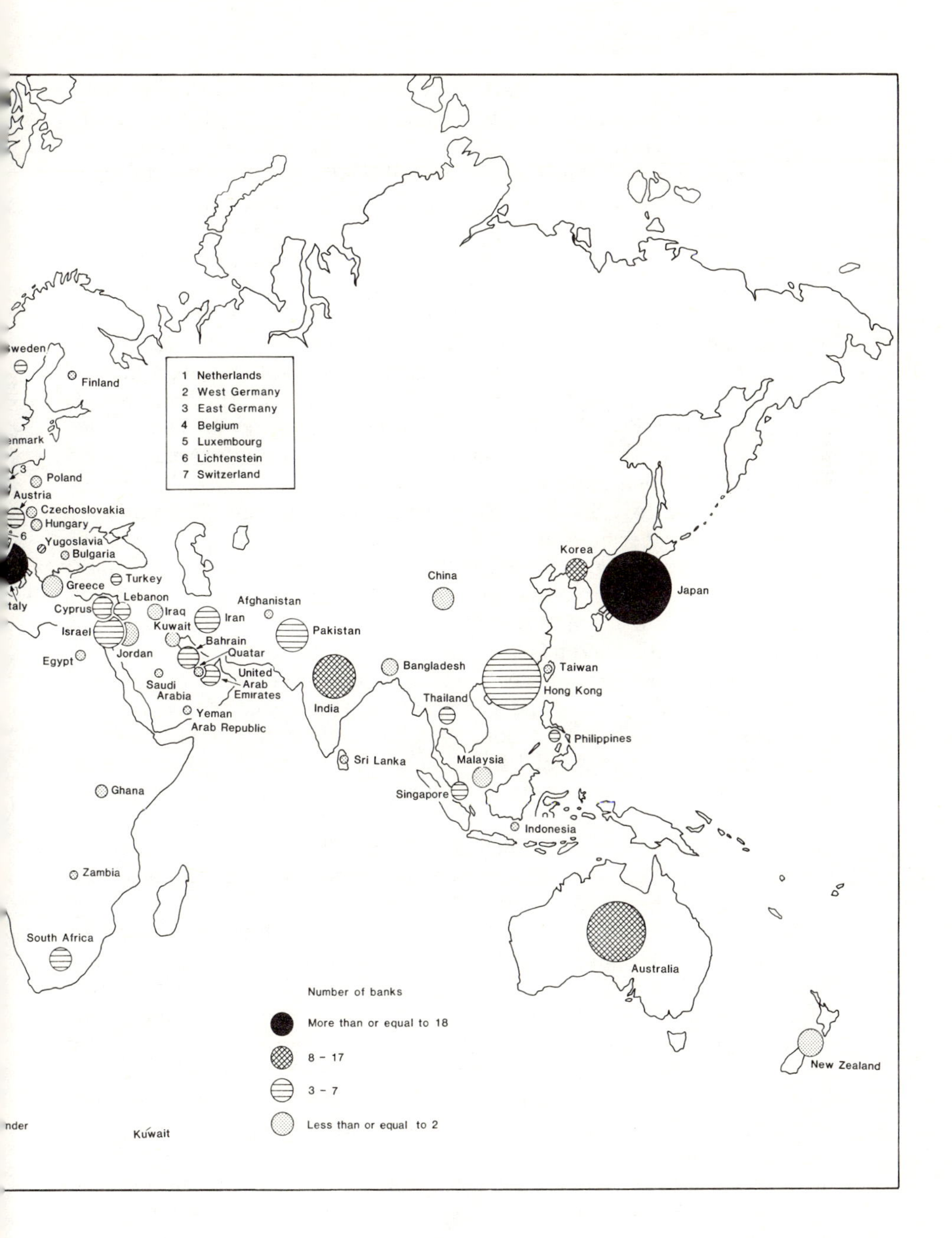

Sweden
Finland
Denmark
Poland
Austria
Czechoslovakia
Hungary
Yugoslavia
Bulgaria
Greece
Turkey
Italy
Cyprus
Lebanon
Iraq
Israel
Kuwait
Iran
Afghanistan
Pakistan
Bahrain
Quatar
Jordan
Egypt
United Arab Emirates
Saudi Arabia
Yemen Arab Republic
India
Bangladesh
China
Korea
Japan
Taiwan
Hong Kong
Thailand
Philippines
Sri Lanka
Malaysia
Singapore
Indonesia
Ghana
Zambia
South Africa
Australia
New Zealand
1 Netherlands
2 West Germany
3 East Germany
4 Belgium
5 Luxembourg
6 Lichtenstein
7 Switzerland
Number of banks
More than or equal to 18
8 - 17
3 - 7
Less than or equal to 2
Kuwait

for services provided from external rather than internal sources. Although large organizations can meet many of their requirements through the internal transfer of goods and services, it is becoming more difficult to retain the services of highly skilled personnel who may only be called infrequently to provide a particular input. Producer service firms can provide the same output to a wide range of clients at lower unit cost and large organizations can reduce their administrative overhead by simplifying their staffing arrangements.

Information is now one of the key requirements of modern enterprises. Since the range and type of information required is both volatile and increasing, organizations need to place those parts of their activities which most need to interpret information at locations where they are best able to receive, assimilate and act upon it. For example, confidence in the quality of information received, or in the process of negotiating a contract depends upon close interaction with sources, clients, or suppliers, and although telecommunications can now be used to expedite such contact, face-to-face interaction is still considered very important. Travel to a face-to-face meeting costs time which could profitably be expended at other meetings, so that users of diverse information such as the headquarters or regional offices of service and manufacturing corporations tend to seek similar locations as closely together as possible, along with the more specialized producer services such as advertising, computer services, or management consultants.

Several examples have been given in this chapter of the distribution of producer services which results, and the way in which the coincidence of different types of services in similar locations gives rise to corporate complexes. These are major nodal centres in which the servicing of corporate or divisional headquarters by a wide variety of high-quality and specialized business, professional and financial services is the *raison d'être*; such complexes do not necessarily focus on core areas such as the New York or London metropolitan regions, they can incorporate also more extensive intra-regional networks such as those in New England or part of Quebec.

Despite the possibility of modifications caused by cultural boundaries of the kind cited in Canada, corporate complexes exert considerable influence on the inter-urban and inter-regional exchange of producer services in a way which leads to a reinforcement of their comparative advantage. Branches or plants controlled from a headquarters in the corporate complex are less likely to purchase producer services (or indeed most other services) from the immediate locality. Locally controlled enterprises are more likely to do so but the empirical evidence from studies in Canada, Britain and Switzerland, for example, shows that a significant proportion of their more specialized producer service requirements are purchased from the nearest corporate complex. In this way access to producer services is not uniform, and this in turn prejudices the ability of firms in cities and regions with economic problems to adapt or innovate in ways which allow them to compete on an equal footing with firms nearer to, or part of, corporate complexes.

The trend towards multinational operation of producer services also reinforces established corporate complexes because of the agglomeration

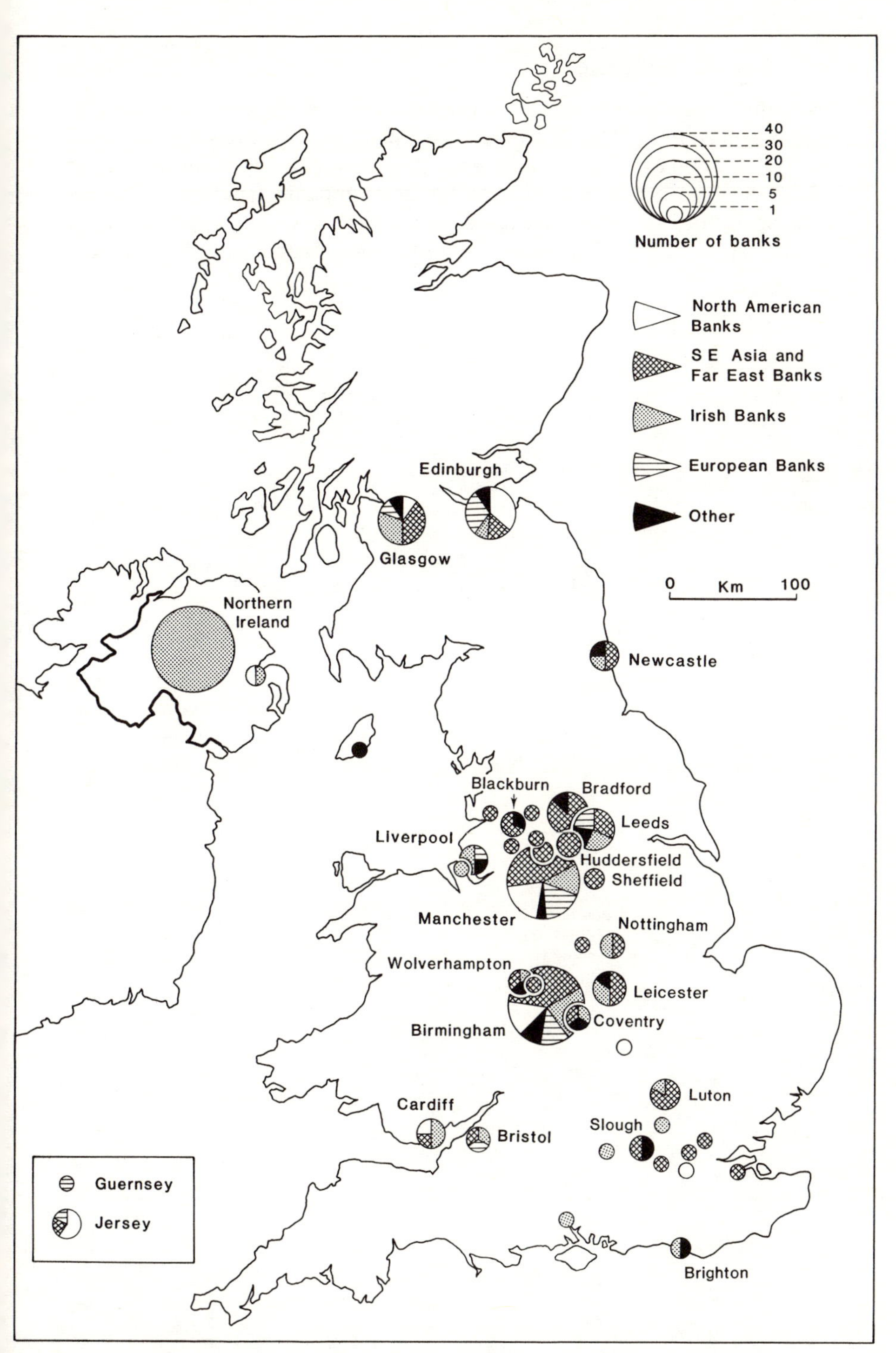

Figure 7.8 Location of branches and subsidiaries of overseas banks in British cities outside London, 1983
Source: As figure 7.6.

economies which these provide for organizations starting up in a new market in a foreign country. Almost all the foreign banks and business services which have established offices in Britain since 1960 have located in central London, while British property, insurance and banking companies which have extended their operations overseas have usually concentrated on the opportunities in Sydney, Hong Kong, Singapore, Frankfurt, San Francisco, or New York.

CHAPTER 8

Locational dynamics of producer and consumer services: competition between CBD and suburbs

INTRODUCTION

Cross-sectional analysis of service industry location provides a useful insight into its distribution and has identified some spatial and other inequalities. Such an approach is justified in relation to the way in which the major theories of location, particularly central place theory, are also essentially static and cross-sectional. It is generally considered difficult to incorporate a temporal component into these models. Yet it will be apparent from Chapters 6 and 7 that private and public services (consumer- or producer-oriented) operate within an environment which demands virtually constant appraisal and reappraisal of the existing location of establishments whether they are part of a major multi-site operation or represent a small, independent concern with just one point of contact with customers and suppliers.

Temporal analysis is, therefore, an essential part of the interpretation of service industry location. It is probably fair to say that temporal trends in the location of services and the processes underlying them have not been widely explored; those studies which have been undertaken tend to be sector-specific and, from a geographical standpoint, reflect a particular interest among urban and economic geographers, for example, in the causes and consequences of location change by retailing, wholesaling and, more recently, office-based service activities (Kivell and Shaw, 1980; Davies, 1976; Dawson, 1983a; Ashcroft, 1981).

The essence of locational change by service industries has been the differential performance of inner urban (including CBD) locations when compared with suburban areas and locations beyond. This is conveyed clearly by some data included in a comparative study of service employment growth in inner and outer areas of six British conurbations (Moore *et al.*, 1981) (Figure 8.1) If service employment in 1952 is taken as 100, there was a steady increase in all the inner areas until the mid-1970s, followed by stability and subsequent decline. Between the mid-1960s and 1977 more than 20 per cent of the service jobs in inner Merseyside had been lost, a rate somewhat faster than that for inner Tyneside. A small decline also occurred in London and Manchester while service sector employment remained stable after 1965 in inner Birmingham and inner Glasgow. Contrast this with the situation in the outer areas of the conurbations, all of which experienced an expansion of their service employment between 1952 and 1977, notably after the appearance of the negative trends in the inner areas. As we shall see later there is a connection between

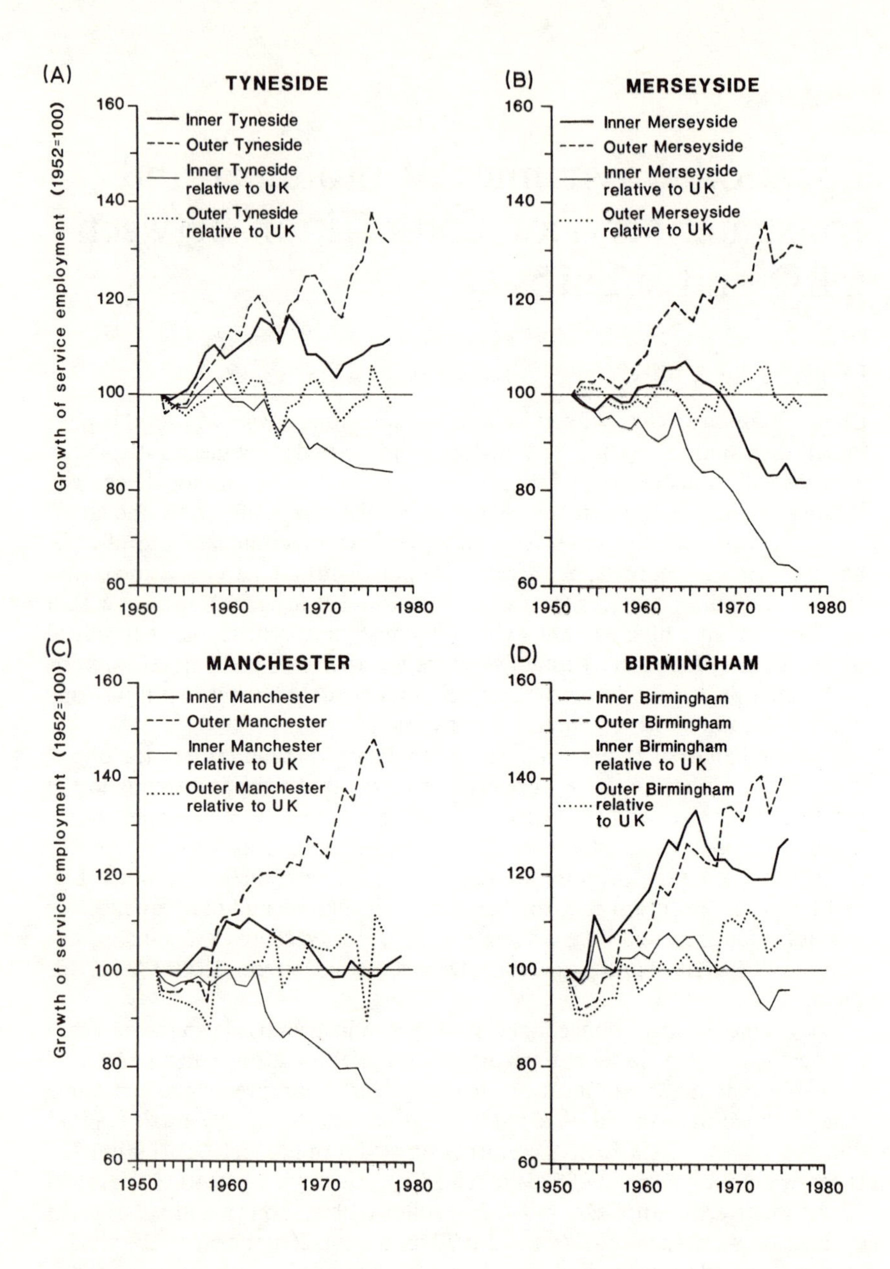

Figure 8.1 Growth of service employment in inner and outer areas of four British conurbations: (A) Tyneside, (B) Merseyside, (C) Manchester and (D) Birmingham
Source: Compiled from Moore *et al.*, 1981, tables 5–8.

events in the inner and outer areas. Overall the increase has been of the order of 30–40 per cent in the outer areas of all the conurbations.

Relative to the national rate of expansion in total service employment, the inner cities also performed poorly (Figure 8.1). The difference is as much as 40 per cent in inner Merseyside. Only towards the end of the 1952–77 period, however, had the outer areas managed to match the national growth rate for service employment, and in Birmingham and Clydeside the rate was ahead of the national rate. When disaggregated into private and public sector services, it becomes apparent (Table 8.1) that the more favourable performance of the outer cities is the product of a rapid expansion of private sector service employment which increased by 60 per cent in Tyneside and by 54 per cent in outer Birmingham. The picture is not uniform, however, in that private service sector jobs decreased to a limited extent in outer Manchester and outer

Table 8.1 Growth of private and public sector service employment in the inner and outer areas of the British conurbations, 1952–76

Conurbation		*Year*				
		1951	*1961*	*1966*	*1971*	*1976*
PRIVATE SERVICES						
Tyneside:	inner	100.0	113.2	118.4	97.2	99.2
	outer	100.0	110.3	133.8	133.8	162.2
Merseyside:	inner	100.0	109.2	107.1	86.6	78.7
	outer	100.0	99.3	113.0	13.3	76.5
Manchester:	inner	100.0	113.7	105.5	85.9	69.9
	outer	100.0	110.3	111.7	95.1	94.1
Glasgow:	inner	100.0	108.5	106.7	91.8	86.0
	outer	100.0	109.0	126.6	121.8	130.1
Birmingham:	inner	100.0	114.8	121.1	103.1	101.8
	outer	100.0	122.1	142.0	139.7	153.5
London:	inner	100.0	110.4	106.0	94.6	88.3
	outer	100.0	109.6	117.2	112.2	121.4
PUBLIC SERVICES						
Tyneside:	inner	100.0	100.3	100.5	101.6	112.0
	outer	100.0	101.3	119.3	145.1	174.8
Merseyside:	inner	100.0	97.0	93.7	89.7	86.0
	outer	100.0	89.2	98.2	104.4	71.0
Manchester:	inner	100.0	102.8	106.0	110.9	119.1
	outer	100.0	97.0	93.7	89.7	89.5
Glasgow:	inner	100.0	103.9	104.8	104.1	114.9
	outer	100.0	103.1	118.0	139.0	155.3
Birmingham:	inner	100.0	107.7	118.9	127.0	133.6
	outer	100.0	117.3	134.5	155.5	175.6
London:	inner	100.0	99.7	99.8	100.6	102.4
	outer	100.0	108.8	113.0	125.9	134.5

Source: Moore *et al.*, 1981, table 7.

Merseyside. The only source of growth in the inner areas was public sector employment, which grew at a relatively faster rate than the national average, but this sector also expanded in the outer cities and thus supplemented the more favourable performance of private services in those areas.

A CONCEPTUAL CONTEXT

Before examining the locational dynamics of specific service activities, it may be helpful to conceptualize the circumstances which affect the locational dynamics of service activities. Some reference was made in Chapter 7 to changes in organizational structure and the way in which they enhance the trend towards centralization of producer services. Such changes are part of a suite of factors internal to service activities which may be distinguished from a second suite of factors which can be considered external. An alternative way of viewing these factors would be to classify them as microscale features, such as organizational changes within banking or insurance, for example, or macroscale features derived from events and trends within the national economy or in consumer tastes and preferences.

Either way, these forces for change in the location of services are more, or less, directly controllable by those affected, internal (endogenous) factors more so than external (exogenous) factors. Simmons (1964) makes a distinction between controlling forces on location such as population distribution and income, and modifying forces such as the introduction of new technology which may ultimately allow alternative locational configurations to be introduced because, for example, they offer the prospect of improved economies of scale or better customer access. In so far as services are, by definition, concerned with being located at the right place at the right time to meet the needs of private consumers or other business enterprises, it seems reasonable to suggest that external or controlling variables are more important agents of locational change than internal variables. Failure to respond to external stimuli may generate a pattern of relict locations which are unattractive to consumers or will give both new and existing competitors an opportunity to make location decisions which lead to the demise of the unresponsive organizations. In this sense failure to respond to internal forces for change may be less catastrophic in relation to the survival of enterprises and, therefore, less important as determinants of locational change.

External influences on location change

The principal external factors with potential consequences for the location of service firms are economic conditions affecting the structure of the economies of which they are a part, the distribution of demand (for which population and/or the distribution of industry could be a surrogate), the technological environment which is increasingly encroaching upon service sector activities, trends in international trade in services rather than in manufactured or primary products, or the degree of institutional control on the behaviour of private and public corporations. All of these factors are outside the direct control of service sector enterprises but they most certainly impinge upon their organizational, operational and locational strategies. A recent example is the decision by one of Britain's 'big four' banks (Barclays) to embark on a fundamental restructuring of its branch network in order to cater more effectively for the different needs of corporate and personal customers (*The Times*, 28 June 1983). Over sixty full-time branches and eighty part-time branches will be closed, and the remaining 2900 branches divided into a hierarchy from 100 key locations down to 750 locations offering only limited personal services.

The state of national economic health will be reflected in the level of wages and the volume of disposable income which can be expended on discretionary requirements. This may not only be reflected directly in the growing turnover of retail, education, or recreational services (whether provided by private or public organizations), but also in increased demand for credit, more extensive use of various methods of saving, more investment in stocks and shares and greater expenditure on health or personal insurance. Such demand will stimulate the development of financial services such as creditbrokers, savings banks, building societies and other forms of financial management which will either respond *in situ* or by introducing new points of contact with their clients, or reorganizing their existing establishment networks. It has already been demonstrated (Chapter 3) that there is a correlation between the level of per capita income and the proportion of services in the economy, and the influence of national economic performance is therefore very important. The significance of this for location is that rising incomes are equated, for example, with changing levels of mobility, different kinds of client expectation, or shifts of fashion to which service industries need to adapt; one part of this adaptation encompasses new or additional locations.

Perhaps the single most important external factor with immediate consequences for the locational dynamics of services is the distribution of clients whether identified in relation to the location of population (consumer services) or of employment (producer services). A well-documented feature of population change in advanced economies is the decentralization of population from the older urban cores to the newer suburbs (Champion, 1983; Lawton, 1974; Hall *et al.*, 1973; Richardson, 1982; Berry, 1973). Between 1951 and 1971 Greater London's population fell by almost 750,000 as a result of voluntary outmigration, population displacement consequent upon slum clearance, and the more rapid growth of smaller existing and new urban areas such as the new towns, in the surrounding metropolitan region. The 1981 Census shows that the move to the suburbs is accelerating; every large city in the UK suffered substantial population losses between 1971 and 1981. The population of Glasgow decreased by 22 per cent, Manchester by 17 per cent and Liverpool by 16 per cent. Most of these changes were the result of population outmovement from the inner cities, and some of the highest population increases have occurred in largely rural areas. All this has been accompanied by macroscale shifts such as that from northern to southern Britain (Office of Population Censuses and Surveys, 1981), or from the north-east to south in the USA (Perry and Watkins, 1977). The distribution of employment has also undergone changes which mirror those revealed in population statistics (see, for example, Spence and Frost, 1983; Fothergill and Gudgin, 1982; Berry, 1975; Berry and Cohen, 1973; Goddard and Spence, 1976; Erickson, 1983).

Service industries are participating in the decentralization of employment from, and within, the major cities (for an example of an early study see Cuzzort, 1955), although the trend has – until recently – been much more strongly developed for manufacturing than for services as a whole; the locational changes by the latter have been selective as to the type and function of the participant activities. On Merseyside between 1971 and 1975 both the CBD and the inner city lost service sector jobs (Figure 8.2) in relative and absolute terms, an 8 per cent decrease in the former and 7 per cent in the latter (just over 6500 jobs for each area). But in a third zone, the outer suburbs, which had 28 per cent of the service sector jobs in 1971, the absolute number of jobs increased by 9 per cent and the share of all service jobs on Merseyside to more than 31 per cent. All MLHs except the utilities (gas, electricity and water) and transport and communications recorded an absolute increase in the outer suburbs, particularly professional and scientific services and miscellaneous services. The former also increased in the inner suburbs together with insurance, banking and finance which, rather surprisingly, decreased by 7 per cent in the CBD (Figure 8.2). It is notable that consumer services have grown substantially in the outer suburbs, while activities with a mixed or producer role show smaller absolute and proportional changes. In a more rigorous analysis (using the negative exponential model) of the relationship between distance from the CBD, the density of office floorspace and its functional attributes Smith and Selwood (1983) show that distance decay is strongly developed for the services examined in Columbus, Ohio. The result of this is to sustain the dominance of inner over outer suburban areas of Columbus as the location for almost all types of service function, even though it is demonstrated that absolute decentralization of office space has been dominant in Columbus since 1970.

Until relatively recently the impact of technological developments on service industries, whether relating to the equipment available within service establishments or to the means of communication between them, were not a major locational force. The advent of advanced telecommunications, the silicon chip and the associated miniturization of computers, together with the increasing ability to arrange for office machines which may be substituting for direct labour inputs electronically to communicate information, has begun to revolutionize the structure of service enterprises and the locational opportunities available to them. At present illustrations of the possibilities offered by the new technology are certainly appearing with increasing frequency, although the spatial consequences of the widespread adoption of, for example, point-of-sale charging to bank accounts are a matter for speculation rather than fact (Jones, 1984). Nevertheless, the impact of technology on the dynamic of service industry location is inescapable and will undoubtedly become more significant during the next decade and beyond.

The growing international component in service activity (see, for example,

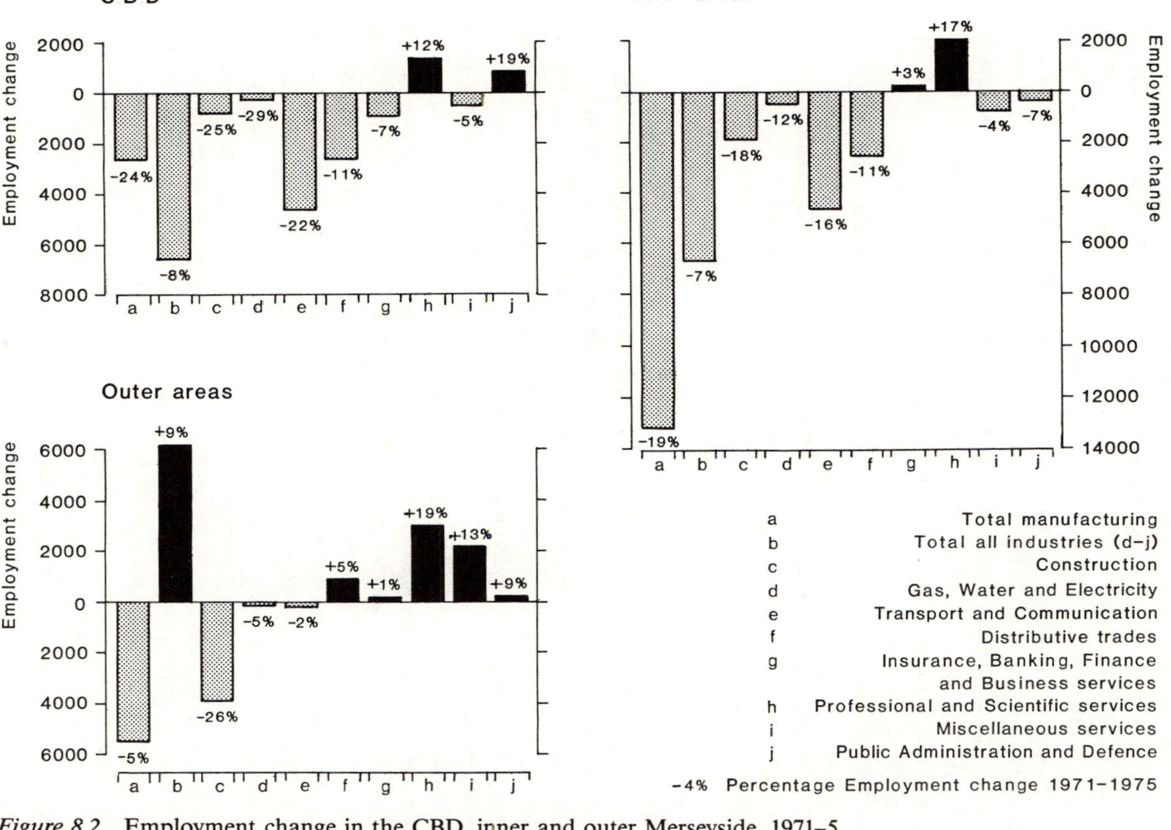

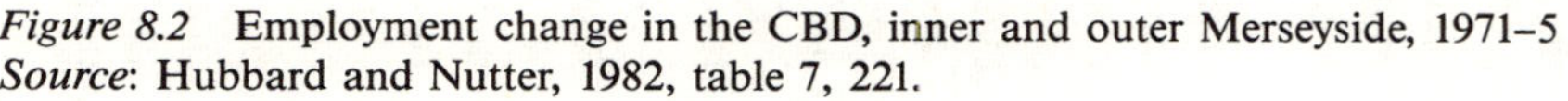

Figure 8.2 Employment change in the CBD, inner and outer Merseyside, 1971–5
Source: Hubbard and Nutter, 1982, table 7, 221.

OECD, 1983) is not unrelated to the opportunities created for greatly enhanced communications, using satellites rather than land-lines and lines under the sea. The rise of international trading in stocks and shares through exchanges in London, Paris, Tokyo, or Hong Kong has undoubtedly been encouraged by the speed with which information about share prices can be relayed between exchanges and buying and selling undertaken at the most advantageous times. Naturally those countries or individual cities with good access to such advanced technologies have attracted existing and new services specializing in foreign transactions. In contrast to the trends in the location of population and employment, the internationalization of certain services may be a force for centralization rather than dispersal and thus contribute to the divergence between the general trend and that for selected services in most cities and regions.

Institutional factors impinging upon the location of service activity may take negative as well as positive forms. On the negative side controls can be used to restrict the volume of construction in order to discourage expansion or to encourage services to consider alternative locations. Restrictions can be placed on the number, density, or spatial coverage of branch establishments; in the USA, for example, banking has until very recently been an intra-state activity controlled by federal and state laws which constrain the geographic areas within which banks can operate (Wels, 1984; Syron, 1984). The structure of banking in the USA is, therefore, very diffuse, with almost 14,500 commercial banks of which the largest five only hold some 19 per cent of all deposits compared with 76 per cent in France and 78 per cent in Canada – which by comparison has only eleven commercial banks (Syron, 1984). Federal tax regulations prohibit cash deposits across state lines, so that rationalization of bank services and opportunities to take advantage of economies of scale are limited. The extent to which banks can branch within states also varies widely from no branching at all to statewide freedom.

On the positive side legislation enacted by local, state or national institutions may be used to attract or to stimulate service activities in the interests of a more diversified local economy or as a means of strengthening the tax base of the city. Financial incentives, easy availability of land, or assistance with retraining local labour are the kind of positive methods which can be used. The incidence of such institutional influences on service industry location is largely outside its control; a failure on the part of the service sector to respond may lead to changes, but the influence of political considerations may well be more significant than the reactions of the service sector at any particular time. The ability of service industries to respond to external institutional factors will rest upon their knowledge of the limitations or incentives in operation and, in general, they are often less aware of these than of technological or of demand-related factors.

Internal influences on location change

The internal factors contributing to the locational dynamics of service industries principally arise from the evolution of their organizational, structural and operational characteristics. Such evolution has been partially generated by internal forces but also encouraged by external factors such as the influence of competitors' activities and the need to counter any negative consequences by undertaking internal adjustments (Davies and Kirby, 1980; Dawson, 1979). An example of the effects of competition has been the demise of the small, independent retailer, at the expense of multiples and co-operatives (Dawson and Kirby, 1979; Scott, 1980; Dawson, 1983b). Faced with increasing operating costs, lack of capital investment and poor locations, often in inner city areas with declining population and customer potential, the number of independent shops (many of which are small) in the UK has decreased from 450,000 in 1950 to 230,000 in 1981. Market share has been reduced from 64 to 39 per cent during the same period (Dawson, 1983b). By 1978 multiples controlled some 42 per cent of total retail sales (Livesey and Hall, 1981; Akehurst, 1983); in 1980 large companies controlling over 100 retail outlets attracted 41 per cent of food sales and 38 per cent of clothing and footwear sales. There is also evidence – although limited by comparison with studies of manufacturing – for growing concentration in retail distribution: 'groceries-based companies are the "engine of change" lying at the heart of, and driving forward, the revolutionary changes now beginning to appear in retailing' (ibid. 170)

A key feature throughout the emergence of service industries (and indeed industries in general) is the distinction between the small firm and the large business organization (Taylor and Thrift, 1983). The latter has become a major form of operation in both the private and public sectors with private large organizations often multinational, and because they are becoming increasingly diversified and cover large geographical areas, they must confront internal problems of control. One response has been to reorganize into a multidivisional structure with separate headquarters which are ultimately responsible to regional or central headquarters. In order to remain profitable these large enterprises must endeavour to provide existing services on the most competitive basis possible, while also trying to identify and develop new services or new markets which may, in turn, require the start-up of additional operations in new locations or adjusting the locational attributes of existing activities. Some of the consequences have been illustrated in Chapter 7 with reference to the expansion and diversification of producer services. Taylor and Thrift (1982) suggest that a new type of large business organization, the global corporation, has now emerged as multinational corporations have lost their competitive edge because they have become much like each other. The global corporations – most of which incorporate service functions among their affiliates – are in a continual process of locational selection and reselection, in an endeavour to efficiently circulate the vast amounts of corporate capital involved in such ventures.

Location is itself part of the internal structure of a firm as well as part of the interface with external factors. Indeed the location problem for a firm is how to find a balance between what is best in relation to internal requirements and external factors; some imbalance between the two is inevitable. Lloyd and Dicken (1977) suggest that this imbalance only becomes significant for location when the stress created crosses a tolerance threshold which may relate to either internal or external factors, or a combination of both. Thus diversification through acquisition and the creation of a multi divisional organizational structure may create surplus premises (as the functions of the acquired firm are transferred to other parts of the organization, for example) while also creating a need for new divisional headquarters suitably positioned to exert efficient control over the establishments and personnel within the division. In the case of smaller firms the tolerance threshold may be crossed when rental increases following lease expiry are unacceptable; when there is inadequate space available for further expansion; or conversely, when a decline in business creates surplus space which it cannot be afforded to retain. Stress also may be created by locational changes made either by competitors or by other businesses upon which the firm is dependent. Changes in the centre of gravity of professional and business services within CBDs have been noted, among others, by Bannon (1972) in Dublin and Sim (1982) in Glasgow.

Locational change need not result from unacceptable levels of stress, firms can take adaptive action while retaining the same location because this avoids the uncertainties associated with a change of location such as the cost or the ability to retain valued clients. Adaptation *in situ* also allows a firm to retain the prestige which may be associated with its current address; this is valued by professional services such as attorneys, architects, consulting engineers and private medical practices, and by financial institutions such as banks, insurance companies and stockbrokers (Vernon, 1960; Armstrong, 1972; Cowan *et al.*, 1969; Gad, 1979). There is, therefore, a good deal of inertia incorporated within the existing location patterns of service industries, and this is probably enhanced by the evidence for lower mobility among small firms (of which there are a disproportionate number in services) (Keeble, 1968; Location of Offices Bureau, 1975). However, some types of services, especially those in the consumer group which are very sensitive to demand, will be less able to treat location change as a comparatively rare requirement. Many retail, hotel, entertainment, leisure and social services are in this category, while it seems that producer services, many occupying office-type premises, are least likely to make a relocation decision when the stress tolerance threshold is crossed.

Laggards, intermediaries and leaders

The small firm is numerically dominant within the service sector (Mitchell, 1980) and, in common with large business organizations, can be divided into 'laggards', 'intermediaries' and 'leaders' (Taylor and Thrift, 1983). The balance between these types of small or large firm has implications for location patterns and the potential for change. The laggards, many of which are small independent consumer service establishments, experience a high turnover rate in which replacement rather than relocation is commonplace and grow very slowly. The latter further reduces the possibility of locational changes or expansion into additional premises which are part of the same organization. The laggards are also unlikely to have any direct dependence upon large service organizations; they will be very likely to have indirect dependence through, for example, the expenditure of income on purchases by employees of large organizations, although this is unlikely to figure prominently in their location decisions. Intermediate small service organizations are, however, more dependent on large organizations especially through subcontracting and franchising. The latter is notable in retail services and causes a high rate of replacement as existing franchise-holders fail to meet targets, default on the operating standards imposed by the franchise-holder, or additional franchises are granted in new geographical areas. In retailing the system franchise is most common; the franchiser sells a way of doing business as well as the name associated with that business. The best example is the fast-food trade in the USA or Australia where a small number of very large franchisers, some of which are now multinational, control large numbers of small firms.

Leaders are the third segment among small service firms; characteristically they are volatile activities at the forefront of innovation adoption and the development of new services. Birth and death rates are high as many of these firms are started by determined entrepreneurs who often, for example, identify gaps in existing markets and concentrate on trying to fill them. The interaction between the leaders and large service organizations is less strong than that for intermediate small firms except where the latter endeavour to achieve takeovers in order to obtain control over a new innovation, or to make available levels of capital support for innovation development which are not accessible to small firms on their own. Hence few small leader service firms will themselves evolve into large business organizations; the probability is further reduced by the problems which they face in obtaining finance to allow an adequate response to larger contracts (Johnson, 1978), their vulnerability to the effects of fluctuations in demand and their limited ability to survive or be profitable because of their dependence on one service line rather than the multi-product base of larger service organizations.

The location effects of diversification

Channon (1978) has shown how the largest 100 service industry firms in the UK have followed diversification strategies between 1950 and 1974. The evolution of the Lex Service Group which started as a small garage and car distribution group in London (Figure 8.3) illustrates the process of diversification by acquisition. This has not only extended the geographic coverage of the car distribution network, but has also brought under the group's control a variety of other service industry activities. The trend has been most evident among financial services, beginning with clearing banks and insurance companies and affecting other financial services since the late 1960s. Diversification has involved the development of overseas markets, the development of international services which compete with North American commercial banks, and a more flexible attitude towards the services offered in competition with other British financial institutions. Recent proposals to introduce computer automation in the UK into the Post Office's network of 22,000 main and sub post offices offers the prospect of banks, building societies and most other types of financial institutions becoming much more accessible to their customers than their own network of branches currently permits (Smith and Rodgers, 1984). Financial

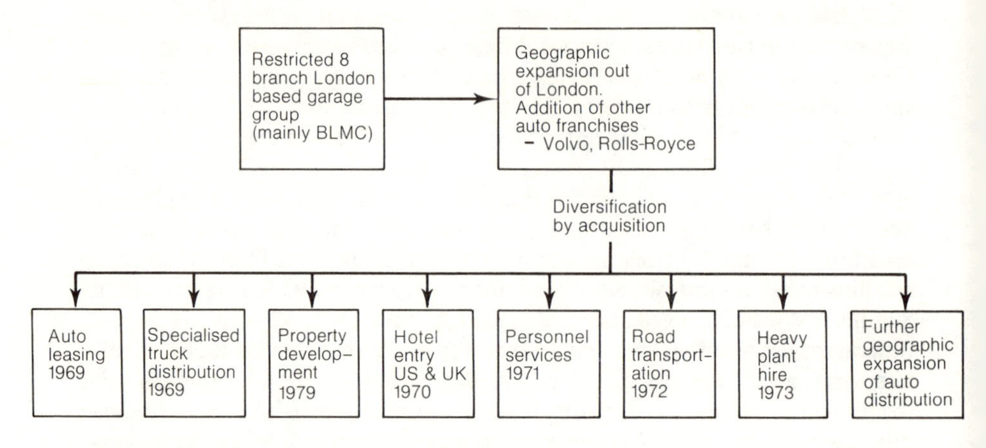

Figure 8.3 Evolution of the Lex Service Group, 1945–74
Source: Channon, 1978, figure 10.2, 263.

groups without their own branch networks will also be able to compete more directly, supplemented by further development of automatic cash dispensers and direct debiting of bank accounts from shop tills. Diversification has, therefore, become a strategy for survival within which locational change will inevitably have a part to play.

The hotel and leisure industry also comprises firms which have diversified into several areas of leisure activity, ranging from cinemas to multi-purpose commercial sports centres. Shipping companies have acquired property, publishing and other interests in order to counter their poor financial performance for much of the postwar period (Channon, 1978). Retailing firms have revealed variable degrees of diversification but in almost all cases this has not resulted in the acquisition of, or merger with, non-retail services. Changes in the geographic coverage of the large multiple-store groups have stabilized and competition now takes place through enhancement of the retail services offered and, as with financial services, the breakdown of the long-standing demarcation between companies on the kinds of goods available from their stores. In contrast to financial services diversification in the retail sector has rarely involved expansion into international operations; the limited attempts to do so have had limited success since consumer tastes and preferences seem to be more inelastic than those of corporate executives using producer services supplied by companies from another country.

Diversification has been accompanied by an increase in multi-divisional organization, especially since the mid-1960s when the demand for management and organization consultants to advise on corporate restructuring greatly increased. By 1970 some 25 per cent of the firms in Channon's (1978) study had introduced a multi-divisional structure and this had increased rapidly to 37 per cent by 1974. The process of diversification has largely been manifest among private services: public sector companies are limited in what they can do by the legislation used to establish them, although this has not protected them from a need for strategic changes of locational significance. Central government departments are similarly restricted and the creation of new functions and withdrawal of existing services is the principal cause of organizational and locational change.

LOCATIONAL CHANGE BY RETAIL SERVICES

Much of the literature tracing the processes and outcome of service industry locational dynamics arises from studies of the distributive trades, especially retailing (see, for example, Kivell and Shaw, 1980; Dawson, 1974, 1980, 1983a; Schiller, 1971; Scott, 1970; Simmons, 1964; Davies, 1976, 1979, 1984). Part of the distribution system involves the transfer of goods from the manufacturer to a warehouse from where bulk is broken and smaller batches are forwarded to retail outlets for purchase by the consumer (Davies and Kirby, 1980; McKinnon, 1983; Firth, 1976). This intermediate stage in the distribution process may be operated by the retail organizations themselves or by separate independent, voluntary, or cash-and-carry wholesaling organizations acting as middlemen. Wholesaling has itself been undergoing some important changes but the limited official statistics together with a paucity of research, at least by retail geographers, makes analysis difficult (Davies, 1976; McKinnon, 1983; Lord, 1984). Storage wholesaling, whereby customers such as small grocers travel to warehouses to collect goods for sale in their own establishments, or receive delivery of bulky goods such as coal or timber, has a limited range of locational alternatives and is often found in low-cost, but accessible, inner city areas outside the CBD.

Cash-and-carry wholesaling, where customers are attracted by lower prices made possible by bulk purchasing by the wholesaler, provides less elaborate display and customer assistance, but easy and free parking (Thorpe and Kirby, 1971; Lowe, 1983). The food trade has been at the forefront in the development of these warehouses, with do-it-yourself (DIY) supplies (Hornik and Feldman, 1982), modular furnishing and domestic electrical appliance suppliers also operating in this way. Do-it-yourself services nicely illustrate the influence of external forces on the organization and location of retailing. In the USA, for example, Hornik and Feldman show how a number of social and economic factors have encouraged the DIY movement (see ibid.). These include the growing disparity between the cost of services and income; between 1972 and 1978 the cost of housing increased by 106 per cent and the wages of skilled construction workers by 131 per cent but disposable personal income increased by only 73 per cent. There is, therefore, considerable incentive to substitute self-help for 'buying in' expensive services such as car maintenance, electrical work, and home improvements and decoration. The availability and quality of services has deteriorated because of a shortage of trained personnel, while attitudes towards helping themselves on the part of those who normally buy in decorating and similar services has changed. Most cash-and-carry warehouses are located in or adjacent to larger urban areas, usually on industrial/warehouse estates at suburban locations with good access to motorway and other high-grade transport routes (Thorpe and Kirby, 1971; Watts, 1977).

Location of retail warehouses

Distribution from warehouses owned and operated by retailers has increased rapidly during the last ten years in parallel with the growth of multiple food, consumer durable and department-store groups (Firth, 1976). Their warehouses act as collection points for the range of goods from diverse manufacturers and from which a suitable number of stores can be serviced, primarily using road transport (Boswell, 1969; Walters, 1975; McKinnon, 1981). Economies of scale and control over the timing and routing of distribution to stores are important reasons for the emergence of this type of warehousing. A map of the recent development of warehouses of this kind in Figure 8.4 clearly reveals an association between location and the potential for rapid assembly and distribution to large areas of the country (McKinnon, 1983). Much of the recent growth of warehousing at Warrington, Birmingham and Basingstoke can be accounted for in this way and most has taken place in the semi-urban rings adjacent to the major metropolitan areas. As warehousing has become more consumer orientated and/or operated by large retail organizations it has become relatively dispersed throughout the urban system but with some concentration at certain favoured locations (see Figure 8.4). This is far removed from the highly localized pattern of warehouses serving relatively small geographical areas typical of earlier periods when independent retailers were predominant or, in the case of the major break of bulk-points at ports such as Liverpool and London, large inner city warehouse complexes which have since become outmoded, underused, or demolished in the wake of containerization and the changing geography of port functions.

Warehousing and related activities comprise a less visible artefact of change in the distributive services than retailing, which has experienced structural and locational changes which are especially manifest in advanced economies. The structural changes which have been taking place involve a shift away from independent to multiple and co-operative organizations supplemented by department stores, mail-order firms and retail markets (Davies, 1976). It seems, however, that this is not an inexorable trend; it is now generally accepted in the USA that multiples (chains) with ten to fifty stores are best placed to be successful, thus ensuring a more stable position for the independent retailers (Rogers, 1983). The halt in the demise of the independent and small chain groups has been encouraged by the availability of space freed by larger and perhaps long-established multiple organizations which have reached the end of their life cycle, by the expansion of franchising which favours smaller operators, the entrepreneurial environment which is prominent in the USA and the growing opportunities for specialist retailers who provide the personal service and range of goods not available in most multiple outlets.

Nevertheless, there has been a clear trend since 1945 towards larger-scale economies using ever-larger units of operation, beginning with supermarkets which were introduced in the USA in the early 1950s and, as with most other innovations in retailing, rather later in Western Europe during the early 1960s (Zimmerman, 1955; Davies, 1976). Later the supermarkets were replaced with

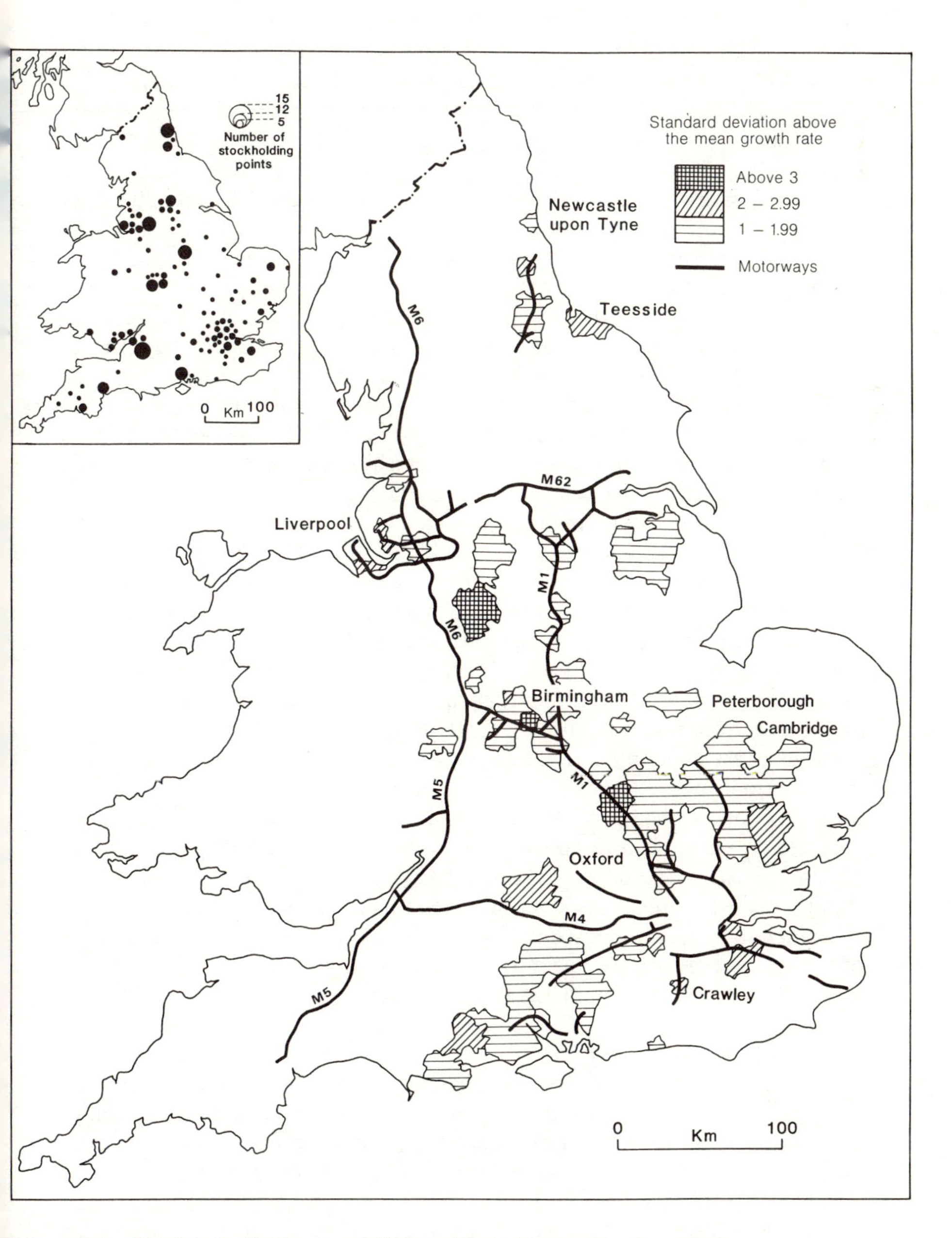

Figure 8.4 Districts in England and Wales with rapid growth of warehouse floorspace, 1974–82, and number of stockholding points (inset)
Source: McKinnon, 1983, figure 2, 393; figure 3, 395.

superstores (>2500 m^2) (Thorpe, 1977), then hypermarkets (>5000 m^2) (Parker, 1975) and discount warehouses during the early 1970s (Davies and Kirby, 1980). Hypermarkets have their origin in France rather than North America and have set the standard for the marketing of food and non-food goods under the same roof (Beaujeu-Garnier and Bouveret-Gauer, 1980; Dawson, 1983a). It is noticeable that convenience goods retailers have led the way in the pursuit of greater economies of scale, but as the distinction between grocery and non-food outlets has broken down (partially as a result of the process of concentration mentioned earlier), this role is now less significant.

These structural changes in retailing must be seen against a backcloth of social and economic trends which imposed considerable direction on the locational trends revealed by retailing. However desirable from the perspective of the retail organization, the search for economies of scale in larger operating units – especially the appearance of superstores, hypermarkets and retail warehouses – would not have been feasible without the increases in personal mobility permitted by the growth of car ownership. Such stores are more widely spaced and permit fewer but more comprehensive (in terms of volume and range of purchases) shopping trips for which a car provides the desired flexibility. Improvements in the provision of high-standard intra- and inter-urban highways have permitted capitalization of the enhanced mobility bestowed by private cars as well as proving a magnet to the location of retail development. Transport improvements have also undermined the accessibility advantages traditionally held by the CBD which, in conjunction with the differential growth of population between central cities and suburbs and the outmigration of population from major towns and cities, has experienced greater competition from alternative retail locations. In addition, there has been a major expansion of female activity rates, from 43–52 per cent between 1970 and 1981 in the USA (Rogers, 1983), which has created a demand for more late-night shopping, particularly in the convenience goods trade. Consumer tastes and preferences have also become more sophisticated, with the ambience of the shopping environment, its comfort, cleanliness, safety as well as mix of speciality and department stores high on the list of shopper priorities. Many of these requirements are easier to meet in new shops or shopping complexes rather than in older, established retail locations.

The shopping centre industry

Perhaps the best-known response to changes in the demand for retail services, increasing competition and the opportunities offered by technological change has been the growth of a shopping centre industry specializing in the development of new retail complexes to replace existing retail centres or to create completely new facilities (Dawson, 1983a; Davies, 1976; Rogers, 1983). The origins of planned centres can be traced back to experiments during the late nineteenth century in the USA but momentum was not really gained until the early 1920s, when the evolving suburban transport systems of large metropolitan areas, such as the tramways in Los Angeles, offered opportunities for the development of ribbon shopping centres in the vicinity of important intersections (Hoyt, 1933; Applebaum, 1932). These centres provided direct competition for the traditional concentration of retail services in the CBD and by 1929, for example, there were already sixteen suburban shopping centres in Chicago (Hoyt, 1933, cited in Dawson, 1983a), increasing to thirty-three by 1933. Convenience stores dominated these strip shopping centres which relied on the attraction of consumers from the adjacent highways; centres comprising comparison and convenience retailers, often much larger than strip centres, came rather later. It has been suggested that the first example of a comprehensively planned out-of-town shopping centre was the Country Club Plaza, Kansas City, which was built in 1923 (Kelley, 1956).

There was no true equivalent of this kind of retail development in Europe, although small clusters of retail services were an integral part of most substantive private and local authority suburban housing schemes, of which there were many, during the period of rapid urban expansion in the UK between 1920 and 1939 (Burns, 1959). The pull exerted by highway-oriented locations was less evident than in North American schemes; developers and local authorities were mainly interested in generating high profits or regular rents from retailers with an effective spatial monopoly within residential areas in circumstances where personal mobility was still comparatively limited by comparison with the USA. Larger suburban shopping centre developments on the North American scale did not appear in Britain until after 1960, and even then at a lower frequency and on a smaller scale than elsewhere in Europe, particularly France (Dawson, 1983a; Alexander and Dawson, 1981).

Planned shopping centres are essentially equated with advanced economic systems; in less developed countries urbanization has been rapid but it has not been accompanied by increases in per capita income among a large enough segment of the population to sustain large new shopping centres. During the 1970s a limited number of centres have been provided, either targeted at selected high-income areas or groups, such as in Mexico City or Nairobi (see ibid.) or as an integral part of development plans for new housing areas. While the level of such activity will likely grow during the rest of this century, the overall impact on the location of retail services will remain limited by comparison with North America, Australasia, or Western Europe.

It is common for analysts to classify planned shopping centres: the 'standard' classification distinguishes between regional, community and neighbourhood centres (McKeever, 1957; Urban Land Institute, 1977) but Dawson (1983a) suggests that this does not do justice to the current diversity of shopping centre types when European and Third World examples are considered. He has, therefore, proposed an extended classification of shopping centres (Table 8.1) within which general-purpose, free-standing planned centres are but one of six types. Theme centres, usually composed primarily of speciality stores, and factory outlet or 'off-price' centres could also be added to the classification in Table 8.2 (Rogers, 1983).

By 1980 there were 22,050 planned shopping centres in the USA alone, a threefold increase since 1964, and these accounted for some 42 per cent of

Table 8.2 A classification of retail centres

Group	*Centre type*	*Location*	*Size(m²)*
FREE-STANDING CENTRES	Super-regional	Intra- or inter-urban	100,000
	Regional	Intra- or inter-urban freeway intersection	50,000
	Community	Urban highways at intersections	20,000
	Neighbourhood	Local highways at intersections	5,000
CENTRES IN EXISTING SHOPPING DISTRICTS	Infill	Near to peak land value in retail area	2,500
	Extension	Close to core replacement	15,000
	Core replacement	Centre of CBD	40,000
MULTI-USE CENTRES	Centres, new towns Downtown megastructures	New community CBD	40,000
ANCILLARY CENTRES	Hotel associated Office associated Transport associated	CBD/large office/hotel	3,000
SPECIALIST CENTRES	Purpose-built In recycled centres	High-income areas	6,000
FOCUSED CENTRES	—	Intra-urban highway junctions	10,000

Source: Dawson, 1983a, tables 2.4 and 2.5, 26–27.

total retail sales (Table 8.3). It is anticipated that this share will increase further to 50 per cent by 1990. As Table 8.3 shows, planned shopping centres cover a wide size range in which small centres (<20,000 m^2) account for one-half of total sales and over 85 per cent of all centres. At the top of this hierarchical classification is the super-regional centre with at least 100,000 m^2 of gross leasable floorspace. Such centres are mainly located in major metropolitan areas in sectors with largely middle- and high-income households and have more than three, and often six, department stores in addition to at least 100 other shop units, incorporating some other services such as cinemas and ice-rinks. Super-regional centres often contain more shopping space and related facilities than established towns and represent serious competition for the custom of an almost exclusively mobile clientele. Regional shopping centres

Table 8.3 Shopping centres in the USA, number and sales, 1980

Size (m^2 in thousands)[1]	*No. of centres*	*%*	*1980 sales ($m.)*	*%*
>1–<10	14,586	66.2	118,677	30.8
>10–<20	4,420	20.0	81,203	21.0
>20–<40	1,695	7.7	58,646	15.2
>40–<80	839	3.8	58,457	15.2
>80–<100	242	1.1	26,984	7.0
>100	268	1.2	41,534	10.8
Total	22,050	100.0	385,501	100.0

Note: 1 Gross leasable area; estimation from square feet in original tabulation.
Source: Rogers, 1983, table 1, 10.

are more numerous and comprise at least 40,000 m^2 of covered shopping, a catchment of at least 100,000 (preferably twice this figure), between one and three 'anchor' department stores which attract other retailers and play a prominent part in the internal organization of retail services in planned centres, parking space for more than 4000 cars and a location immediately adjacent to major highway (usually motorway) intersections (Figure 8.5). Typically US regional shopping centres occupy free-standing greenfield sites but elsewhere land shortages and parking controls (Japan) or the policies adopted by local planning authorities (the UK, Australia) have modified development patterns (Dawson, 1983a). There is only one strictly 'regional' shopping centre in Britain and this was opened in 1976. The Brent Cross shopping centre in north-west London has almost 80,000 m^2 of retail space, two department stores and over 3000 parking spaces. European priorities have been to encourage integration between old-established shopping areas and new shopping centres, thereby controlling the conflicting demands for the scarce resource of land in countries such as the Netherlands, Belgium and the UK.

New shopping centres in the range 10,000–30,000 m^2 are classified as community centres; they have smaller catchments (25,000–100,000), more limited parking provision (1000–1500 vehicles), smaller sites, a more limited range of retail types and are anchored by discount department stores or catalogue showrooms (Rogers, 1983). Major highway orientation is less influential for the location of community shopping centres (Figure 8.5) which are often developed in association with large new residential extensions of the kind currently taking place around the 'sunbelt' cities (Perry and Watkins, 1977). At the lower end of the hierarchy the neighbourhood centre (3000–10,000 m^2) is the most ubiquitous and found, with variations in detail but not function, throughout West European, North American, Japanese and Australasian cities. The anchor

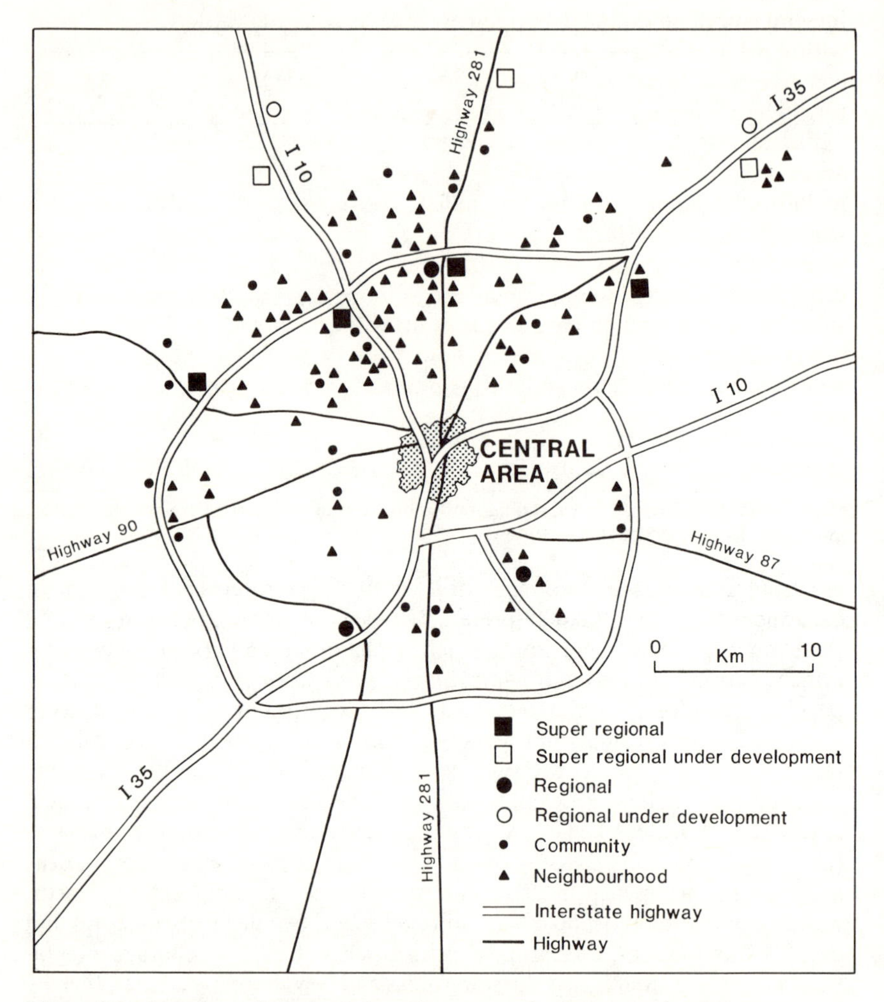

Figure 8.5 Location of planned shopping centres in San Antonio, Texas
Source: Dawson, 1983a, figure 22, 25.

stores may be a supermarket and/or a drugstore, in a group of stores mainly concerned with providing convenience goods and personal services within a catchment of beween 8000–20,000 population. Parking is much more limited than at regional or community centres, especially in European cities where many neighbourhood shopping centres are former village centres subsequently absorbed by the tide of postwar suburban expansion. Because neighbourhood centres primarily service local needs (i.e. within five to ten minutes' walking distance or five minutes' driving time), they are often associated with public services such as primary schools, health centres and district libraries. This pattern of development is best seen in new town neighbourhood units, where service centres which combine private and public activities are provided as an integral part of an overall development plan. Such centres tend to be embedded within residential areas but Dawson (1983a) notes that in the USA commercial developers favour neighbourhod centres at intersections on major distributor highways separating major residential areas.

Planned shopping centres have arisen in response to the extension of urban areas but pre-existing retail service centres are also subject to extension and replacement (Table 8.2). The possibilities are diverse and local circumstances will influence the detailed possibilities. Infill and extension becomes necessary where increases in local demand require a response to the risk that trade will otherwise be diverted to other retail areas. Extension is most common among suburban centres which have not kept up with the growth of population in newer suburban extensions. It may not always be possible to respond in this way, however, because of limitations imposed by local planners to conserve land for other purposes or to limit the encroachment of retail functions on established residential areas. Infill then becomes attractive (if the space is available) and may involve the introduction of a new centre behind existing shop frontages but with access to the principal shopping thoroughfare. But perhaps the most interesting centre type in this group is the estabishment of new retail facilities in CBDs.

Impacts on established retail centres

The growth of planned, and primarily suburban, shopping malls during the 1960s and 1970s not only took place in response to changing demand, but also at the expense of long-established shopping centres, especially the CBDs (Muller, 1981; Black, 1978; Smith, 1972; Sternlieb and Hughes, 1981, 1983). The causal link is, however, more difficult to measure than is commonly assumed (Kivell and Shaw, 1980; McDonald, 1975) but retail sales in the CBDs of US cities certainly began to decline during the early 1960s, and the decline has been accelerating during the 1970s (Table 8.4). Between 1972 and 1977 CBD sales in large metropolitan areas (sales exceeding $850 million per annum) declined by 2 per cent, equivalent to 46 per cent when adjusted for inflation (Muller, 1981). In general, there is no difference in the trend according to city size or region, and while in some cities CBD sales have increased, the rate of growth has been two or three times slower than for the central city or SMSA as a whole. Using data for a much longer time period (1954–77) for over ninety SMSAs grouped into regions, Robertson (1983, 323) concludes that retailing has experienced 'massive absolute decline' regardless of location. As a proportion of SMSA sales CBD losses were highest during the 1970s and there was no significant difference between sunbelt and frostbelt cities. Although the rates of decline were most rapid in smaller SMSAs, the CBD share of overall SMSA retail sales remained higher than in the larger metropolitan areas.

This contrasts with the strength of the CBD for retail services in cities in the UK, at least between 1961 and 1971. However, the proportion of total turnover accruing to central area retailers decreases with city size, so that in 1971 those in the 100,000–250,000 size range generated 45.7 per cent of total turnover in the central area compared with 24.9 per cent in cities with a population

Table 8.4 Changes in CBD, central city and SMSA retail sales: selected US cities, 1972–7

SMSA	*CBD*		*Central city*		*SMSA, % increase (1972–7)*
	Sales[1] *(1977)*	*Change (%)*[2]	*Sales (1977)*	*Change (%)*	
Chicago	932	9.1	8179	23.6	48.0
Boston	463	7.3	1832	12.8	35.8
Minneapolis	324	14.0	1265	26.2	62.2
Tucson	261	−2.8	1381	58.3	68.1
St Louis	209	14.2	1373	18.1	55.0
Milwaukee	176	7.9	1926	39.2	55.4
Salt Lake City	171	19.1	884	47.6	71.0
Albuquerque	140	−39.1	1434	77.3	77.4
Kansas City	91	−22.3	1703	37.2	58.5
St Paul	91	−12.9	905	35.2	62.2
Omaha	69	−15.7	1357	43.9	57.1

Notes: 1 In $m., 1980 values.
2 1972–7.
Source: Muller, 1981, table 2, 185.

exceeding 1 million (Kivell and Shaw, 1980; see also Thomas, 1975; Schiller, 1977). It will be shown in Chapter 9 that institutional intervention in the development of urban areas has played an important part in sustaining the prominent place of the CBD in the location of British retail services despite the demand by retail companies and the property development industry to locate new facilities in the suburbs or greenfield sites outside the main built-up area of large cities. The future of the city centre as a retail location is better assured than has been the case in North America but since the mid-1970s there has also been more interest there in reinvestment and rehabilitation of CBD retailing.

In order to explain this 'rediscovery' several theories can be advanced; the growth of regional and super-regional malls has relegated the CBD to equivalent status and since it remains at least as (if not more) accessible regionwide as suburban centres its development potential can be assessed accordingly. Many of the early regional shopping centres are in need of modernization and do not offer the range of merchandise or environmental assets of newer centres further away from downtown areas; this prompts assessment of alternative development strategies in which the CBD is included. Downtown economic development agencies anxious to retain the tax base of a city will encourage CBD development because some of the highest land values are found there (partly because other services such as office-based activities still demand downtown space) (Bies, 1977). Federal initiatives towards improving central city environments following the riots in some large cities such as Los Angeles (Watts) in 1968 have also encouraged the CBD redevelopment lobby (for example, the Economic Development Act 1970 and the Community Development Act 1977). There is a reduction in the outmovement of population combined with a concern for more energy-efficient urban forms in which public transportation plays a more prominent part (a number of US cities such as Washington, DC, San Francisco, Baltimore and Philadelphia have invested in rapid-transit systems which serve and cross downtown areas). Hence it would seem that accessibility advantages, together with some deterioration in the competition from older regional shopping centres in the inner suburbs, provide downtown areas with an opportunity to divert demand from within their primary trade areas.

The latter argument follows from the widely recognized (Spink, 1981) fact that 60–70 per cent of regional mall sales come from the population within ten minutes' driving time; CBDs can also orientate their retail services towards more diverse markets, such as downtown employees, tourists, conventions, entertainment and special retail functions. Other services, therefore, have a supporting function in the development of new retail space which must replace often functionally obsolete and decaying structures in prime locations; such schemes may be as large and complex as suburban regional malls and most large American cities now have schemes planned or in progress (De Vito, 1980; Spink, 1981). Several cities (Seattle, Portland, Denver and Philadelphia) have created pedestrian malls in an effort to upgrade the retail environment. Others have undertaken complete redevelopment of large areas such as Quince Market (Boston) or Fisherman's Wharf (San Francisco). Rehabilitation of parts of the CBD having historic significance (Pioneer Square or Pike Place Market in Seattle) has also been undertaken. But Robertson's (1983) results contradict the notion that CBD retailing is being revitalized and that the CBDs of the sunbelt cities are more viable than those of the frostbelt cities.

One reason for the continuing importance of the CBD as the location for retail services in British cities is the inertia caused by redevelopment following the damage caused by wartime bombing between 1940 and 1945. During the years immediately after the war there was no clear indication that suburban locations would expand in a way which would create direct competition for retail services with established city centre areas. All the cities in this position, such as Swansea, Plymouth, Coventry and Southampton, quickly decided to try to retain the prestige, image and commercial vitality bestowed by the highly centralized urban structure of the prewar years. Retail services would play a leading role in the rejuvenation process since they (together with employment in office buildings) were at the leading edge of the interface between city centres, and their immediate urban environs and the regions beyond. At the same time, the opportunity was taken to introduce innovations which eased the interaction between retailers and consumers; the most important was the pedestrian precinct, first introduced in Rotterdam in the Lijnbahn in 1948, where traffic was effectively excluded and space freed for less congested, safer, healthier and more relaxed use of retail and other services. As Davies (1976, 178) suggests, 'the precinct concept ultimately flourished in the 1960s as the cornerstone of new shopping developments that are best described as the "in-town" equivalents of the North American "out-of-town" centres'.

City centre precincts have undergone substantial changes in design and scale since the early, open-air, examples in Britain's new towns such as Hemel

Hempstead and Harlow. There has been a trend towards enclosed malls, much like their North American counterparts but lacking the regularity in design and functional structure (for a full discussion see Bennison and Davies, 1980). The city centre is the general location, within it the location of new precincts does not follow a set pattern. Local morphology, the size and shape of land parcels zoned for retail use in an environment shared with competing uses, the proximity of parking facilities and public transport termini, the size and objectives of each new precinct and the kind of retail services likely to want space in a new precinct are some of the variables to be considered. Some of the largest precincts, such as the Arndale Centre (Manchester) or the Victoria Centre (Nottingham) are added to existing retail areas with which they are connected as directly as possible in a way which encourages pedestrian circulation between the new and the established areas. Alternatively, new city centre precincts such as Eldon Square (Newcastle upon Tyne) and St George Centre (Preston) have been 'embedded' into the established retail infrastructure. Between 1967 and 1977 some 250 centres have been constructed in British city centres, and although many have been successful, some have been less than satisfactory.

While representative of the changing needs of retail services and new concepts in city centre redevelopment, city centre shopping precincts will themselves cause changes in the location of retailing and other services within their general environs (Table 8.5). Bennison and Davies identify social, environmental and social impacts, both positive and negative, of new shopping centres in city centres (see ibid.). Although increases in consumer disposable income have created a demand for new retail floorspace, much of the provision within new precincts is occupied by retailers relocating from their previous establishments in other parts of the same central area. National multiple stores may vacate premises in another part of the city centre, especially if the new centre is likely to change the 'centre of gravity' of pedestrian flows. Other retail services, such as local independents or specialist retailers, may be unable to afford the higher rents demanded in new centres but will wish to move as near as possible, thus vacating other premises in the process. If they fail to respond, they may suffer from reduced turnover and eventually face closure. On the positive side the premises vacated sometimes act as a low-cost 'seedbed' for new retail services, especially stores which provide goods in response to current fashion and to new but not yet well-established technology, or may serve a very selective segment of the market for which a central location is not essential. This diversifies the retail infrastructure of the city centre and enhances its status as a regional shopping centre.

LOCATION CHANGES BY PRODUCER SERVICES

Dispersal of consumer services at all scales of analysis is a well-developed and widely analysed phenomenon but location change has also occurred among producer services (Stanback, 1979; Cohen, 1977; Armstrong, 1972; Fuchs, 1983; Daniels, 1975; Alexander, 1979). By comparison with consumer services, however, producer services are highly centralized activities with, as we have seen,

Table 8.5 Positive and negative impacts of retail renewal in city centres

Economic		*Social*		*Environmental*	
Positive	*Negative*	*Positive*	*Negative*	*Positive*	*Negative*
Adds new stock	Reduces old stock	Allows for efficient shopping	May favour car-borne shoppers	Modernizes outworn areas	Changes traditional character
Accommodates larger modern stores	Discriminates against small independents	Provides new shopping opportunities	May limit choice to stereotypes	Reduces land-use conflicts	Creates new points of congestion
Increases rates and revenues	Increases monopoly powers	Provides more safety	Creates new stress factors from crowds	Scope for new design standards	Intrusive effects on older townscapes
Creates new employment	Changes structure of employment	Provides more comfort and amenities	Attracts delinquents and vandals	Provides weather protection	Creates artificial atmosphere
Improves trade on adjacent streets	Reduces trade on peripheral streets	Concentrates shopping in one area	Breaks up old shopping linkages	Leads to upgrading of some streets	Causes blight on other streets
Enhances status of central area	Affects status of surrounding centres	Potentially greater social interaction	Becomes dead area at night	Integrates new transport	Causes pressure on existing infrastructures

Source: Bennison and Davies, 1980, 14.

disproportionate representation in the largest urban centres. Thus much of the evidence relating to relocation and redistribution is derived from studies which have focused on large metropolitan areas such as New York (Armstrong, 1979; Quante, 1976), Seattle (Daniels, 1982), Sydney (Alexander, 1978), Manchester (Damesick, 1979), Washington, DC (Fuchs, 1983), Toronto (Code *et al.*, 1981) and London (Daniels, 1977). In most of these and other studies producer services are not examined on an industry or MLH basis, but by using office floorspace, office employment, or corporate headquarters as a surrogate. The connection is a strong one but certainly far from perfect. Indeed Gottman (1983, 72) urges caution because geographers have perhaps too readily adopted the office 'as the spatial expression, relatively easy to survey and measure, of the geographical environment of service work'. With this caveat in mind and using office activities as surrogates for producer services, location changes have been promoted by a mix of push–pull factors which partially overlap with those that have influenced location change decisions by consumer services.

Relocation of office-based producer services

In certain respects office-based producer services are victims of their search for centralization, in that such a limited number of pressure points have been created both within national urban systems (see Chapter 7) and within individual urban areas. Negative externalities have inevitably followed from such concentration, including rising rents and other property taxes, increases in salaries reflecting competition for specialized as well as more routine staff, and escalation of fringe benefits and ever-lengthening journeys to work as the population in metropolitan areas has become more dispersed. Traffic and pedestrian and other types of congestion within city centre office, retail, restaurant and other areas is commonplace and, among larger organizations, fragmentation of premises (due to inability to expand within their principal buildings) has pushed up operating costs and reduced efficiency. There is also evidence (Quante, 1976) that the general social and economic decay within the CBDs of large Ameican cities, reflected most acutely in crime statistics, has also encouraged corporations to consider alternative locations in order to be able to attract the kind of staff that they require.

The way in which such externalities are perceived by service firms will be partially determined by the information which they have about the alternatives. A comparison with suburban locations, for example, reveals the prospect of lower accommodation costs because land and development costs will be lower, together with an opportunity to obtain most staff at lower wage rates, improved accessibility to labour markets (especially to part-time female labour), prospects of less-congested, more comfortable working conditions attractive to employees and possibly leading to lower staff turnover, less absenteeism and greater daily reliability in arrival and departure times and improved productivity. At the inter-urban level in the USA the shift of population from the frostbelt to the sunbelt, and the 'quality of life' factor, has encouraged longer-distance relocation of some producer service functions (Cohen, 1977).

Table 8.6 Location of corporate headquarters in selected SMSAs, 1958 and 1975

SMSA	*No. of corporate headquarters*		*Change 1958–75 (no.)*
	1958	*1975*	
New York	142	104	−37
Chicago	50	44	−6
Pittsburgh	23	15	−8
Los Angeles	17	21	4
Philadelphia	17	13	−4
Detroit	16	13	−3
Cleveland	15	17	2
St Louis	14	12	−2
San Francisco	13	9	−4
Boston	7	12	5
Minneapolis–St Paul	7	12	5
Dallas	6	6	—
Cincinnati	4	3	−1
Kansas City	3	1	−2
Atlanta	—	5	5
Washington, DC	—	2	2
Denver	—	2	2
Total SMSAs	334	291	−41

Source: Armstrong, (1979), table 3.15, 87.

Starting with Goodwin's (1965) paper on the location of management centres in the USA, extensive use has been made of the *Fortune* 500 lists to chart the changing geography of headquarters location (see, for example, Stephens and Holly, 1981; Semple, 1973; Burns, 1977; Armstrong, 1979). Similar data is also available for British companies (see, for example, Evans, 1973; Goddard and Smith, 1978) and Japanese companies (Abe, 1984; Nagai and Myaji, 1967).

Changes in the spatial distribution of corporate headquarters may result from physical relocation of corporate headquarters and changes in the performance of corporations (assets, number of employees and sales) which lead to their exclusion from the top 500 or 1000, or as a result of mergers and takeovers which cause headquarters to be centralized at one location (Table 8.6). For these reasons too much should not be inferred from Table 8.6, although other census or *County Business Patterns* data for employment shifts confirm the underlying trend towards outmovement of producer service activities from their traditional strongholds (Daniels, 1984). Thus between 1958 and 1975 the New York SMSA lost thirty-eight corporate headquarters (27 per cent), Pittsburgh lost eight (35 per cent), Detroit lost three (19 per cent) and Chicago lost six (12 per cent). Conversely, Boston gained five (71 per cent increase), Houston gained ten (1100 per cent), and Los Angeles gained four (24 per cent). The majority of these changes have taken the form of suburban rather than inter-city relocations; only six out of forty relocations from New York by headquarters of the top 500 involved the sunbelt cities, the majority stayed in New Jersey, Connecticut, or suburban New York, less than one hour from Wall Street (Stephens and Holly, 1981). The scale of headquarters operations in some cities may have increased but the rank order is much the same as it was three decades ago (Cohen, 1979).

The relocation of corporate headquarters is probably less important than differentials in regional and urban growth processes between the emerging south and west and the slower-growing north-east of the USA (Semple, 1973; Stephens and Holly, 1981; Rees, 1979; Semple and Phipps, 1982). Indeed Stephens and Holly note that

> the head offices of industrial corporations display a marked conservatism in their locational behaviour. Just as the urban hierarchy displays a tendency toward long-term stability, so also does the distribution of corporate influence. (Stephens and Holly, 1981, 298)

If the latter is measured in terms of the assets, sales and number of employees of corporations controlled from headquarters in the older and well-established corporate complexes such as New York, there was no significant change between 1955 and 1975. In this way corporate complexes have, if anything, been strengthened if as Cohen (1977) has observed modern corporations have become increasingly dependent upon the external support provided by advanced business and professional services for which there are relatively few suppliers available (see also Stanback and Noyelle, 1980). Most of the companies providing these crucial strategic and other inputs are located in a small number of 'international' cities such as New York, San Francisco, Paris, London and Tokyo. There are, therefore, significant backward linkages between relocated corporate headquarters in suburban Connecticut and specialized services in Manhattan (Table 8.7), or alternatively, a company such as General Motors which is officially headquartered in Detroit maintains a large executive office in New York which keeps in touch with financial markets and operations (Cohen, 1979).

In common with the dynamics of retail service location the degree of suburbanization by producer services differs between North American and other cities. The tradition of centralization is still a powerful influence on location in Sydney (Alexander, 1978), Edinburgh (Fernie, 1979), Wellington (Davey, 1972) and Leeds (Facey and Smith, 1968). It may well be that below a certain threshold of size de-concentration is inappropriate because any negative externalities in the CBD are outweighed by agglomeration economies which are more difficult to replicate in the suburban centres where office-based producer services can be expected to locate. This may be the situation in Australian metropolitan areas which have recently been trying to identify suitable strategies for the location of suburban offices (Daniels, 1985b). Retention of a centralized pattern of location is also encouraged by institutional factors, both private and public in origin (see Chapter 9), which are also present, but less powerful, in North American cities. There is also less scope for suburban areas (which

Table 8.7 Links between relocated corporate headquarters and other producer services, New York and Houston, 1975

Type of relocation and producer service used	*Utilize New York firm or mostly New York*	*Non New York firm(s)*	*No information*
Suburban relocations from New York since 1965 (24)			
Bank lender	22	—	1
Pension fund administrator	11	1	12
Investment banker	24	—	—
Law firm	21	1	2
Accounting firm	19	5	—
Out-of-city relocations from New York since 1965 (10)			
Bank lender	8	1	1
Pension fund administrator	5	—	5
Investment banker	10	—	—
Law firm	4	4	2
Accounting firm	4	5	1
Firms headquartered in Houston (9)			
Bank lender	4	6	—
Pension fund administrator	3	5	1
Investment banker	7	—	2
Law firm	1	6	2

Source: Cohen, 1979, table 4, 13.

are often incorporated within the same municipal jurisdiction) to provide tax and other incentives which are attractive to developers of suburban office space as well as to the firms moving in. The competition to attract economic development among suburban municipalities or counties around central cities in the USA is intense and generates wide variations in tax and other incentives.

While there may be international differences in the volume of location change by office-based producer services, Alexander (1979) concludes that the underlying reasons for change are much the same everywhere. He has compiled broadly comparable information relating to reasons for office moves from fourteen different studies undertaken during the late 1960s and 1970s. With one exception, a study by Quante (1976) of corporate relocation in New York, all the studies show that lack of space for expansion (an internal factor) is high on the list of reasons, followed by high rents and related costs (11/14) and expiry of lease (9/14). The latter is the most frequently occurring external factor, followed by congestion (8/14).

A similar pattern emerges in a recent study of 142 firms, mainly producer services, in the Washington, DC, area which had made location decisions between 1981 and 1983 (Fuchs, 1983). Over 75 per cent of the firms had relocated an existing office, 17 per cent had begun operations for the first time and the remaining 10 per cent were new branch offices. Among the firms which had relocated 46 per cent cited expansion needs, 20 per cent referred to the effects of rental increases, 13 per cent to the expiry of leases and 9 per cent to the importance of consolidating the operations of the organization. The impact of these push factors does not necessarily involve shifts from central to less central locations. There is substantial evidence (for example, Alexander, 1978; Daniels, 1982; Damesick, 1979; Goddard, 1967; Bennett, 1980) which shows that short-distance locational change within the same area of the city is more common than inter-area or inter-suburban movement (Figure 8.6). Migration from the centre of Toronto is both numerically smaller and less clearly focused on selected suburban centres than intra-suburban relocations. Between 1960 and 1977 in Toronto only 12 per cent of the total recorded moves (845) were made by office firms originating in downtown (Code *et al.*, 1981). Indeed it seems that the relative importance of the central core of Toronto as a point of origin for migrant firms has declined betwen 1960 and 1970. Only 36 per cent of the forty-seven firms in Fuch's study (1983) which moved from within the District of Columbia relocated to suburban areas outside, while most of location changes in the inner and outer suburbs also took place within those areas.

An assessment of the factors influential in the relocation decision quickly reveals the rationale for short-distance locational change. The search for suitable locations starts with an areawide search (Daniels, 1982; Fuchs, 1983), followed by a more detailed second stage involving individual sites or buildings within the short-listed area(s). At this stage accessibility, which takes many forms (Table 8.8), emerges as the prominent consideration, although its importance does vary in relation to type of service and area. Hence in the Washington, DC, study 83 per cent of the business services cited proximity to business clients, and 71 per cent of the R&D establishments and 60 per cent of consultant services, although analysis by area shows that the significance attached to this factor decreases from 95 per cent in Washington, DC, to 14 per cent in the outer suburbs. In the latter highway accessibility, which to some extent is a surrogate for contact with clients or other parts of the organization elsewhere in the Washington area but especially in Washington, DC, was most frequently mentioned by firms (31 per cent). It is easier to retain access to business associates by minimizing relocation distance or establishing new firms in areas where contact potential is known to be high.

But the physical linkages between office firms could be breaking down in response to the effects of technological innovation and as their association with particular functional clusters in the CBD is also declining (Edgington, 1982a). This has created the opportunity for greater dispersal of office functions within the CBD, a process also documented by Pritchard (1975), Bannon (1972) and Goddard (1967). It may well be that the pre-eminent influence of local linkages in the location of producer services is declining, thus allowing a greater opportunity for decreasing centralization of these services both within CBDs and beyond them. This possibility is supported by the results

Table 8.8 Significance of accessibility factors for location of producer services, by firm type and area: Washington, DC, 1982

Factors	*Government representation*	*Research and development*	*Activities*				*Areas*			*Total*
			Financial, insurance and real estate	*Consultancy services*	*Business services*	*Sales/ marketing*	*District of Columbia*	*Inner suburbs*	*Outer suburbs*	
Proximity business associates	79	71	22	60	83	46	95	79	14	55
Highway accessibility	11	14	27	13	8	9	—	8	31	17
Near metrorail	32	7	11	—	16	—	20	18	5	13
Proximity to executives' homes	11	11	7	—	24	28	3	8	19	11
Access to employees' homes	11	25	15	7	—	—	5	13	14	11
Easy access to building	11	7	—	13	—	5	8	5	6	6
Proximity to old office	5	7	—	33	—	—	8	5	5	6
Near airports	5	—	—	13	8	18	3	5	8	6
Access to other areas	5	11	4	—	—	18	—	3	11	6
Access to CBD	—	—	22	—	—	—	5	5	3	4
Near secondary routes	—	7	—	—	8	—	—	—	6	3
Totals	19	28	27	15	12	11	40	39	63	142

Note: Proportion of respondents is classified by activity and area citing each accessibility factor.
Source: Fuchs, 1983, tables 24 and 26, 83–84.

of a comparative study of office location in five Dutch cities in which it is shown that 'the range of locational tolerance for offices is in fact not limited to traditional districts in or near the urban core' (De Smidt, 1984, 121; see also Van Dinteren, 1984). There are also other facets of location choice by producer services (see p. 225) which seem to contradict the importance which has always been attached to the effect of agglomeration economies.

The pattern of suburban office locations generated by the need for accessibility does not only involve intra-city requirements, but may also reflect special needs such as contact with business associates outside the city. Hence international airports act as magnets for relocating and new suburban producer services (Bennett, 1980; Hoare, 1973; Aschmann, 1976; Code et al, 1981). In a comparison of office activities in the vicinity of Toronto's international airport and in the Don Valley area (see Figure 8.6) of the city Bennett (1980) found some evidence for statistically significant influences on the spatial structure of office activities which could be attributed to the airport. Firms in the Don Valley were much less intensive users of air travel than suburban offices in the vicinity of the airport. Most of the firms attracted there were small to medium-sized organizations representing engineering consultants, technical services, transportation services and, the largest group, branch offices of foreign-owned manufacturing firms. Bennett also notes that 80 per cent of the twenty-five head offices located in the airport area are foreign-controlled with a large complement of professional staff engaged in orientation functions. Nevertheless, the attraction of airports only affects a small proportion of all suburban firms but in Toronto it was found that recently formed firms and movers were more likely to be attracted to the airport area and it is, therefore, likely that the influence of this factor on the pattern of suburban office development will increase in the future.

The office park is the ultimate expression of the role of the suburbs as a counterpoint to the diseconomies of CBD and even conventional suburban office nodes. There are at least fifteen office parks adjacent to the I-287 corridor in Morris County (New Jersey) most of which have been developed during the last five years (Figure 8.7) About twenty miles from downtown Manhattan the County Department of Industrial and Economic Development stresses the international reputation of the area because more and more companies (14 per cent of the *Fortune* 500 are represented) discover they can operate successfully there, the county tax rate is the lowest in New Jersey, it has an AAA (exceptionally high) business credit rating, modern new interstate highways (I-80 and I-287), excellent park facilities and a very good labour pool and educational facilities. Identification of office parks is complicated by the diverse terminology used by developers and real-estate brokers; titles include commercial parks, office centres/parks, corporate campus, business centre, office complex, executive campus, research park and professional village. Irrespective of the label, geography occupies a prominent place in the promotion of office parks, with emphasis on environment (rolling countryside, trees, expansive lawns, streams and lakes) and travel times to Manhattan and labour (usually expressed as 'within minutes', nearness of regional shopping malls,

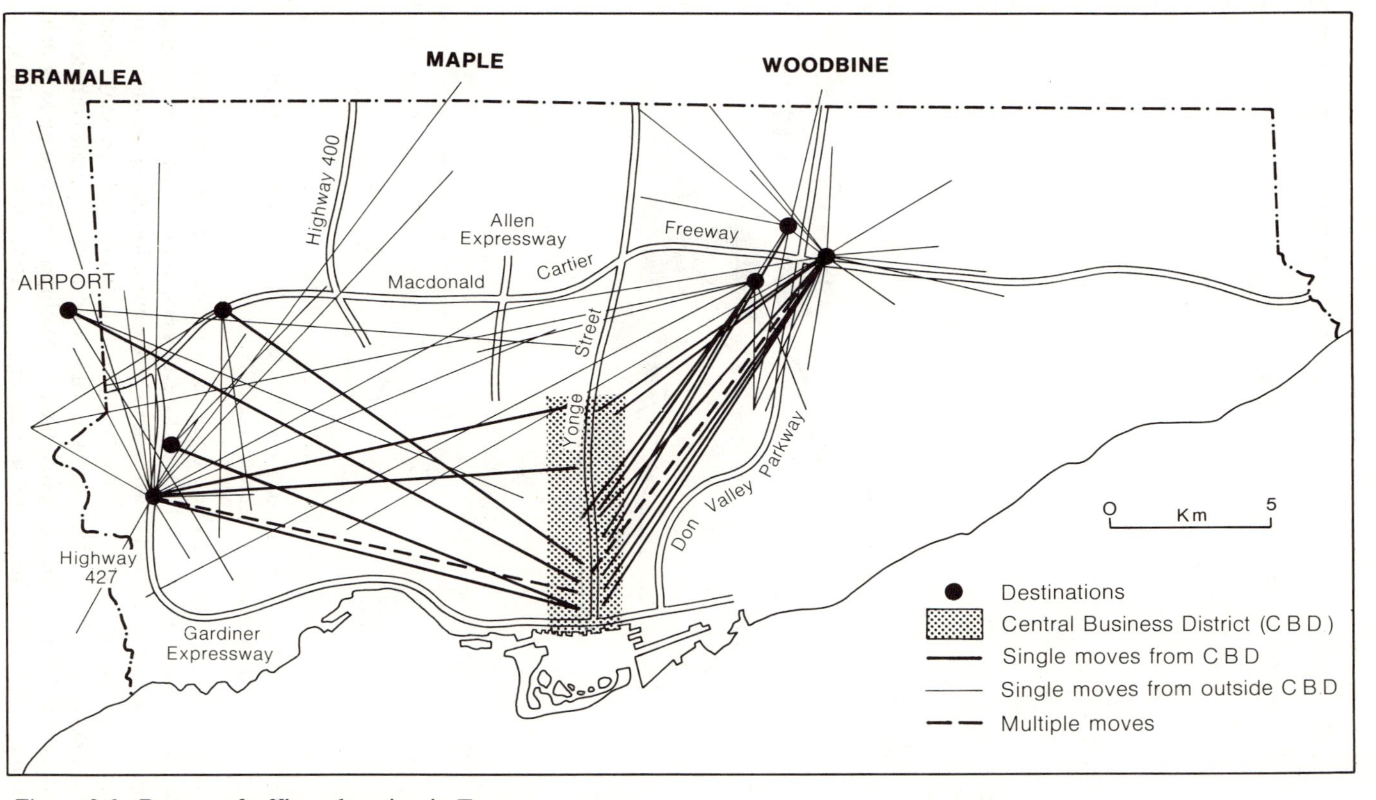

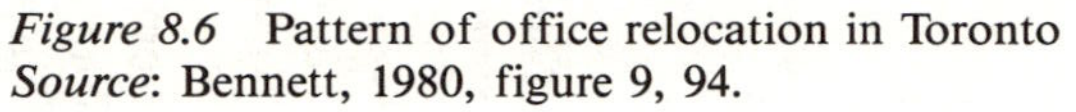

Figure 8.6 Pattern of office relocation in Toronto
Source: Bennett, 1980, figure 9, 94.

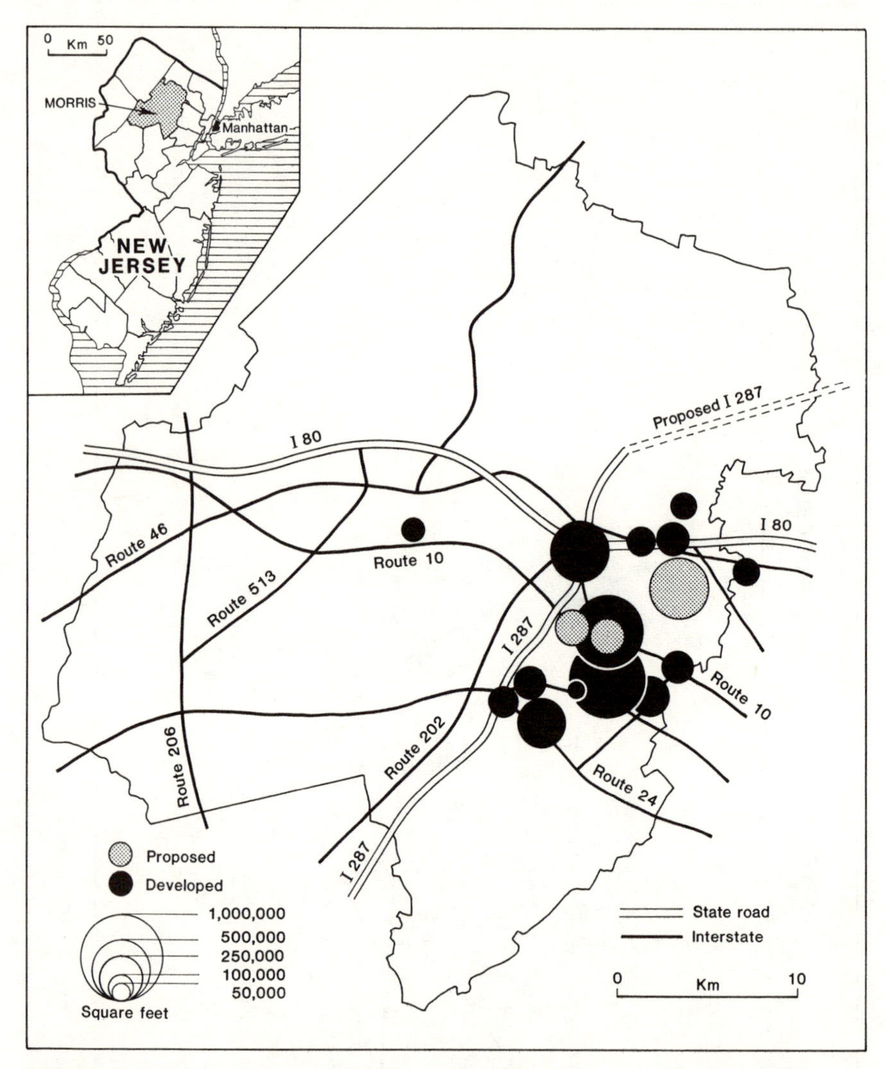

Figure 8.7 Existing and proposed office parks in Morris County, New Jersey, 1983
Source: Compiled from data in *Black's Guide*, 1983.

restaurants and cultural facilities, and access to interstate and other major highways). In addition, many office park developers, as well as those providing offices in secondary or specialized centres, draw attention to the relationship between their sites and other national or regional corporations (especially those from the *Fortune* 1000 list); this suggests that second-level corporate complexes (outside the CBDs) are emerging in large metropolitan suburban areas which will prove attractive for further suburbanization of producer services and give rise to a pattern of concentrated de-centralization.

Table 8.9 Some positive and negative consequences of the suburbanization of producer services

Economic		*Social*		*Environmental*	
Positive	*Negative*	*Positive*	*Negative*	*Positive*	*Negative*
Reduces operating overheads (rent/ rates)	Induces lower-density space use, reduces operating savings	Reduces average journey to work times for CBD and inner city recruits	Increases journey to work times of suburban recruits	Improved building design	Extensive unimaginative ground-level parking lots
Availability of cheaper labour	Increases competition in local labour market	Increases private transport use for journey to work	Reduces public transport use for journey to work	Landscaped development	'Sterilization' of large areas of suburban land at low-use intensity
Stimulates female activity rates	Extends labour catchment areas	More comfortable/ pleasant working conditions	Limited access to other services during working hours	Upgrades existing suburban commercial centres	Increases congestion in established suburban centres
Diversifies suburban business complex	Enhances differentials between development prospects of suburban areas	Increases residential amenities	Rising cost of accessible suburban residential areas	Reduced pedestrian/ transport conflict	Visual intrusion of large areas of parked vehicles
Increased suburban growth prospects	Mainly routine functions located in suburbs				

Suburbanization of producer services has generated both positive and negative effects with reference both to firms and to their employees (Table 8.9). Most of the positive economic consequences relate to the firm, in particular the lower overall operating costs compared with more centralized locations. But these advantages may be immediate rather than long-term effects because of the tendency for costs of suburban locations to rise as competition pushes up wage rates and steep increases in rents occur when leases become renewable. The producer service firm is less concerned with the positive or negative social consequences of moving to the suburbs, although the positive aspects such as easier journeys to work or a more settled workforce makes for better productivity and lower overheads arising from the need to acquire replacement or temporary staff. Again, unless certain limitations are placed on the location of the office space occupied by producer services (see Chapter 9), the social benefits enjoyed by employees are short-lived as congestion, longer journeys to work, or rising housing costs follow from the growing demand for suburban office locations. This also suggests that we should not assume that the dispersal of consumer and producer service activities within cities is an inexorable process. There are checks and balances, the relative importance of which changes over time, which suggest that constant re-evaluation of this assumption would be prudent. The pervasive influence of technological change (see Chapter 10) or the intervention of local and central government agencies also promotes a climate of uncertainty about projecting locational trends on the basis of experience during the last fifteen or twenty years.

Nevertheless, there is enough evidence available about intra-urban location change by office-based producer services during the last twenty-five years to allow the construction of a descriptive model (Figure 8.8). It is assumed that prior to 1960 a highly centralized location pattern focusing on the CBD was typical of producer services. Between 1960 and 1969 when the disbenefits of CBD location – combined with an accelerating growth and diversification of producer services – highlighted suburban areas as possible alternative locations, a low-density dispersed pattern of locations, often comprising single buildings, emerged. Such development could be considered exploratory since the requirements of the market had yet to be tested or properly expressed and the office space industry had to take a calculated chance in the absence, in the case of European countries with well-developed planning systems, of any guidance from development or strategic plans. The demand from producer services for suburban office space became more clearly structured after 1970 and centralization, in search of agglomeration economies similar to those in the CBD began to appear at intersections between suburban freeways or public/private transport interchanges in European metropolitan areas (London, Paris and Amsterdam, for example). Since 1980 suburban office centres have been consolidated for reasons similar to the original attractions of the CBD for producer services and have introduced intra-suburban location changes by firms which may have originally moved during the 1960s, for example, to a dispersed development. It is notable that office parks do not fit at all easily into this schema. The reasons will remain unclear until more is known

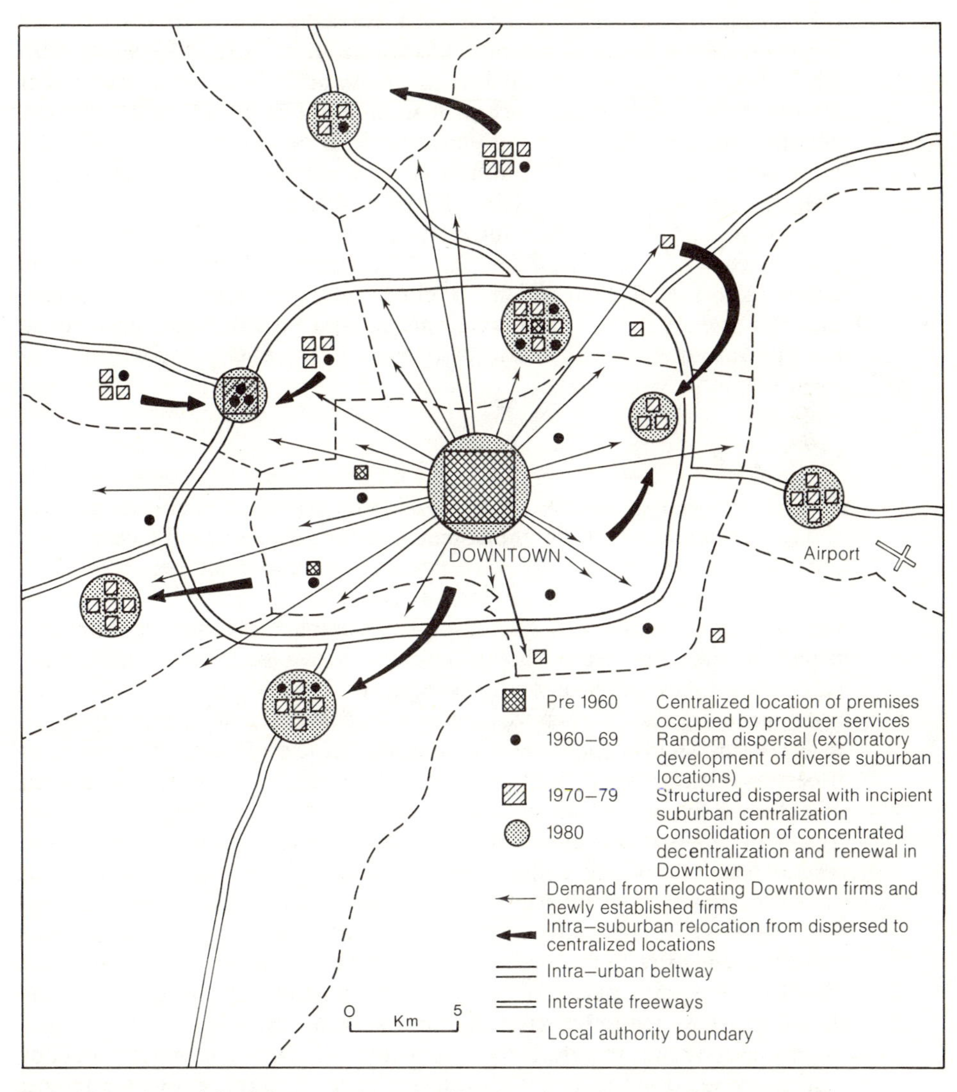

Figure 8.8 A model of the evolution of office-based producer services location within metropolitan areas

about the occupants of office space in office parks and, in particular, the attributes of their linkages with other activities both within and outside the parks. In many ways office parks offer an environment which is the antithesis of that pertaining in the suburban office centres developed along conventional lines, that is with a morphology and density much like that of the CBD.

There are some parallels between this model and that proposed by Erickson (1983) for the evolution of the suburban space economy in which much more comprehensive employment data for the period 1920–80 for several of the older

US metropolitan areas has been used. In particular, the second stage in the Erickson model, dispersal and diversification, coincides (although somewhat lagged in time) with the random dispersal phase of the producer service model. The third phase, infilling a multi-nucleation, in the Erickson model coincides with the incipient centralization and consolidation phase (post-1970) of the more sector-specific model. A 'follow-the-leader' syndrome may also be operating (see ibid.) whereby the risk of failure is perceived to be minimized by moving in to the suburbs which have already proved to be satisfactory to firms already having taken the plunge. Throughout the period since 1960 the CBD has also attracted new office development but mainly by way of replacement and rehabilitation of outdated premises, a process which has become more prominent since 1980 in association with the more general revival of CBDs for other service functions.

SUMMARY

There have been several references in earlier chapters to the ways in which changes in employment, organizational structure and location have affected service industries. But the main thrust of the presentation has been an analysis of producer and consumer service location patterns within a static, cross-sectional framework. This chapter has demonstrated that such an approach is partial, first, because of the influence of factors such as population and employment redistribution on the location of demand for services, and secondly, a requirement that service industries keep their location constantly under review in an environment where adaptation, anticipation, or innovation are necessary for survival and growth. In other words, there are sets of internal and external factors which together cause many service industries to make location changes whether in an absolute or differential sense. Some firms are more responsive than others and can be characterized as leaders rather than intermediaries or laggards. Temporal analysis of location is, therefore, an essential part of the interpretation of location. Many of the changes which have taken place involve shifts in retail or warehouse as well as some producer services from the centre of cities to the suburbs and beyond, or differential growth of services between outer and inner urban locations.

This chapter has, therefore, focused on the empirical evidence for location changes in relation to employment in services as a whole and with reference to the specific examples of retailing, warehousing and office-based activities many of which are producer services. In common with other parts of this book the examples given are inevitably selective, and this should not be taken to mean that the processes illustrated are not also affecting other services such as transport or the various educational and health services. In addition, the emphasis on the locational dynamics of service industries in urban areas, especially the changing place of the CBD for the location of services, is to overlook the changes which have also been taking place in rural areas. Here centralization within selected rural service centres is quite the opposite to the dispersal within urban areas.

It is suggested that the changing distribution of service industries within cities, particularly producer services, has followed a sequence beginning with centralization in the CBD through widespread suburban dispersal and on to concentrated dispersal in a small number of well-located suburban centres. The latter stage has been associated with a resurgence of the CBD. Such a sequence may also be applicable to other services such as retailing. The descriptive model does call into question whether it is reasonable to expect a continuation of the outmovement which has dominated the dynamics of service industry location during the last twenty years. The answer depends to some extent on the public policies and institutional factors which are considered in Chapter 9 and the degree to which technological considerations, which are now becoming more prominent in the service industries, are allowed to exert their potential effect on the operation, organization and scale of typical service industry establishments (see Chapter 10).

CHAPTER 9

Public policies, institutions and the location of services

INTRODUCTION

The processes manifest in the distribution of consumer or producer services have been largely explored within the context of geographical space dominated by private sector location decisions. A competitive environment ensures that service firms are anxious to provide the best possible facilities for their clients; they will endeavour to do so from those locations where viability is both assured and likely to provide the largest margins. It has already been shown how efforts to reduce uncertainty when making a location decision have created a certain 'conservatism' about where to locate, with the result that certain established agglomerations of producer services or particular types of location for consumer services have been able to perpetuate their initial advantage. In short, the market has tended to foster the development of an imbalance in the spatial distribution of service industries.

It would be misleading, however, to assume that this is the outcome of purely market influences, such as the distribution of corporate clients or the spending power of shoppers, because in most economies where the market is allowed to exist, it does so within a framework of public and institutional initiatives and guidelines which operate at the national, regional and local scale. Hence the imbalances in the spatial distribution of economic opportunities which have emerged in the wake of changes in industrial structure, and which are often biased towards particular regions, have long been recognized as a legitimate target for intervention by public agencies in the UK, France, the Netherlands and Ireland. Such intervention attempts, for example, to bring about some redistribution between favoured and less-favoured regions and to persuade service industries to expand in unfashionable areas, or seeks to limit the scale, density, appearance, or location of service activities within individual metropolitan areas.

RATIONALE FOR PUBLIC INTERVENTION

Advocates of some kind of public intervention in the location of service industries would make their case on a number of grounds. First, the market has created undersupply and a limited range of consumer services in less prosperous cities and regions or in the inner cities; secondly, access to services such as public transport or medical facilities in rural areas has been heavily curtailed because of the high costs of provision in extensive areas with low-density populations; and thirdly, the evolution of marketing practices among distributive services, in particular, has created a heavy demand for 'greenfield' and similar sites in and around the edges of urban areas at the expense of established retail service centres and/or vacant warehouse districts within existing built-up areas. Countries such as the UK, Netherlands and Belgium where land for urban development is in relatively short supply are especially affected by this problem. A fourth reason arises from the possibility – since more research evidence is still needed – that the concentration of producer services in a limited number of centrally located corporate complexes perpetuates the difficulties of readjustment confronting peripheral or disadvantaged cities and regions; the presence of producer services will not only provide alternative employment, but also provide the infrastructure necessary to both attract and stimulate dependent industry and employment. Finally, large-scale agglomeration of certain types of services passes on costs to the community and individuals in the form of congestion, higher land prices, escalating rents, or wage rates which have become unacceptably high. By invoking policies which are justified in the cause of the public interest, it is hoped that some of the spatial and related socio-economic inequities resulting from market mechanisms will be ameliorated. The extent to which the public interest is exerted does, of course, vary from country to country and, where appropriate, from one urban area to another. But even in the USA some planned limitations on the location of service activities, especially within urban areas, are now accepted as a legitimate weapon for use by publicly appointed local agencies such as planning commissions.

The latter is typical of public intervention at the local scale; land-use plans, zoning ordinances, or development permits are used to guide services to the most appropriate areas with reference to population need or more general development strategies. Control and intervention of this kind may be both positive and negative, but usually the latter, and has been used in some countries such as the UK for at least fifty years, although service industries have merited special attention only comparatively recently by comparison with other economic activities within the purview of local public agencies. Policies devised at the local scale might also involve tax and similar financial incentives designed to encourage selected activities, an approach used most frequently in circumstances where the traditions of public intervention in the market process are less acceptable or less well developed, such as in the USA.

Local attempts to influence the location of services are generally more direct, however, than those made by agencies operating at the regional scale. Certainly, in Europe policies emanating from regional agencies have largely been expressed in plans and strategies to which services make a contribution rather than acting as a specific target. The regional plans produced in Britain during the 1960s and early 1970s (see, for example, South-east Economic Planning Council, 1967; Northern Region Strategy Team; 1977) all make reference to service industries, often in the context of office development, but the implementation of their proposals has been difficult because of the absence of regional agencies with statutory planning and other powers; it is left to local agencies to interpret regional policies for service industries and to integrate them, if they see fit, within the plans for their areas of jurisdiction.

The third scale at which public policies are employed is at the national level. Most of the significant initiatives during the last twenty years have been introduced at this scale and will be the principal focus here. National policies are often targeted at regional problems but are in a somewhat different category to policies devised within regions, in that administration of the former is usually undertaken by central, rather than regional or local, government agencies using guidelines and rules which apply with equal force (in theory at least) to all the industry sectors or to all the areas of the country specified for their operation. Marquand (1979) has divided national policies which are relevant to service industries into three groups: policies which are directed specifically at service activities; policies which incorporate services along with manufacturing; and policies which exclude services either indirectly or explicitly.

LATE GESTATION OF NATIONAL POLICIES FOR SERVICE INDUSTRIES

The significance of Marquand's classification is that it indirectly draws attention to the gestation of national policies for service industries; historically service industries have not been considered worthy of public intervention and concern about imbalances in regional economic development engendered a policy response couched exclusively in terms of reviving or diversifying the manufacturing sector. The motivation to replace manufacturing job losses with more of the same, if sectorally different, employment was perhaps understandable given the relatively limited proportion of services in overall employment in the 1930s, except in special cases such as port economies. In areas with a tradition of employment in manufacturing it would also be easier to draw upon the available labour.

During the years immediately following the Second World War such concerns continued to hold sway and in most advanced economies this remained true until the late 1960s and early 1970s, when policies which more specifically incorporated service industries first began to appear; examples are the investment grants available to industry under section 7 of the UK Industry Act 1972 and similar grants available in the Netherlands and Germany; relief available on interest payments for commercial loans in Italy, Denmark and Belgium; or removal grants for workers and/or premises in the UK and the Netherlands. In all these examples the terms of assistance available are on the same basis as those available to manufacturing, or are available only to certain specified service activities. It is assumed, therefore, that service industries require the same kind of assistance, or will respond to the same kinds of restriction on growth in specified areas of a country, as manufacturing activities. It may in fact be more important to assist with the labour rather than the capital needs of services since their dependence upon high-cost equipment is less, in relative terms, than that of manufacturing; an absence of suitably trained or skilled personnel, on the other hand, will be much more problematical. Thus some countries, such as the UK, have training and retraining programmes which include the acquisition of skills which allow individual workers to make the transition from blue- to purple- or white-collar employment because they are more immediately marketable to firms in the service sector.

Consensus about the value of national policies which are specific to service industries has developed in a pedestrian and tortuous way. None the less, there have been (and in some cases remains) a number of public policies aimed directly at services (Table 9.1). These have been most comprehensively reviewed in relation to the EEC and the UK by Marquand (1978, 1979; see also Friedrich, 1984). There are esentially two approaches: first, restrictions on location in overcrowded areas in order to divert growth elsewhere or to promote relocation to areas of underprovision; and secondly, a variety of financial inducements such as investment, removal, or training grants (Table 9.1) which create conditions favourable to retaining services already established in areas of need or which will persuade others to move away from the areas of overconcentration. Such approaches might, of course, be used independently or in association. Most of the EEC countries do not employ public policies directed at specific service sector activities; the main exception is the tourist industry which has been the target for specific measures in France and the UK.

Table 9.1 National and regional policies specific to service industries, EEC countries

Country	*Type of policy*
France	Investment grants, tax relief, initial grant for jobs maintained, land-use planning policies, infrastructure aids (tourism), planning controls in congested areas, information agencies
Ireland	Investment grants, tax relief, rent relief, training grants, grant towards current costs of R&D
Netherlands	Investment grants, tax relief, grant towards current costs, land-use planning policies, information agencies, state financial participation
Germany	Loans, assistance with transport cost
UK	Investment grants, initial grant for jobs maintained, planning controls in congested areas, information agencies, training programmes, land-use planning policies, agencies with special powers, state financial participation

Source: Marquand, 1978, appendix 8.1 tables I and IV.

PUBLIC POLICIES FOR THE REDISTRIBUTION OF SERVICES

Some of the diseconomies associated with the trend towards concentration of certain services, especially those which occupy office space in major metropolitan areas, have been cited in Chapter 8. The costs of agglomeration may lead organizations to rationalize their operations or undertake structural changes, which may involve relocation of parts of the organization, designed to minimize the effects of negative externalities. But to focus on internal problems of agglomerated service activities is to overlook the costs which they are also imposing upon other services as well as other economic activities. High costs of land assembly for the construction of office buildings, escalating rents, congested highway and public transport systems, long journeys to work, or high housing costs are not just experienced by each service sector organization and its employees, but by its very presence in a large agglomeration, it is responsible for similar overheads faced by other organizations in the vicinity. The costs of maintaining public services such as transport or telecommunications are also affected by the ebb and flow of large numbers of workers at the beginning and the end of each working-day, or of telephone traffic which peaks during the first half of each working-day, for the rest of the time capacity far exceeds demand and results in inefficient and costly utilization of manpower and resources.

It is factors of this kind that motivated public policy initiatives in France, the Netherlands and the UK from the mid-1960s onwards (Grit and Korteweg, 1976; Beaujeu-Garnier, 1974; Bateman, 1976; Daniels, 1975; Burtenshaw *et al.*, 1981; Bannon, 1978). Two approaches have been used: first, restrictions on development which is occupied by services in certain designated areas; and secondly, the introduction of dispersal policies for service functions under the direct control of the public sector, that is civil service activities (Toby, 1973; Thorngren, 1973; Civil Service Department, 1973; Bannon, 1973; de Smidt, 1985). Perhaps the most closely analysed example of the former has been the sequence of attempts in the UK between 1963 and 1979 to impose controls on the development of office space in certain areas of the country (Daniels, 1975, 1982; Manners and Morris, 1981) in the belief that this would divert both demand for office space and the supply of space to the areas not covered by the office development permits (ODPs) required in, essentially, London and the south-east region for any development larger than 300 m^2 (later raised to 1000 m^2).

For a period during the late 1960s it was almost impossible to obtain ODPs for new or replacement office space in London and this limited the opportunities for expansion by office-based services, pushed-up rents because of the declining vacancy rates and caused organizations actively to evaluate the costs of their existing operations and ascertain whether they could be undertaken more cheaply but just as effectively elsewhere. They were reminded of comparative costs such as rents, labour costs and working conditions by the Location of Offices Bureau, which was established in 1963 and provided free and impartial information to clients about the alternative locations outside London. The Bureau also sponsored research into various aspects of relocation which helped to monitor the effectiveness of the dispersal policies with respect to both the firms involved and the office employees affected.

The records maintained by the Bureau provide an invaluable guide to the effectiveness of a policy which limited office development and service industry expansion in one area of the UK in an attempt to divert economic activity elsewhere. The records compiled by LOB statistics are only broadly indicative of the effectiveness of ODP policy (because not every firm that had thought of or actually moved out of central London consulted the Bureau) and it remains a matter for conjecture whether the dispersal of jobs or firms would have taken place anyway as a result of the operation of normal market forces; recent evidence (see pp. 236–8) suggests that these can be just as powerful in influencing service industry location decisions. Nevertheless, the Bureau acted as a catalyst at a time when planning policies in areas outside London supported the attraction of decentralizing organizations and market forces were favourable. Between 1963 and 1977 a total of 4227 firms which were considering moving an estimated 349,878 jobs consulted the Bureau; less than one-half (2026) of the firms eventually decided to relocate all or part of their operations, and these represented 145,155 jobs (Location of Offices Bureau, 1977). Small to medium-sized firms (less than 100 employees) comprised the majority of movers (80 per cent), although only 25 per cent of the jobs; and 414 firms with more than 100 employees moved 75 per cent of the jobs.

The predominance of small firms is partly a reflection of the response by office-based services to the pressures generated by the ODP policy (Table 9.2). Prominent among those deciding to relocate were offices representing distributive trades, insurance and professional and scientific services many of which were small firms. In terms of jobs insurance accounted for 20 per cent of the total, followed by banking and finance (10 per cent) and transport and communications (10 per cent). Moves by manufacturing offices comprised just 34 per cent of the total, with engineering and electrical goods and chemicals and allied industries the major contributors. It is revealing to compare the number of jobs actually moved with those which firms decided not to move (the ratio is included in Table 9.2). Business, professional and miscellaneous services (i.e. predominantly producer services) have ratios well below 1.0; for every one job moved between one and two jobs remained in central London. This contrasts markedly with mixed services, such as banking and finance and insurance, which have ratios exceeding 1.0, and for every three insurance jobs moved only one was eventually considered ineligible. This provides a crude indication of the attachment of producer service firms to the central London corporate complex, leaving mixed services to relocate, often partially, those functions least dependent upon agglomeration economies and largely engaged in planning and progammed contacts (routine functions) rather than orientation activities. Earlier data (Location of Offices Bureau, 1975) indicates that some 37 per cent of the relocations up to 1975 were partial, 47 per cent were complete moves and the remainder involved the creation of new branches or the retention of a small 'front' office in London.

The functional selectivity of service industry relocation may give rise to spatial polarization between decision-taking and routine functions. But even less acceptable within the context of ODP policy objectives would be a limited degree of spatial redistribution. In order to achieve the objective of improving the mix of economic activities in the provincial regions and cities there would need to be long-distance relocation to places more than 100 miles from London

Table 9.2 Office functions dispersed from London, by industry sector, 1963–77

Industry sector	*Firms and jobs moved*				
	No. of firms	*%*	*No. of jobs*	*%*	*Ratio*[1]
Primary industries	4	—	31	—	0.04
Manufacturing industries	706	34	54,558	34	1.06
Construction	58	3	6,612	4	2.06
Service industries	1,281	63	98,345	62	1.14
Gas, electricity, water	3	—	565	—	0.10
Distributive trades	258	13	11,163	7	1.06
Insurance	248	12	31,385	20	2.90
Banking and finance	88	4	15,906	10	1.41
Professional/scientific	211	10	8,894	6	0.75
Business services	136	7	6,385	4	0.57
Miscellaneous services	116	6	5,637	4	0.83
Trade associations	99	5	2,286	1	0.57
Transport and communications	117	6	15,915	10	1.11
Public administration	5	—	209	—	0.63
Unknown	6	—	299	—	0.12
Total	2,055	100	159,845	100	1.11

Note: 1 Number of jobs moved divided by the number of jobs represented by firms which eventually decided against relocation.

Source: Location of Offices Bureau, 1977, from table 8, 38.

and involving large numbers of firms and jobs. The reality has fallen far short of this; up to 1977 some 82 per cent of the firms which moved did so within the south-east region, the majority choosing locations less than twenty miles from central London, that is within the metropolitan area (Location of Offices Bureau, 1977; Pare, 1981). However, organizations moving larger establishments have tended to move further than smaller firms because they are better able to absorb the transitional costs such as payments to assist staff relocating with their office or the costs of moving equipment, records and furniture. Therefore, only 72 per cent of the decentralized jobs remained in the south-east.

This should not be taken to mean, however, that the required long-distance movement of jobs occurred. Of the 43,988 jobs established outside the south-east, almost 40 per cent went to the south-west; Bristol, a prominent growth centre for office space throughout the 1970s is only some 115 miles from London. There may also be a 'quality of life' component, in that the environmental and related attributes of the south-west will be perceived as superior both by decision-makers and those employees they wish to encourage to move with them. The distributive trades, insurance, and banking and finance moved an above-average proportion of jobs outside the south-east, but 49.3, 52.1 and 32.7 per cent respectively were moved to the south-west (Location of Offices Bureau, 1975). Business services also moved a disproportionate (relative to the proportion of firms) number of jobs to provincial locations with more than 83 per cent going to the northern region. The contribution of these location shifts to diversifying the economic base of provincial areas and stimulating further economic development has almost certainly been limited by the tendency for firms, in all industrial sectors, to disperse routine office functions, which evidence suggests (see Chapter 7) generate fewer local linkages than headquarters or regional establishments which include decision-making functions. There may be short-term multipliers associated with, for example, the construction of office space to accommodate the demand from migrant services (Yannopoulos, 1973; Ashcroft and Swales, 1982a) but long-term multipliers are more limited.

Even if it is assumed that the LOB was only aware of 50 per cent of the actual number of office jobs dispersed (see Hall, 1972), the annual rate of movement by private sector organizations was of the order of 21,000 jobs or 2–3 per cent of the total number of office jobs in central London. The ODP legislation did not limit the use to which space vacated by decentralizing firms could be used and the vacuum created was readily filled by the expansion of those offices remaining in central London or by inmigrants such as foreign banks (see Chapter 8). Thus the net decline in central London employment has been much smaller than the above figure would suggest; between 1966 and 1981 office employment in the City of London, for example, fell steadily from 267,330 to 220,000 (approximately 3000 per annum) (*The Times*, 8 April 1983). Estimates produced by the Greater London Council (Weatheritt and John, 1979) indicate a 9.8 per cent decrease in central area office employment between 1961 and 1976, a 7.2 per cent increase in the inner area (outside the centre) and a 47.2 per cent increase in the outer suburbs. Also it cannot be assumed that the central London decrease is attributable solely to outmigration; job losses, especially in clerical work, caused by improvements in productivity associated with increased use of information and office technology will also have played a part throughout the 1970s, together with a decelerating growth of office-based service employment generally (see Gershuny and Miles, 1983). The average floorspace per office employee has been steadily rising in areas such as the City and the accommodation requirements of modern office equipment rather than the labourforce may explain the upward trend. The changing ratio of managerial, administrative and professional workers to clerical workers will also generate demand for higher space standards per employee.

Uncertainty about the precise reasons for the decline in office-based services in central London is symptomatic of a wider concern about whether the ODP initiative or the creation of the LOB was really necessary. The debate began in a climate of uncertainty about office employment growth in London (Evans, 1967; Manners and Morris, 1981), moved on to arguments about the effects of restricting the supply of new and replacement office space on spiralling rents, and the effects of the freeze on business rents between 1972 and

1975, and ended with a debate about the failure of the policy, together with the LOB, to persuade enough firms to leave the south-east completely, or on the other hand, that the Bureau was too successful and should reduce its effort (Location of Offices Bureau, 1976). As a result of the latter, the terms of reference of the LOB were changed in 1977 to embrace, first, more explicit promotion of office employment in inner urban areas (both London and provincial), and secondly, to attract international manufacturing and service firms to locate office functions in the UK (Location of Offices Bureau, 1978). The services of the Bureau were now available to firms throughout the country rather than to just those in central London. Following the election of a Conservative administration which was committed to reducing public intervention, cutting back on public expenditure and reducing to a minimum the number of quasi-autonomous agencies such as the LOB, its prospects and the ODPs became very uncertain; both were abolished in 1979.

SERVICE MOBILITY WITHOUT PUBLIC INTERVENTION?

It is interesting to observe the subsequent behaviour of office-based services, whose location decisions have since only been subject to local control on supplies of office floorspace through the planning permission system. The removal of controls and persuasion to relocate would be expected to encourage renewed expansion in central London, yet many companies have actively considered and undertaken moves from the capital during the period since 1979. Market forces provide the explanation, in that prime office buildings in central London are among the most expensive in the world; £30–£40 per sq. ft is not uncommon in the City, which attracts higher rents than locations in the West End. Thus a secretary may cost an employer £6000–£7500 per annum before a salary is paid and a company occupying a small building of 10,000 sq. ft may have to face total charges (rent, rates and services) of up to £500,000 per annum. This compares with rents of less than £7 sq. ft for prime office space in Glasgow, Liverpool, or Leeds and combined rents and rates charges which are only two-thirds of the *average* of those prevailing in London. At a time of national economic recession companies have looked closely at the relative costs of operating in central London as against other metropolitan or provincial locations. Large organizations such as Imperial Chemical Industries, International Business Machines, Blue Circle, Chemical Bank, Rank Xerox and Commercial Union Assurance have either relocated all or some of their central London offices or are in the process of evaluating alternatives. Commercial Union produced a plan in 1982 involving the loss of 1200 jobs and relocation of 1500 staff from its headquarters in the City to Croydon, aimed at saving the group £20 million per annum (*The Times*, 15 September 1982). Another insurance company, the Prudential, announced reorganization of its regional and sub-divisional offices in 1982 and its plans to create a new centre outside London to handle all the processing of underwriting claims and accounting for its Holborn and City office business (*The Times*, 24 October 1982). Rank Xerox have developed a 150,000 sq. ft campus-style complex at Marlow in Buckinghamshire, about thirty miles from central London. Such large moves by prominent organizations can be expected to attract publicity but it is reasonable to assume that smaller organizations have also been moving out of central London for similar reasons.

While the market may be promoting a surprising continuation of the activity encouraged by earlier public policies, the resulting redistribution of service employment remains locationally selective. The south-east continues to be the principal destination or the area favoured for the location of new enterprises. Towns such as Reading, Swindon, Poole, Milton Keynes and Brighton have received much of the overspill; one particular area to benefit from this is the so-called 'western corridor', extending from Hammersmith in inner west London to Bristol (Figure 9.1). Most of the growth centres within the corridor are inside the south-east with the exception of Bristol and Swindon. Since the late 1970s this zone has attracted computer software and other companies involved in

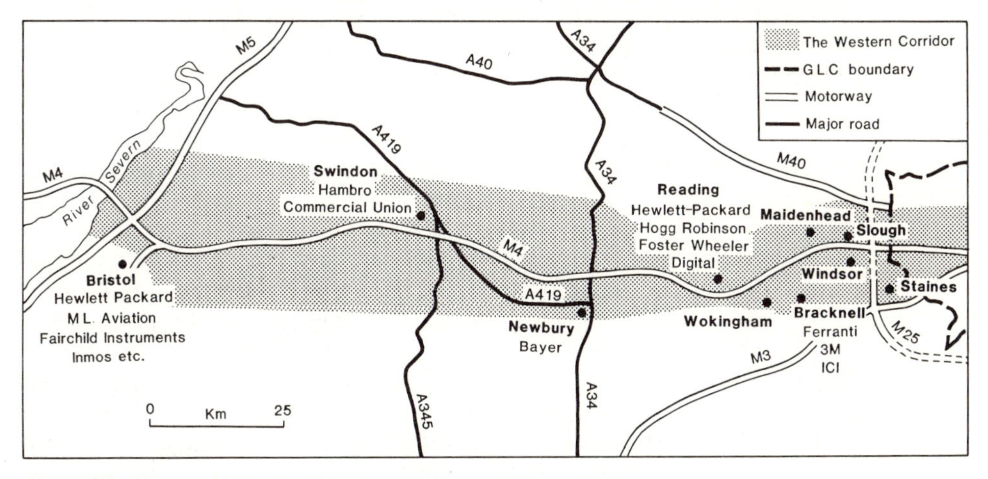

Figure 9.1 The 'western corridor', London to Bristol, 1983
Source: Redrawn from a diagram in *The Times*, 30 June 1983.

microtechnology and this, together with good access to lines of communication (rail, motorway and Heathrow airport), greenfield sites in an attractive environment and the availability of skilled labour, has stimulated the steady emergence of a complex of service and manufacturing activities (Hall and Markusen, 1985).

It would seem that with or without direct public policies for services, private sector activities will not be easily persuaded to move further than is really necessary or to regions which are not perceived to offer an acceptable quality of life. The UK is not alone in this experience. The Netherlands has encouraged dispersal from the Randstad to provincial cities in the north and east of the country but with very limited effect (De Smidt, 1984, 1985). From the late 1960s France has also pursued policies directed at private sector services with similar objectives to the UK and using a combination of restrictions on development through a permit system, a tax on development which becomes more punitive towards the centre of Paris and an agency promoting relocation, initially for the large provincial cities designated as 'métropoles d'équilibre' and later the medium-sized provincial cities (Burtenshaw *et al.*, 1981; Philippe, 1984). Just as in the UK, this kind of intervention has met with limited success because of the ability of the development industry and office firms to adapt to or find ways around the limitations on their locational preferences. The concentration of office employment in central Paris has diminished but it is hard to ascribe this to the effects of the public policy instruments, and as in Britain the change has favoured locations within the Isle de France region rather than substantial inter-regional movement of services.

REDISTRIBUTION OF PUBLIC SECTOR SERVICES

The alternative to efforts to influence the location of the private sector is to seek to modify the location of public sector services. Since they are under direct central government control, they should in theory be more easily manipulated in the interests of regional development priorites. Policies for relocating civil service departments and their staff have been introduced in Sweden (Thorngren, 1973; Swedish Government Decentralization Commission, 1978; Swedish Ministry of Industry, 1982) and the Netherlands (Toby, 1973; De Smidt, 1983, 1985). Despite some enforced dispersal of some government departments during the Second World War, most of the administrative functions of the civil service in the UK, in the late 1950s, exhibited a pattern of distribution much like that which was giving rise to concern in the private sector. This provided an opportunity for central government to work by example and two initiatives, one in 1963 and the other in 1973, were taken to plan and implement a programme of civil service dispersal in addition to locating new departments outside London (Civil Service Department, 1973). The principal benefit to the areas chosen for the dispersed departments would be the creation of badly needed employment together with the growth-inducing effects of large civil service departments, which would attract or induce the growth of private sector office-based services. Although the evidence supporting this hypothesis remains elusive, any growth-inducing effects would be limited if high-order decision-making functions were not included in the dispersal programmes. In general, routine work has most commonly been relocated or diverted to provincial locations; the Post Office Savings Bank (Glasgow), Department of Health and Social Security (Newcastle upon Tyne), Vehicle Licensing Centre (Swansea), Department of Employment (Runcorn), Passport Office (Newport and Peterborough) and Manpower Services Commission (Sheffield) are just a few of the dispersed departments which are mainly engaged in executive and routine work, leaving high-level decision-making functions in headquarters departments in London.

Figure 9.2 Alternative programmes for the dispersal of civil servants, 1973; and actual programme, 1977–88
Source: Civil Service Department, 1973; cited in Daniels, P.W., *Service Industries: Growth and Location*, London, Cambridge University Press, 1982, figure 31, 80.

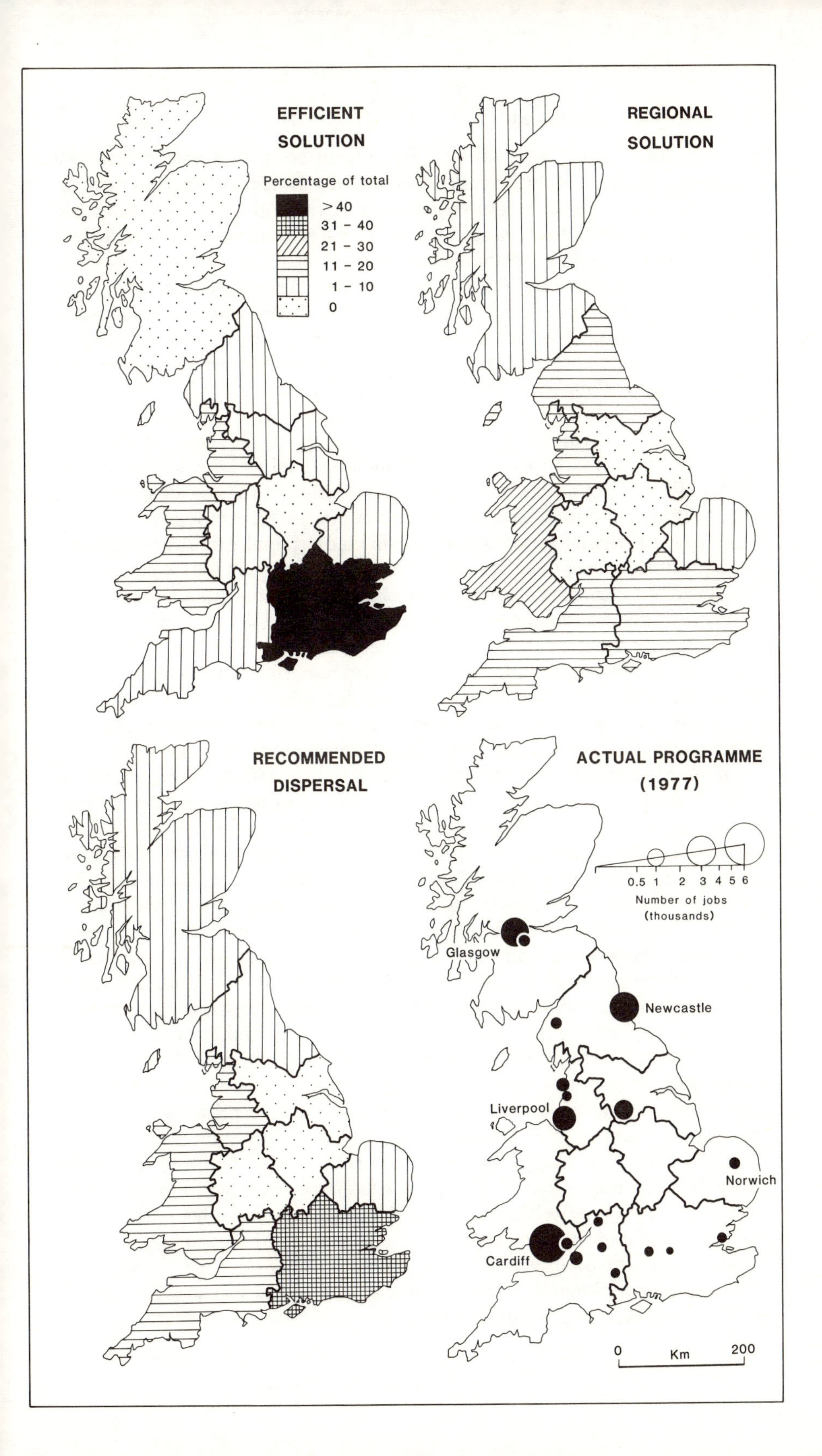
EFFICIENT
SOLUTION
Percentage of total
>40
31 - 40
21 - 30
11 - 20
1 - 10
0
REGIONAL
SOLUTION
RECOMMENDED
DISPERSAL
ACTUAL PROGRAMME
(1977)
0.5 1 2 3 4 5 6
Number of jobs
(thousands)
Glasgow
Newcastle
Liverpool
Norwich
Cardiff
0
Km
200

Certain departments were also excluded from consideration for dispersal because of their pivotal role as decision-making units together with the high communication costs, far outweighing any savings in accommodation and related costs at the decentralized locations, associated with the face-to-face interaction which their personnel require.

Therefore, the report by Sir Henry Hardman (Civil Service Department, 1973) found that the most efficient solution for the redistribution of 31,000 headquarters civil servants would require more than 40 per cent to remain in outer London and the south-east (Figure 9.2). Even the programme recommended in 1973 would retain some 35 per cent of the civil servants within the south-east. The programme which was finally agreed favoured the regions, particularly certain major provincial cities, in particular Cardiff, Newcastle upon Tyne, Glasgow, Sheffield and Liverpool, to which large relocations would have an immediate impact on construction and related industries (Ashcroft and Swales, 1982a, 1982b) and provide a firmer foundation for the development of self-sustaining provincial office centres of the kind originally mooted by Wright (1967). But for reasons already alluded to, the commitment to this programme has waned – a feature enhanced by the increasing opposition of senior civil servants to dispersal plans for their departments (see, for example, Pacione, 1982). Similar reluctance has been encountered in the Netherlands, even though the distances involved are much smaller, and in Sweden which has a programme for dispersing civil service work from Stockholm to the northern regions (Sundqvist, 1975). The resource costs and benefits of government dispersal programmes are also uncertain; over a ten-year period the cumulative net costs of moving the Manpower Services Commission to Sheffield were negative but decreasing during the first six years and changing to substantial net benefits thereafter. But similar estimates for much longer-distance dispersal from London to Scotland show increasing cumulative net costs for the first six years and continuation of higher costs than benefits throughout a ten-year accounting period (Parliamentary Committee on Scottish Affairs, 1980).

The accumulated evidence clearly suggests that nationally directed public policies aimed at redistributing service activities on the basis of restraint on growth in specified areas which will compensate for decline in other sectors or will have growth-inducing effects have not been successful. Furthermore, there will be limits to the extent to which such redistribution can continue; the surplus in one region can only be reduced to some threshold level beyond which further outmigration would be unacceptable. The growing volume of protest, whether justified or not, against the promotion of private sector dispersal from London or of civil service dispersal from Stockholm because of the effects on the local economy is symptomatic of much louder protests if public policies were wholly effective. The prospect, therefore, is that the gap between need and supply may well remain in provincial areas, leaving them the problem of how to respond. Before turning to a specific example based on the approach used by the City of Liverpool, it is necessary to consider briefly the other dimension of national public policy directed specifically at services, the use of financial incentives to assist relocation.

REDISTRIBUTION THROUGH INCENTIVE

In 1973 a Service Industry Grant Scheme was introduced in Britain to encourage the growth of office and service employment in areas designated for special attention (the Assisted Areas). More recently the title has been changed to the Office and Service Industries Scheme (OSIS) which provides special grants to office and service industry undertakings, such as administrative offices, R&D laboratories, marketing departments and central training establishments, which create additional employment in the Assisted Areas and serve a wider than purely local market (Department of Trade and Industry, 1983). The last mentioned criterion possibly gives undue emphasis to the export role of the eligible services at the expense of those which may be import-saving and have an equally genuine choice of location between Assisted Areas and elsewhere (Marquand, 1979). The effect of this limitation may be to reduce unnecessarily the opportunity for creating a mix of local services which might prove attractive to firms and projects eligible for other kinds of regional assistance. Three types of grant are available: a negotiable grant to employers of up to £8000 for each job created in the Assisted Area, a fixed grant of £2000 for essential employees moving with their work up to a maximum of 30 per cent of the jobs being increased and a contribution to the cost of outside consultants commissioned to determine the feasibility of establishing a particular activity in the Assisted Areas (Department of Trade and Industry, 1983). These are the maximum grants available in the Special Development Areas; the grants for each additional job, for example, are lower in the Development and the Intermediate Areas where the economic problems are relatively less severe.

The level of assistance can be illustrated with an example of an accounts department which an insurance company outside the Assisted Area wishes to locate in a Special Development Area. At the time of the move seventy-five jobs will be involved of which twenty will be essential transfers from the present location. Within three years the total staff is expected to increase to 110. The maximum grant available to the employer for each job created is £8000 but negotiations indicate that £6500 per job would be adequate to allow the company to proceed with its plans. Hence the total grant would be 110 × £6500 = £715,000. One-half of this would be paid one year after the first job was created and the balance after a further two years, provided that the target of 110 is reached. In addition, all twenty essential employees moving with the office would receive a tax-free removal expenses grant (this group accounts for less than 30 per cent of the total number of jobs created). The total cost of these grants would be 20 × £2000 (£40,000), giving a total overall cost to the OSIS scheme of bringing about the relocation of £755,000. Although substantial, it seems that this level of financial assistance is not attractive and does not outweigh more practical considerations, such as market potential or minimum annual turnover, as determinants of the final location decision (Daniels, 1985c).

The response to the scheme by service industries has been negligible, at least by comparison with manufacturing (Table 9.3). The cumulative number of offers of assistance finally accepted under the OSIS scheme is less than 7 per cent of the total and equivalent to the same proportion by value (approximately £58 million). The estimated new service employment is just 28,615 out of a total for all industries in excess of 420,000. There are considerable variations between regions in the number and proportion of jobs created. The north-west

Table 9.3 Offers of OSIS assistance accepted, 1972–83, by regions and assisted areas of Britain

	No. of grants	*%*	*Value (£m)*	*%*	*Estimated new employment*	*%*
REGIONS						
Scotland	57	3.6	7,882	3.4	2,646	2.4
Wales	89	7.9	9,471	8.1	4,350	7.0
North-east	80	7.9	10,293	8.2	4,490	7.8
North-west	171	8.2	14,257	7.7	8,514	9.0
Yorkshire and Humberside	94	7.0	8,146	11.4	6,174	9.6
South-west	16	5.1	419	2.2	383	3.1
East and West Midlands	22	5.3	7,500	3.4	2,058	10.6
ASSISTED AREAS						
Special Development Areas	180	6.9	23,402	5.8	8,795	5.2
Development Areas	139	7.1	19,874	11.3	7,435	7.2
Intermediate Areas	210	6.4	14,693	7.6	12,385	8.5
Totals	529	6.7	57,969	6.9	28,615	6.8

Source: Department of Trade and Industry, *Industrial Development Act 1982: Annual Report*, London, HMSO, 1983.

and Yorkshire and Humberside have attracted almost one-half of the new service jobs, while Scotland has attracted just 2646 jobs or 2.4 per cent of the total jobs created (see Table 9.3); there is almost an inverse relationship between need and the proportional contribution of new service jobs (see also ibid.). Hence they accounted for some 5 per cent of all new jobs in the Special Development Areas but 8.5 per cent in the Intermediate Areas. More offers of assistance were accepted in the former but the value of each grant was lower than in either of the other two types of Assisted Area. Care must also be taken with interpretation of the employment estimates; it is not known, for example, how many of the jobs lost to the non-assisted areas as a result of OSIS are subsequently replaced, how many of the new service jobs in the Assisted

Areas displace pre-existing jobs in the same or related activities, or to what extent they absorb previously unemployed labour rather than those already employed elsewhere in the local economy. Therefore, Marquand (1983) suggests that the only way to evaluate policies such as OSIS is in terms of whether they create gains in national employment and output rather than just moving jobs and output around between regions.

USING LOCAL INITIATIVES

Attempts to influence the distribution of service activities through the mechanism of national policies and initiatives has not only proved universally difficult, but also has often overlooked the particular local needs of the areas which are supposed to benefit. By the end of 1983, for example, only four major firms had accepted support from the OSIS scheme in Liverpool and offers of support exceeding £50,000 only amounted to £400,000 between 1972 and 1982 (City of Liverpool, 1983). This is hardly a substantial response to the deep-seated economic problems of the city which has Special Development Area status and has suffered major job losses during the past fifteen years (see, for example, Gould and Hodgkiss, 1982; City of Liverpool, 1982). Total employment in Liverpool has fallen by an estimated 89,000 jobs between 1971 and 1981; service industries declined by 17 per cent (37,000 jobs), although 90 per cent of the loss was concentrated in the transport and communication (which includes a substantial producer function) and distributive trades. The decline of the latter is linked with the loss of population from the city. Forecasts of employment change from 1981 to 1986 (City of Liverpool, 1982) suggest a continuation of the downward trend across all sectors (Table 9.4), although in proportional terms the service sector will become more important with some growth in producer services (but insufficient to cover for continuing job losses from blue-collar services).

To some extent the opportunities for local producer services to expand in Liverpool or for branches of external organizations to be located there depends on the availability of suitable accommodation, mainly office space in free-standing buildings in the CBD. One of the difficulties which has beset attempts to attract more 'growth' services has been the shortage of new, modern office space; while the property development cycle (see pp. 247–50) has affected development in Liverpool, the peaks and troughs have been much flatter than elsewhere. The acknowledged reason for this is the 'image' of the city as perceived by the largely London-based property development industry, the financial institutions upon which a growing number of development schemes depend for support and the agents responsible for marketing office space. The city also has an ill-deserved reputation for prolonged labour disputes in the construction industry and building costs, which are supposedly higher than in any of the other major British provincial cities (City of Liverpool, 1983).

Whatever the merits of these arguments, the development rate in Liverpool has been slower than in other provincial cities in the Assisted Areas even though office vacancy has been as low as 2–3 per cent since the mid-1970s. Since this

Table 9.4 Employment in Liverpool, by sector, 1971–86

Industry sector	*1971*	*1981*	*1986*	
			high	*low*
Primary	411	50	42	42
Manufacturing	110,611	68,166	58,275	48,509
Services	215,353	180,505	174,931	148,601
Blue collar				
Gas, electricity, water	3,551	2,187	1,917	1,614
Transport and communications	52,531	35,318	28,129	25,489
Distributive trades	47,599	31,000	27,314	24,486
Miscellaneous services	29,088	28,500	30,703	26,560
White collar				
Insurance, banking, etc.	18,361	14,500	13,445	11,518
Professional, etc.	50,019	48,000	48,482	38,934
Public administration	17,755	21,000	24,941	21,000
All industries	346,094	256,387	241,965	204,130

Note: The 1981 employment figures are estimates and the high and low forecasts for 1986 are based upon several assumptions, including levels of public spending, no increase in national employment levels, the likelihood that unemployment will not fall, and no major new economic regeneration programmes.

Source: City of Liverpool, 1982, tables 1.2 and 6.2.

has failed to get the development companies to provide additional space (whether new or refurbished), the city council has embarked upon a policy which, it is hoped, will stimulate the local market. It is also hoped that the renewed availability of modern, well-located buildings will help both to retain growth services already established in central Livepool and to encourage others, especially producer services, to move into the city. Other long-term benefits of this policy include the generation of additional rates income, refurbishment and/or redevelopment of old office buildings, and the receipt of ground rent on buildings owned by the local authority. There are two features of the policy: first, financial support, in the form of rent guarantees, has been made available where appropriate to encourage advance office building and to invite bids from developers for sites owned by the city council and for which a ‘planning brief’ is available. To some extent the local authority is, therefore, sharing some of the risk with the development companies. Secondly, taxes are not levied on new office space until it is leased, and the case for exempting from rates those buildings which have been previously occupied but are temporarily empty for refurbishment is considered on merit (see ibid.).

While it is too early to assess the yield to the local economy, it is certain that since the early 1980s there has been an upsurge of office development activity in central Liverpool. A number of key sites in prime locations with first priority for development (including planning permission) and immediately available (especially in the Moorfields area) are now occupied, or are in the process of being occupied, by new office and mixed commercial development. How far this activity is attributable to the city's initiatives as opposed to a market response in circumstances where long-term vacancy rates have been well below the national 'norm' is uncertain. Some of the new space already available remains unlet and pre-letting is exceptional. It also remains to be established that the new or refurbished space is taken up by new or inmigrant services rather than horizontal movements by established city centre organizations upgrading space as opposed to expanding their labour requirements. On the positive side the local initiative is helping to enhance the appearance of the office district, thus making it more attractive to firms as well as perhaps signalling to conservative investors that prospects are improving. Without these efforts to stimulate development, the general deterioration in the service economy of the CBD, in particular, might have continued and ensured that Liverpool remained caught in the spiral of decay which national policies (including many which have not been discussed here) have signally failed to stabilize.

The kind of initiative taken by Liverpool remains exceptional and has probably been born of particularly difficult local circumstances. In the USA attempts to revive declining CBDs with large-scale and multi-functional, prestigious projects, such as the Renaissance Center in Detroit, have depended mainly on the mobilization of private sector investment. The risks involved in these largely speculative developments are great and all have by no means been successful. Perhaps more immediate benefits to local service economies follow from decisions – all too rare – by large corporations, such as Standard Oil of Ohio, to invest in new major international headquarters in problem areas such as the centre of Cleveland. This should do a good deal to bolster confidence in the long-term prospects of downtown and the market for locally supplied intermediate services.

In more buoyant situations demand from mixed and producer services for both centralized and suburban locations is such that it can profitably be channelled elsewhere in the interests of diversifying inner city economies or to provide an economic base for fast-growing suburban municipalities. Local initiatives may, therefore, be best expressed in the form of planning policies for strategic office centres in addition to the use of fiscal incentives of the kind used in Morris County, New Jersey (see Chapter 8). Two of the best-known examples of the former are City-Nord, Hamburg (Husain, 1980), and La Défense, Paris (Burtenshaw *et al.*, 1981). Both are multi-functional and completely new complexes with good transportation services to downtown and surrounding suburban areas and have successfully attracted the headquarters of major multinational manufacturing companies in the oil, chemical and office equipment industries.

Much more common, however, has been the designation of existing commercial areas as favoured locations for office development in both inner and outer suburban areas; examples include the designation of strategic office centres in London (Daniels, 1975; Damesick *et al.*, 1982) and office centres in Greater Manchester (Greater Manchester Council, 1975; Damesick, 1979). Such centres are particularly important for the location of mixed and producer services serving regional or even national markets. When trying to identify centres suitable for such development, planners must take into account the characteristics of the firms to be located there and the consequences of concentrated development. Offices have significant transport impacts because of the high level of trip generation and, with a significant proportion of the trips to suburban centres by car, associated high demand for parking either within office buildings or on the adjacent (often residential) streets (Daniels, 1980). In North American cities the majority of work trips to suburban office areas or to office parks are undertaken in private vehicles but in European cities as many as one-half of the employee trips at some locations still depend upon public transport; the convenience of strategic centres for access by public transport is thus an important consideration in location choice. Office centres also create a demand for dependent or support services such as restaurants, hotels, office equipment maintenance and supply, printing services and general retailing; the more convenient these are, the better the prospect for a successful attempt to develop a viable centre. Croydon, in south London, is one of the best-known examples of this approach in which local authority foresight during the late 1950s, followed by subsequent initiatives and commitment, has ensured the development of an impressive complex of office-based mixed and producer services many of which have moved some or all of their activities from central London. Examples in other British cities include Edgbaston (Birmingham), Stockport (Greater Manchester) and Bootle (Merseyside), or in Hamburg the Haverstehude and Rotherbaum areas and the Laatzen district in southern Hanover (Burtenshaw *et al.*, 1981).

Enthusiasm for office centre policies which largely reflect local planning priorities and initiatives is tempered by the disbenefits which arise from uncontrolled or inappropriately targeted development. If a centre grows faster than the transport system serving it, there is the likelihood of an overloaded public transport service, parking problems and highway congestion. Further public investment which will benefit private capital will ease these problems but the need to invest can be avoided by placing a limit on the volume of office employment. This can be achieved by careful zoning and judicial use of planning permissions but this can also create difficulties because not all planning permissions ultimately lead to completed development (only 50–60 per cent of the floorspace permitted) and there is always concern that an attractive, but perhaps mistimed, development will select an alternative location and take with it the benefits which the preferred office centre would have enjoyed by gaining. The notion of a threshold size for office centres is appealing but planners find it hard to find the right balance between protecting public interests and meeting the needs of service activities which might be lost to competing locations in other local authorities if unable to locate in a centre which is considered to have reached (or passed) its threshold.

INFLUENCE OF INVESTMENT BY PRIVATE SECTOR INSTITUTIONS

A large proportion of the accommodation used by office-based services is provided on a speculative basis, that is without a firm commitment by the client(s) to occupy an office building when it comes on to the market. Some buildings are constructed on the basis of prior arrangements or are custom-built by the large corporations with the financial resources to invest in their own buildings. In North America, Australasia, Hong Kong, Singapore and some European countries, and in particular the UK, speculative provision of office space is the norm, and this means that national, and especially local, public initiatives and location policies for service industries must take into account the investment decisions made largely by private sector institutions. Patently it is

> unproductive to adopt a policy that satisfies many planning objectives only to find that no developer is willing to invest in the 'chosen' locations. A policy for office development must attempt to understand and take account of the needs and problems of the developers who build the offices. (Greater Manchester Council, 1975, 71–2)

The significance of the development industry should not be underestimated because during the last ten years the involvement of financial services such as banking, insurance and pension funds, which are themselves major users of office space in urban areas, has expanded greatly (Barras, 1979a; Malone, 1981; Daly, 1982; Bateman, 1985).

Provision of office floorspace is a highly profitable activity which was largely dominated until the late 1960s by individual developers who acted essentially as middlemen. The profits are potentially large because of the substantial difference between the development costs and the capital value of a completed office building. The latter is determined by the rental income and the acceptable yield from investment in property; since rents will be highest at prime locations or points of greatest demand, developers will tend to concentrate as far as possible on such locations or attempt to identify as yet undeveloped sites with a potential to meet the requirements of users. As Marriott (1967) demonstrates, once they had earmarked locations in and around city centres where they anticipated that demand for office space existed, or could be created by providing new buildings, the developers assembled (often discretely) the plots of land which together would provide a large enough site for a profitable office development (see also Whitehouse, 1964; Ambrose and Colenutt, 1975). The task might take several years and entail great risks long before profits were made. Once the sites were assembled, planning permission was sought and, if obtained, the design, construction and other services required to produce a finished building would be brought together in a way which would give the developer as much control as possible over the return on the investment. Financial institutions provided bridging and loan finance but were rarely directly involved in the property development process.

The supply of office floorspace follows a cycle of booms and slumps; there have been two cycles in the UK since 1945 and a third is currently in progress (Barras, 1979b, 1983, 1984). Following a detailed scrutiny of the development cycle in London, it has been suggested that it is

> created by the delay of three to four years between the start and completion of a development scheme. Increasing demands lead to shortages of space and rising rents. Only when the first wave of developments comes on to the market, stabilising rents and values, does the level of new starts begin to slacken off, by which time a potential over-supply has been created. Speculative development then drops to a low level until this new supply is absorbed and the cycle can begin again. (Barras, 1981, 10)

The office development cycle for Greater London and central London is illustrated in Figure 9.3 (see also Catalano and Barras, 1980). The slump at the end of the first cycle in the late 1960s saw the demise of the independent property developers who were involved in a number of takeovers which created larger and stronger property development groups which were well placed to take part in the next property boom in the early 1970s. When this boom collapsed following the oil crisis of 1973, the properties owned by several groups

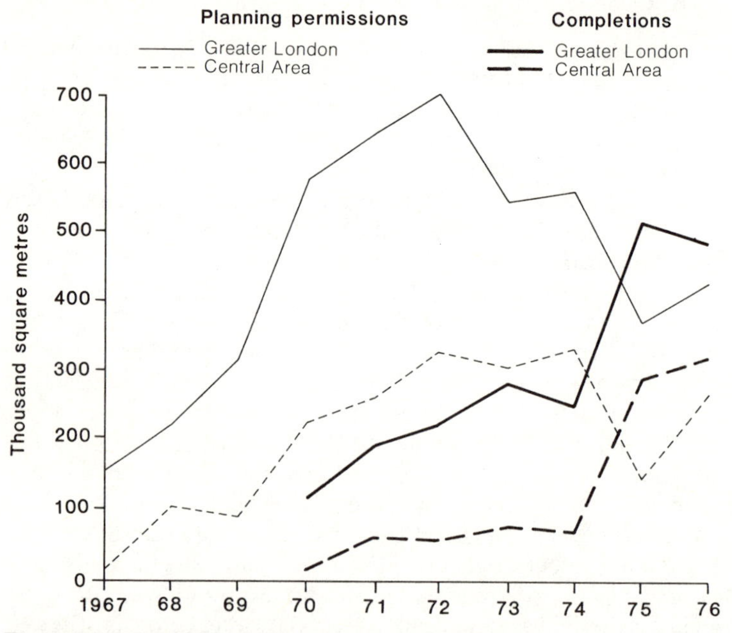

Figure 9.3 The office development cycle in London
Source: Barras, 1981, figure 9, 26.

which had overstretched themselves, including attempts to introduce speculative developments in other European cities such as Brussels, Paris, or Amsterdam (Burtenshaw *et al.*, 1981), became the targets for acquisition or equity arrangements involving the major banks and pension funds, thus broadening the direct involvement of institutions in decisions about the supply of new and refurbished office space (see also Bateman, 1985).

Hence there are now at least four types of capital deployed by four different agents: commercial capital represented by the property company, landed capital

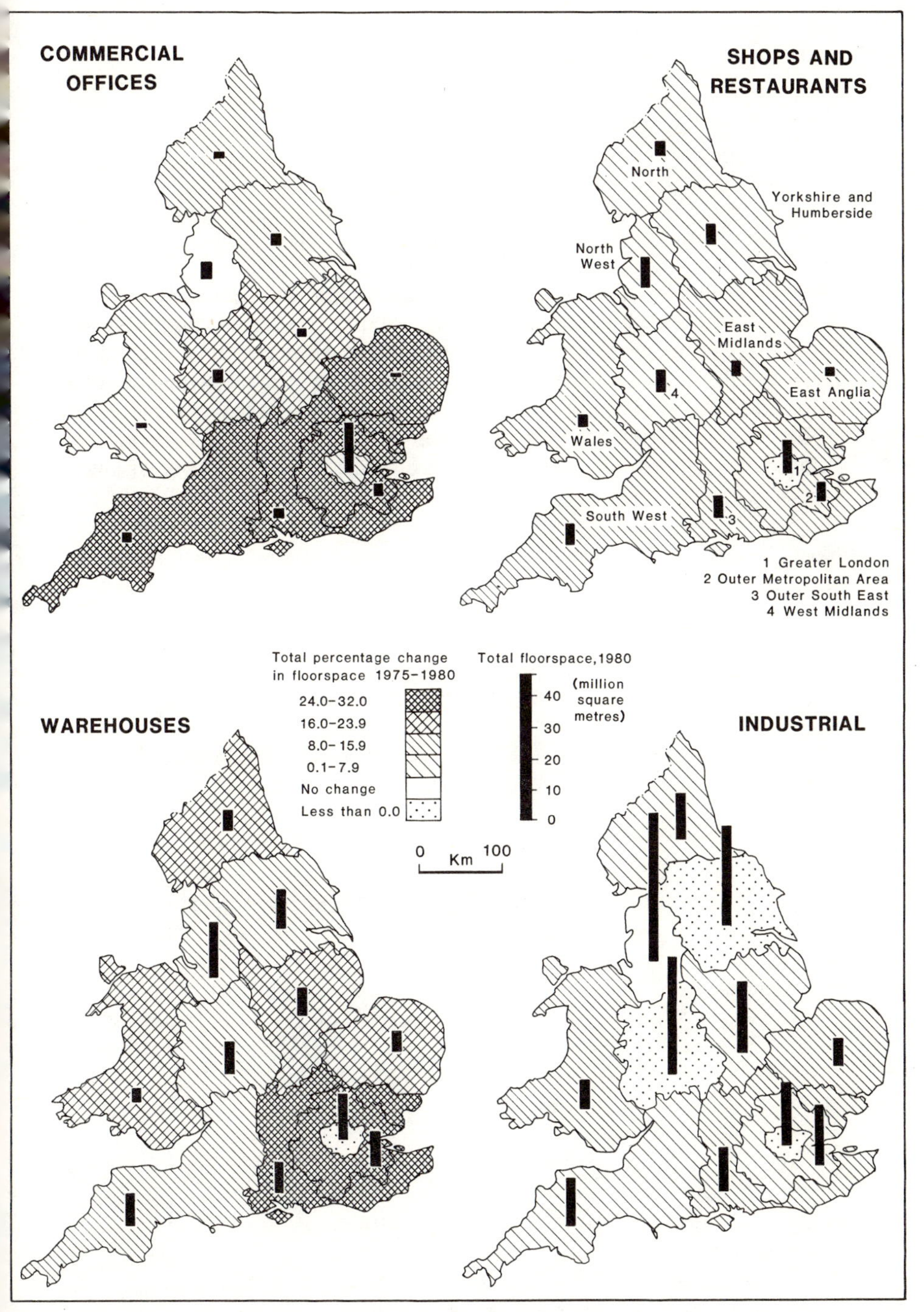

Figure 9.4 Floorspace changes in selected land uses, 1975–80; and total floorspace, 1980, planning regions of England and Wales
Source: Central Statistical Office, *Regional Trends*, London, HMSO, 1981, table 10.9, 118.

or the landowner, financial capital representing the funding institutions and industrial capital in the form of the construction companies (Barras, 1979a). The latter have also been taking an increasingly direct role in office development during recent years. Financial institutions are now an integral part of all aspects of the development process rather than acting solely as a source of fixed interest mortgages in the way typical of the first postwar office development boom. However with the interests of shareholders or insurance and pension fund policy-holders in mind, it is not surprising to discover that financial institutions are very cautious about the locations at which they choose to place investments; this tends to mitigate against local initiatives to encourage office-based services in 'problem' cities or regions such as Liverpool or Glasgow. The regional patterns of change in commercial offices, warehouses, shops and restaurants and industrial floorspace (Figure 9.4) reflects this, especially for the first two types of floorspace. The same organizations are also major service sector occupiers of commercial office space; for example, out of sixty-two known major users of buildings larger than 10,000 m^2 in the City of London (see ibid.), 35 per cent of the floorspace was occupied by banks (twenty-six) and a further 17 per cent by other major financial users (fourteen) including insurance companies and brokers.

The growth of international banking and the recent influx of overseas banks (see Chapter 7) has further strengthened the influence of bank location requirements on office space provision. Boroughs such as Tower Hamlets in inner east London, for example, have found it difficult to attract new office space and employment to replace job losses in manufacturing (Damesick *et al.*, 1982). Yet an inner London borough (Hammersmith and Fulham) which is in the path of the popular 'western corridor' out of London has found that funding institutions and developers find it an attractive investment proposition (McKee, 1981). Hence the 'confidence shown by the financial institutions cannot be exaggerated and the Borough Council has always been conscious of the need to be sensitive to their views' (ibid. 5). Since the actors in the office development process are increasingly representing both the supply and demand sides of the equation while, at the same time, representing one of the few economic sectors to be growing, it is not difficult to see why their influence on the success of public policies for the location of service activities should not be considered lightly.

SUMMARY

The analysis of the location of service industries is incomplete without some reference to the effects of public policies and the role played by the institutions that invest in the buildings and other facilities used by service industries. The need for some kind of public intervention has arisen largely as a result of the highly skewed spatial distribution of office-based producer services. It was suggested in Chapter 7 that the centralization of these services has taken place at the expense of marginal or peripheral cities and regions within national economies. Such has been the imbalance that some countries have used a variety of policy instruments, operated by state and local governments, to try to achieve a more uniform distribution of service activities and related employment opportunities.

Apart from early redistribution of civil service functions prompted by the Second World War, for example, most of the comprehensive efforts to guide the location of service industries began during the 1960s. Most of the policies introduced were negative, that is restrictions were placed on the amount of new or replacement office space which could be provided in certain designated areas or extra taxes were levied on companies according to the number of employees engaged at premises within specifed areas. Such controls were supplemented later with positive policies using financial incentives made available in the areas to which it was hoped that service industries could be persuaded to relocate all or part of the activities; grants to assist with relocation, a number of years of rent-free accommodation, or minimum red tape in connection with planning applications might, for example, be offered. Initiatives of this kind have been introduced in France, the Netherlands and UK, initially by central government and more recently by individual local authorities disillusioned with the effectiveness of national policies.

Such changes are symptomatic of the litany of doubt surrounding the merits of explicit policies for the location of service industries. Of course, some public services over which governments have direct control, such as civil service functions, can be redistributed from capital cities in order to diversify regional employment opportunities. But experience in Sweden, the Netherlands and UK indicates that intent and practice are difficult to reconcile because of the resistance of civil servants, especially those in senior posts, to relocation. In the case of private sector services the record of substantial inter-regional redistribution is poor, although intra-regional shifts in south-east England or the Isle de France have certainly been achieved as a direct result of ODP and similar policies. Indeed such movement has exceeded levels considered acceptable by the source areas and caused a reappraisal of the policies, as happened in London during the mid-1970s. This eventually resulted in the abandonment of the ODP system and the closure of the Location of Offices Bureau. This left incentives as the main instrument of public policy towards service industry location, and perhaps because the level of assistance is not large enough or does not match the actual requirements of service firms, these have not been at all successful.

But even if the incentives were adequate, there is a limit to the number of potentially mobile service firms in the overcrowded areas, so that in recent years increasing attention has been given to ways of encouraging the expansion and retention of service industries already located in the problem areas. This, together with assistance designed to encourage local entrepreneurs to start service businesses, may be more useful than relying on inter-regional movement. The latter is also made difficult in some areas by the shortage of suitable accommodation because of reluctance on the part of development companies and/or institutional investors to provide new office space in, for example, cities where economic prospects are considered marginal – even though, as in Liverpool, office vacancy rates have been low enough throughout the 1970s to have made it reasonable to expect the development industry to want to provide new floorspace. Because there has been insufficient modern office space immediately available for rent, mobile service firms are less inclined to choose Liverpool: the shortage of inmigrant firms is perceived by the development industry as justification enough for not providing new office space. It is very difficult to break out of a cycle of this kind, so much so that Liverpool has recently undertaken office development jointly with development companies, thus sharing the burden of risk, in an effort to boost the attractiveness of the city centre for office-based service or manufacturing organizations.

Institutions are conservative about where they invest, but they are also careful about when they invest in new development. This causes a cycle of development of the kind which has been illustrated for the City of London. Hence at certain times the demand for warehouse or office space exceeds supply, pushes up rents and related costs and may cause service firms to consider moving all or part of their activities to less costly locations. This may also be encouraged by the heavy taxes levied by inner city local authorities trying to sustain various public services with a declining population and employment base. Therefore, despite the demise of the Location of Offices Bureau and of the ODP system in London, the market has continued to create conditions which encourage outward movement from congested areas. However, much as before, this is mainly short distance and has not benefited more distant regions and cities.

CHAPTER 10

New technology and the geography of services

INTRODUCTION

It was suggested at the beginning of this book that services of one kind or another permeate almost every facet of life in the modern world. This has made it all the more surprising that relatively little attention has been given to variations in the spatial patterns of services which 'ought to be the first task of a geography of services. Linking specific services with certain locations and elucidating the reasons for the linkages observed could not fail to improve our understanding of human geography on the one hand and of the socioeconomic functioning of society on the other' (Gottman, 1983, 63). An attempt has been made in the preceding chapters to rectify some of these oversights.

There is little doubt that our interpretation of the existing locational attributes of service industries and the possibilities in the future is governed to a degree by the effects of technological innovations which have already had a major impact on production in the manufacturing sector and have now become highly significant for services. The task of illuminating the geography of services is, therefore, complicated by the rapid diversification of information technology during the last two decades (Coates, 1977) which, it seems, allows a much more foot-loose approach to the choice of location. This may be characterized, for example, as a telecommunications–transportation trade-off (Jussawalla, 1978; Nilles *et al.*, 1976) which is an expression of the extent to which information exchange using telecommunications can provide an alternative to the conventional physical movement of people, goods and, most important to many service activities, information. The prefix 'tele-' signifies, of course, that many transactions, whether associated with producer or consumer services, can now be carried out remotely as if distance no longer existed.

Historically service industries have relied heavily for their successful operation upon the physical movement of people and information. Any technological developments with the potential to render this dependence less significant could have consequences for location patterns. It has already been shown that existing location patterns reflect the historical dependence of many service industries upon various kinds of physical transport. The postal system provides a physical mechanism for information transfer but it has essentially fulfilled the needs of users without changing the locational opportunities available; in other words, the postal system has responded to needs and priced its service in a way which reconciles the wide differentials in the cost of moving mail between locations which are far apart by comparison with those which are close together. Other modes of information transfer such as face-to-face communication have had a more important influence upon location than information exchange via the postal service because the costs of travelling or the value of the participants' time increases with distance and cannot be 'averaged out' in the way achieved by the postal service. The telephone has provided partial substitution (Poole, 1977; Clarke, 1977) for the immediacy of face-to-face meetings but as a means of information exchange it lacks the nuances which are detectable during person-to-person interaction and any agreements reached or promises made will, in a business context, require validation through an exchange of letters, contracts, other documentation, or a face-to-face meeting. In this respect the telephone has been a vehicle for arranging more meetings than acting as a substitute. Most consumer services are provided at some place or point at which it is necessary for users to assemble in order to be able to consume (or otherwise benefit from the use of) them; physical movement of people, vehicles bringing in new stock or departing with deliveries, together with a sequence of trips linking the nodes which comprise each individual's activity matrix, is the consequence of this.

Urban agglomeration economies or the influence of central place principles will become obsolete if ways are found reliably, efficiently and confidentially to provide the services which both producer and consumer activities offer. Telecommunications technology, combined with improvements in the information-handling capabilities of office, retail, and domestically based computer and related equipment, offers these opportunities and it seems essential to consider briefly whether there will be significant spatial consequences at both the urban and regional scale (Goddard, 1971, 1980; Dhillon *et al.*, 1978; Bell, 1979). In many ways telecommunications are the invisible layer of transport supplementing the physical transport links between economic activities or between cities, regions and nations.

The effects of new technology on services will not be confined to spatial patterns: levels of employment, occupational skills and the structure of organizations are also likely to be subject to change (Gershuny and Miles, 1983; Green *et al.*, 1980). Indeed some of these changes will contribute to the spatial effects of new technology (Goddard, 1980; Mandeville, 1983). Information processing is increasingly the means by which firms actualy deliver their services and is of central importance to many business operations (Piercy, 1984). At the same time, it affects all functions and levels of management and invites close scrutiny of company strategies both operational and locational. In particular, information technology provides increasing potential for centralizing management and organizational control (Mansfield, 1984) which may reinforce the discrepancy between the location patterns of consumer and producer services. In the remainder of this chapter attention will be focused briefly on the nature of the new information technology, followed by a short overview of its possible impact on selected aspects of service sector employment and location.

SOME CHARACTERISTICS OF THE NEW TECHNOLOGY

The new technologies available to, and increasingly used by, service industries can be divided into three areas; telecommunications, electronics and computing (Figure 10.1). In electronics perhaps the single most important advance has been in the number of memory bits or switches held on one silicon chip; from just one in the 1960s to well over 100,000 in the early 1980s and still rising (Forester, 1980, 1985; Schaff and Friedrichs, 1982; Heyel, 1969; Braun and MacDonald, 1978). The production of microchips has also become much more reliable, so that equipment failure rates attributable to poorly constructed chips

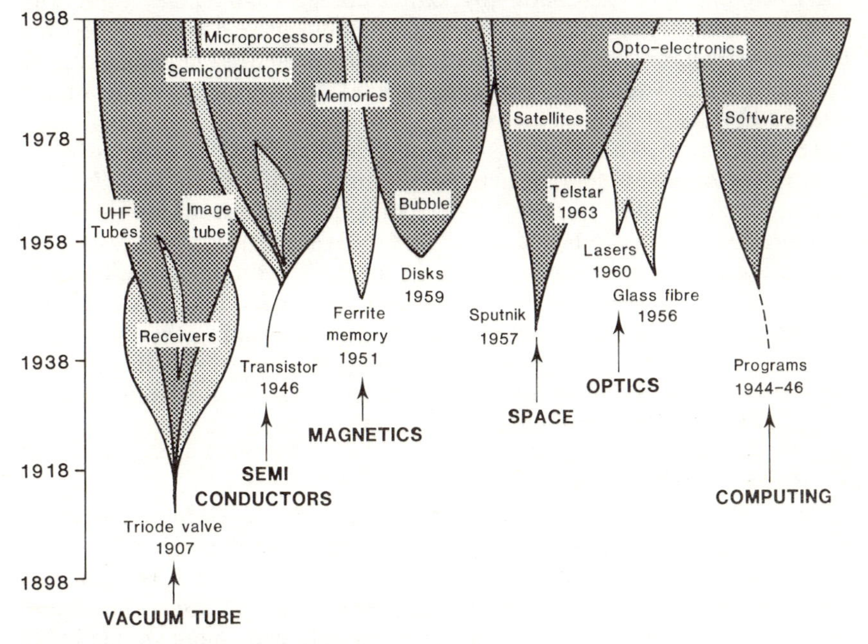

Figure 10.1 Advances in information technology
Source: Danzin, 1983, figure 1.1, 22.

have decreased sharply. The ensuing development of microprocessors has permitted the construction of small, low-cost but powerful desk-top computers and storage and information transfer devices.

In the field of telecommunications not only has the capacity to transmit information increased sharply following the advent of optical fibres (more than 1000 million bits per second is now possible), but the time required to communicate an item of information has been greatly reduced by using satellites as relay stations. There have also been improvements in long-established equipment such as the telephone, telex and the typewriter. Telex has had a reputation of being a slow keyboard-to-keyboard transmission service using bulky, noisy electromechanical terminals which were liable to transmit error-ridden messages printed in upper case only. Modern technology has changed this, much quieter equipment which closely resembles a word processor is now available in telex exchanges, so that the number of connections in the UK alone is around 100,000. The ability of telex to hold its own is a by-product of the growing international scope of business and financial services (as well as of manufacturing); it overcomes the problem of communication between different time zones which the telephone is clearly unable to resolve. Hence about 53 per cent of originating telex traffic in the UK is for overseas destinations (Woolnough, 1983). Also in prospect is a completely new and much faster service known as teletext, a system for conveying documents via data transmission between word processors, for example, which can transmit a 2000-character letter (two pages) in 15 seconds – by comparison with two minutes using conventional telex. Teletext also offers a more comprehensive range of upper- and lower-case symbols and incoming messages can be stored as outgoing messages are transmitted. West Germany, Sweden and the UK now have operational teletext services.

The sophisticated hardware which is now available has limited value to service firms, however, unless its latent power can be harnessed through the software required to make it operate. All the computer-based systems used in offices require software which may be available 'off the shelf' as in the case of word processing or data base management or, for the large range of tasks which are specific in structure, content and objective to individual organizations, require programs tailored to meet specific needs. Software now occupies by far the largest part (as much as 80 per cent) of the cost of new information systems and the highly skilled programmers who write the often long and very complicated programs remain in short supply.

Technological convergence

Prior to 1950 improvements in computers, electronics and telecommunications tended to take place separately from one another but more recently they have been converging and together comprise information technology (Figure 10.2) (Bjorn-Anderson *et al.*, 1982; Martin, 1978). When each of the segments shown in Figure 10.2 could be linked together, services became more vulnerable to technological change. Advances in the speed and memory sizes of computers, for example, were of limited value to multi-branch service firms until it was possible for the computers to communicate with one another through telecommunications devices or for users to 'interrogate' their machines from several remote terminals. Office-based services are especially likely to be vulnerable to the versatility of information technology. Enhanced telephones, centralized and free-standing local computers, printers, voice synthesizers and sophisticated reprographic equipment (Danzin, 1983) all now can use digitized information to reconstruct any kind of message and process and retrieve or consult it irrespective of the distance involved.

Applications of new technology to the office can be classified into at least four groups: mechanization, computer aids, office information systems and decision support systems (Butera and Bartezzaghi, 1983; Driscoll, 1979). Mechanization is most likely to affect routine tasks such as typing, filing and information retrieval and will change both product quality and clerical and

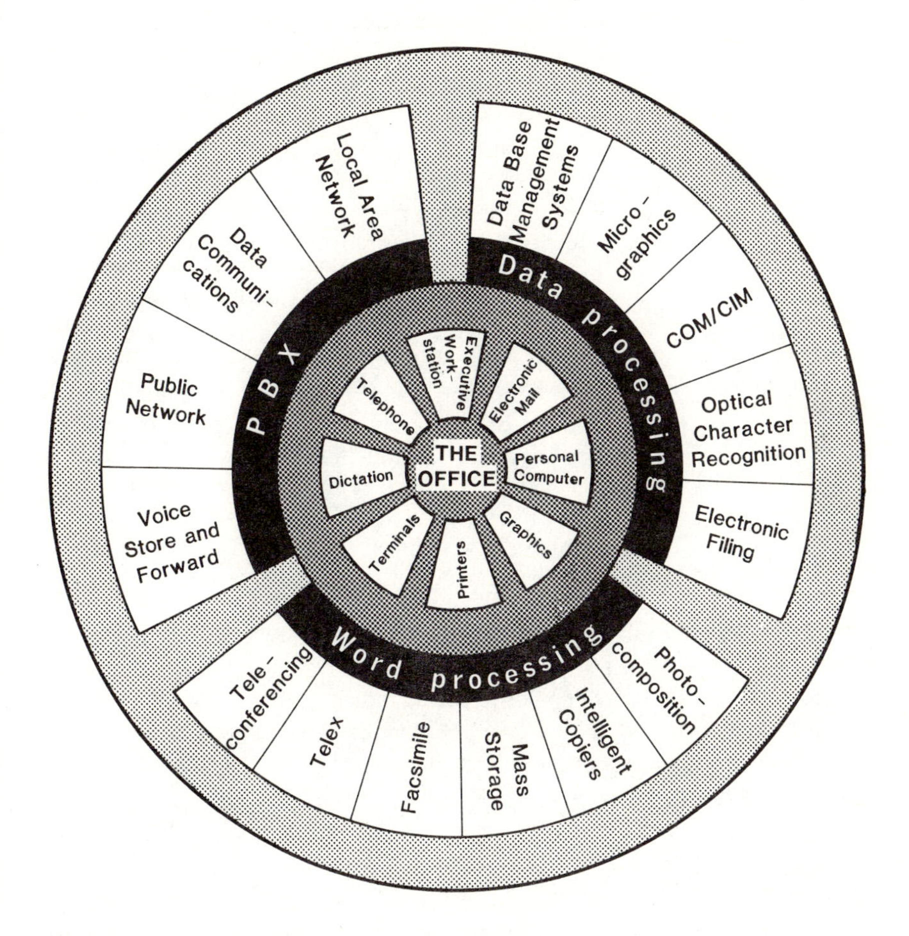

Figure 10.2 Office automation: merging of office equipment, data processing and telecommunications

secretarial productivity. Computer aids will also have an impact on similar areas, together with creating changes in job roles and the communication channels used by managerial and professional workers. Most organizations are confronted by problems of information storage and retrieval, together with applications to effectiveness and efficiency, and flexibility and responsiveness to new client requirements or shifting markets. Office information systems which are computer-based can help to minimize these difficulties and to optimize an organization's response to unexpected events. As information has become more readily accessible and there is more to assimilate decision support systems, such as management simulation, project development and financial planning models, have become useful tools for executive decision-making. These are just some of the options currently available; they are being constantly supplemented by new possibilities.

The component of information technology with the greatest potential for spatial consequences is most obviously telecommunications which can aid productivity and change the way in which individual office workers interact (Webber, 1963; Harkness, 1977; Abler, 1975; Toffler, 1981; Kellerman, 1984). The picture-telephone, for example, helps to overcome some of the inherent disadvantages of being unable to see the person with whom a conversation is taking place. Currently any visual communication between discussants requires travel to meetings, or alternatively, the assembly of participants in special studios. There are nine Confravision studios in the UK but at £100 per half-hour few companies are prepared to cover the expense of using them when the benefits are uncertain (Pye and Williams, 1977). A new video-conferencing system has been introduced by British Telecom and should be affordable by large companies if not every household (Cookson, 1983b). Video-conferencing equipment can be installed in company offices because it uses existing narrow-band communications networks rather than television-capacity lines of the kind used for Confravision. The system can be operated in any meeting-room with normal lighting and a facility for filming documents or objects is also provided. Forty offices in seventeen companies were involved in the British trials in 1984, and although still not cheap, it is believed that the cost can be justified on the basis of the fares, expenses and travel-time saved by the highly paid executives who usually participate in video-conference sessions. Although lowering the demand for air and other modes of travel, it remains uncertain whether Confravision or video-conferencing is an adequate substitute for face-to-face interaction: to a degree it exaggerates any imperfections in an individual's presentation because users have expectations of quality allied to their experience of television. The reactions of participants are also more difficult to assess in a way which can be crucial in complicated business negotiations. On the other hand, remote conferencing leads to more efficient meetings, in that participants are less inclined to draw out discussions in the interests of justifying the time and expense of travelling to the meeting.

Telecommunications networks

It is also possible now for firms to have their own communications networks for exchanging speech, data, telex, teletext, and facsimile instead of relying upon public networks which are often overcrowded. Such networks are especially attractive to companies with offices and manufacturing plants spread over a large area. One such company, Blue Circle, has therefore invested £3 million in its own digital network (Figure 10.3), replacing its previous telephone exchanges which used old-fashioned analogue models with calls mainly going through the public switched network. The new system consists of low-capacity lines converging on six major nodes which are, in turn, linked by high-capacity transmission lines. The design is such that no single exchange acts as a focal point, and if one of the exchanges at Hull, Armitage or Greenhithe fails, calls can be rerouted via the other two. About fifty establishments are linked to the network and the company expects to cut its telephone bill from £1.9 million to £1.4 million a year, to make savings in data transmission costs, to gain an

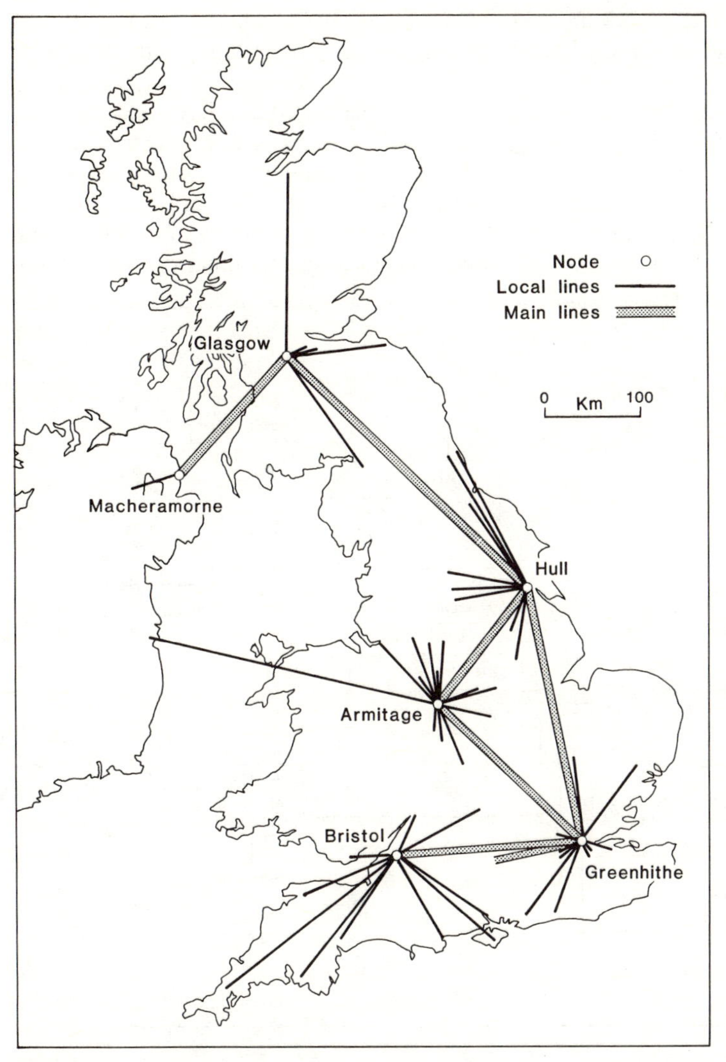

Figure 10.3 Digital network for all communications needs (speech, data, teletext, facsimile, telex) operated by Blue Circle Cement
Source: *The Times*, 7 September 1982.

advantage over its competitors and to give greater customer satisfaction.

Another approach which will be attractive to some businesses and, it is hoped ultimately to private households, is to provide a national network for transferring streams of data at high speed from one location to another. One British company (Mercury) has agreed with British Rail to lay its optical fibre cables along its tracks and so provide a figure-of-eight national network centred on Birmingham with the southern loop serving places such as Bristol, Reading,

London, Watford and Northampton and the northern loop including Stoke-on-Trent, Liverpool, Manchester, Leeds, Sheffield and Derby. This trunk network should ultimately have a carrying capacity of 140 megabits per second per fibre.

The convergence of technologies has also permitted a variety of digital information capabilities to be combined within the same service, particularly for communication within an organization. This requirement has been met by the development of local area networks (LANs) (Figure 10.4) and wide area networks (WANs). A cable is installed in a building or group of buildings and computers, word processors, facsimile machines, data terminals and other kinds

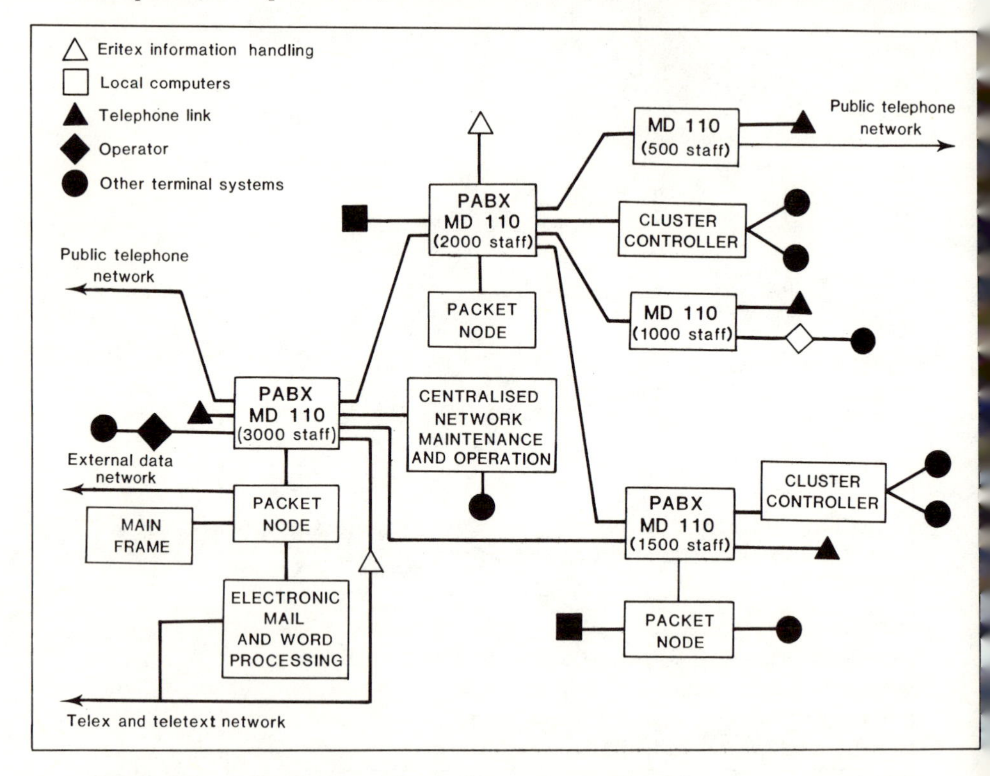

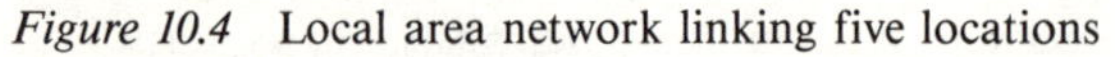
Figure 10.4 Local area network linking five locations

of equipment are linked to it in order to communicate over the cable. Networks can be arranged in a number of ways, including a star configuration in which all links feed through a central point, a 'bus' configuration in which everything is organized in relation to a single cable, or a ring network within which data circulate until they find the correct destination. It is estimated that there were 3000 LAN connections in Western Europe at the end of 1983, rising to 30,000 by the end of 1984 and to 250,000 by 1987.

The so-called office of the future is also a product of the interweaving of formerly separate technologies (Uhlig *et al.*, 1982; Strassman, 1980; Bagdikian,

1971). An office of this kind would operate without the large volume of paper which still dominates the day to day work of almost all office workers, from the most junior clerk to the company president or managing director. It is the need to assimilate, exchange and output information on paper that has created much of the postwar demand for white-collar workers whose typical working-day is far from solely devoted to the task for which they are employed (Figure 10.5). A large proportion of employees' time is spent doing other things, some of them unproductive such as being away from desks, waiting to receive or to return important phone calls, or checking through files to collect a very

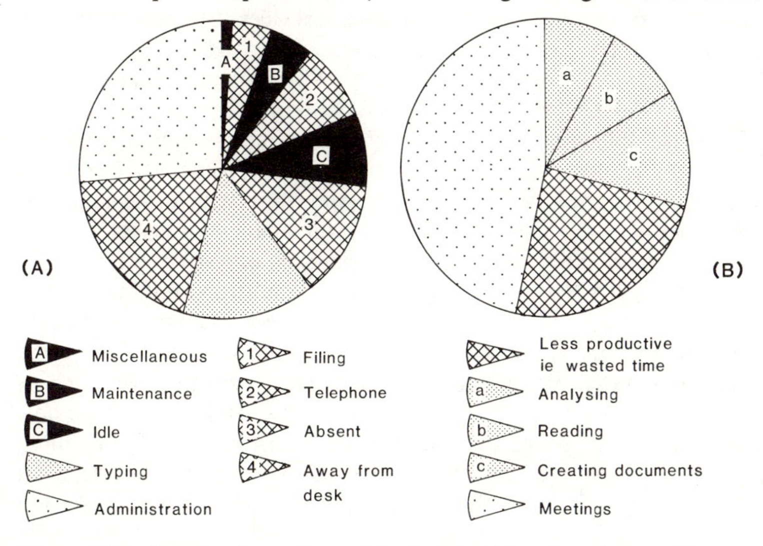

Figure 10.5 Workloads performed by typists (A) and executives (B)
Source: Office Management Information Systems, reproduced as diagram in *The Times*, 14 January 1982.

specific piece of information. In the office of the future all interaction is electronically organized without the need for time-wasting use of paper; each worker occupies a multi-function or executive work-station which consists of a battery of screens, keyboards and phones which provide immediate access to data bases, viewdata facilities, electronic mail for forwarding and receiving memos and messages to colleagues (it can also be used as a personal diary), telephones providing video images of participants, and a portable terminal, so that the user can communicate with the work-station from a remote location such as at home or while away from the office on company business.

Debenhams, a major British retail group, uses a private system similar to that described (Brooks, 1983) in which all sixty-nine of its stores are linked into the network. Among other things messages can be exchanged electronically, so that a manager can receive a note on his visual display unit within seconds rather than expending office, secretarial, or telephone time to complete the task over a much longer timespan. Information can also be stored and retrieved for quick analysis of, for example, the previous week's trade figures or executives can obtain daily information at their work-station without the need to consult bulky computer listings which may be out of date before they become available. It is possible, for example, that one in six office workers will be using multi-functional work-stations by 1990 with one for every three workers by 1995; in the finance industry the adoption rate will be higher with one work-station for every six employees by 1987.

New technology and consumer services

Most of the examples of new technology cited so far have been adopted by, or have a potential for adoption by, producer services. But services which depend upon substantial over-the-counter interaction with customers are also vulnerable to information technology in ways which may also have spatial implications, particularly for travel (National Economic Development Office, 1982). A recent example is the introduction by the Nottingham Building Society of a computerized banking and shopping service called Homelink (Cookson, 1983a). Claimed to be the first such system in the world which is available as a national service, it allows customers to carry out financial transactions and purchase a variety of goods and services using a domestic television set. Customers, initially limited to those investors with substantial deposits, can transfer funds between accounts, check building society and bank statements on the television screen, pay household bills, order and pay for goods, arrange holidays, or order traveller's cheques without the necessity to leave home. It is hoped that within three years 100,000 homes will be linked to the service.

This system is a symptom of the shift towards a cashless society in which, for example, over-the-counter exchanges of cash at banks or retail outlets will become the exception rather than the rule. Cash dispenser machines are already commonplace at banks, thus reducing the need for cashiers. Other technological changes adopted by British banks include electronic funds transfer systems, customer cashier terminals, autobanking, point-of-sale facilities and interbank transfers using systems such as SWIFT (Society for Worldwide Interbank Financial Telecommunications) (Marshall and Bachtler, 1984). Authorization of credit card transactions via electronic terminals in retail outlets is already operating in larger department stores in the USA and such point-of-sale terminals will become more widespread in other countries as the advantages to both retailers and consumers become clearer. These changes, together with systems like Homelink, will alter the pattern of banking; the distinctive services offered by banks, building societies and insurance companies will become blurred as financial supermarkets become feasible. Merrill Lynch, a major US stockbroking firm, now provides in addition to its normal buying and selling of stock service, loans, credit cards, investment and financial management advice, an estate agency and arranges insurance.

Information technology is, therefore, generating fundamental changes in the way in which all types of services whether producer, mixed, or consumer interact with their clients or customers. Inevitably questions then arise regarding the effects of capital investment in information technology on the labour inputs required by service industries. The switch from labour- to capital-intensive production in manufacturing has induced massive labour shedding in some industries: are similar prospects now in sight for the service sector?

INFORMATION TECHNOLOGY AND SERVICE EMPLOYMENT

Until comparatively recently the expectation that technology would threaten jobs in the services sector was considered unrealistic; the office worker in particular was immune from cyclical fluctuations in the demand for labour characteristic of other economic activities in advanced economies. In the past management has concentrated its investment on improving the productivity of blue-collar workers with an average of £15,000 per worker expended on automated equipment compared with just £1200 on each office worker. This is now beginning to change as efforts to achieve efficiency and higher productivity are focused more on white-collar work, where it is believed that the scope for improvement is very large by comparison with that for blue-collar work. Hence, while the number of workers in banking in the USA, for example, has been growing throughout the 1970s, the rate of growth has been slowing down from 4.6 per cent per annum between 1960 and 1973 to 3.2 per cent between 1973 and 1976. Much the same is happening in the UK, where recruitment by banks has declined sharply since 1980 and staff turnover rates have also decreased noticeably as employees endeavour to protect the jobs which they occupy. Staff turnover has also declined in insurance companies from one in six employees in 1979 to one in ten in 1982 (Barber, 1982). All this is occurring even though the volume of banking and insurance transactions is rising, and it suggests that as productivity has been improving by as much as 100 per cent as a result of introducing word processing, for example (Sleigh *et al.*, 1979), a reduction in the rate of recruitment is to be expected.

The place and sector-specific employment effects of information technology on services will depend on diffusion rates (see, for example, Pritchard, 1982; Henize, 1981), trends in the demand for services (especially those provided from offices), the size distribution of the firms adopting the new technology, the effects of mergers and reorganization, the ease with which new technology can be used within different types of services, or the willingness of workers to co-operate with the changes in work practices, job specification, or even job status (see, for example, Coombs and Green, 1981; Green *et al.*, 1980; Kendall, 1979; Bosworth, 1983). The latter is not insignificant in view of the way in which information technology is infiltrating service occupations across the full range rather than primarily the clerical and related areas in the way typical of the adoption of mainframe computers during the 1960s and early 1970s. De-skilling of all white-collar work has become a reality as decision-making is taken over by computers which have been programmed with many of the procedures, knowledge and responses which were formerly the domain of individuals and accumulated over a long period of time (Weir, 1977; Dymmel, 1979; Guiliano, 1982). Yet it is important to note that it remains very difficult to separate the effects of technology from the impact of other factors on employment such as demand for particular services, shifts from full- to part-time work, or the condition of the national, regional, or even local economy.

The size of service firm is important because in smaller establishments, which are much more numerous in services than in manufacturing, labour displacement opportunities are more limited because one individual will be responsible for several tasks rather than one, as would be the case in a large organization with extensive division of labour. It is possible that smaller firms will also be less willing to invest in new equipment, but it may be a necessary strategy for effective competition and survival against larger rivals, and it may also depend upon location (Daniels, 1983c). In order to provide the kind of intermediate service equivalent to that available from firms much nearer to the 'heartland' of the British economy small firms in remote provincial towns such as Carlisle were found to be very receptive to information technology. Small firms also have a lower ratio of clerical to managerial and professional workers and this also makes them less vulnerable; clerical and secretarial occupations have been the most readily substitutable to technology ever since batch processing of insurance claims or the clearance of cheques became feasible on company computers during the early 1960s. In some of the larger financial companies the number of clerical staff displaced by electronic office machines has often run into several hundreds.

The diffusion of information technology will also vary between service industries (see, for example, Gershuny and Miles, 1983). Many consumer services, have been slow to adopt information technology; in some instances the service provided limits the possibilities, while for others the scale of operations such as those of small shops does not justify the capital investment in or use of new technology. Office-type workers are also fewer in most consumer services, although in retailing for example much of the work they now undertake could be replaced by point-of-sale terminals and computer-controlled warehousing and stocktaking, using electronic number codes attached to each product for sale and identified by scanners at the point of purchase. Since they have a much higher proportion of office workers, both professional and clerical, producer services are already major users of information technology and this trend will continue and increasingly impinge upon the more professional occupations as specialized software – which assists architects with design work or surveyors with the estimation of quantities, for example – becomes available. Public services have been slower to adopt information technology, perhaps because of the very large scale of many of their operations, including those such as the Department of Health and Social Security or the Inland Revenue which handles vast quantities of information about millions of clients at local and regional offices throughout the UK. Gershuny and Miles (1983) suggest that such monolithic organizations will probably concentrate on improving their own internal data banks and information exchange by using the new technology before trying directly to improve the service to clients. The employment effects are uncertain but clerical workers, for example, a large proportion of civil service manpower, will be particularly vulnerable.

As these diverse examples might lead one to expect, there is considerable variation among the forecasts of the aggregate impact of information technology on employment. Indeed, 'the effects on employment . . . are vague and in dispute as far as precise figures are concerned' (Marshall and Bachtler, 1984, 447). Bird (1980) provides a useful summary of the diverse forecasts made during the late 1970s. A French study of banking and insurance jobs in France forecast a 30 per cent reduction in jobs during the period 1978–88, while a British study (Sleigh *et al.*, 1979) forecast a modest overall increase in banking employment up to 1985, followed by a reduction if banks do not develop new services or experience more competition from other financial services, such as building societies, which are already providing banking-related facilities. Over 40 per cent of private sector administrative and clerical jobs in insurance, banking and building societies are at risk from new technology in the 1980s (Virgo, 1979), or a 15 per cent decrease in insurance industry jobs (Sleigh *et al.*, 1979) is likely as a shift takes place from routine to higher-grade jobs. But Barras and Swann (1983) conclude, after gathering data from a sample of insurance companies, that computerization has slowed down the rate of employment growth rather than caused employment cuts.

Forecasts of this kind make rather grim reading, especially for the economies of Western Europe which have been creating very few new jobs and stimulating only limited levels of new investment, encumbered by sluggish productivity and unemployment levels twice those of the USA. As was the case in the period immediately following the extensive adoption of mainframe computers, there is some prospect of new occupations which arise directly from the new technology and which did not exist before. But in view of the deeper penetration of information technology into all levels within service organizations (such as the use of personal computers) it is not likely that any new occupations will totally compensate for those displaced because they are outmoded. There can be little doubt that occupational skills will change, particularly among those which have traditionally not been involved in the use of keyboards. Both professional and administrative workers will require keyboard skills in order to fully utilize an integrated work-station, for example, and typists will still be in demand. On the other hand, general clerks, postroom personnel, or telephone switchboard operators will not be required because properly integrated and networked information systems will incorporate these tasks without the need for human decisions. Legal clerks, draughtsmen, accounts clerks and clerical workers in local authorities or hospitals and in the civil service are also highly vulnerable. Information management occupations, such as librarians, data base compilers or information system managers, will become much more central in the day to day work of organizations and be as critical to their survival as good book-keeping or the personal acquaintance of clients must surely have been among many service firms at the end of the last century.

Adaptability and a willingness to consider retraining will become part of keeping a remunerated place in the labour market. Decision-making jobs remain 'safe', and the jobs associated with the design, installation, operation and servicing of information systems will remain, but the majority of the workforce not involved in these two areas will be vulnerable. Part of the adaptation brought about by information technology may also involve a redefinition of jobs and expectations. The increase in part-time and self-employment noted in Chapter 2 is a symptom of this, together with more recent innovations such as job sharing or the removal from workers of the usual contractual obligations of employers, such as providing an office, pension arrangements or a company car, and replacing them with access to the company's intelligence information (marketing, product life, sales guides, and so on) and any other specialist requirements which are available to office-based employees but with the workers now based at home.

INFORMATION TECHNOLOGY AND THE LOCATION OF SERVICES

The combination of information technology and changing job characteristics makes it feasible to perform service industry tasks at a much wider range of locations than in the past. Some observers go so far as to suggest that the home can now become the workplace as well as the base for many interactions with consumer services which will not require any travel (see, for example, Azimov, 1978; Toffler, 1981). The Xerox Corporation has begun to encourage some of its employees to work from home, thus reducing the overheads arising from office leases or fringe benefits but ensuring that their fifty or so 'networkers' who work from home remain competitive while bearing more of the overhead costs. The company estimates that for each £10,000 of full-time labour costs in central London, £17,000 must be added for non-wage overheads such as office accommodation (*The Economist*, 29 September 1984). A British computer services company, F International, employs worldwide 850 freelances, mostly home-based, and has five almost continually workerless regional offices in the UK and similar subsidiary establishments in Copenhagen, Amsterdam and New York. Harkness (1977) has estimated that up to 50 per cent of all workers could operate from home if they chose to use the technology currently available.

Evidence has been cited in earlier chapters, particularly in the discussion of office-based service activities, which suggests that organizations have recognized the possibilities for a more decentralized mode of operation. Both Goddard and Morris (1976) and Thorngren (1973) showed that any of the contact needs of decentralized establishments which could not be fulfilled at the new location, which might be seen as a considerable disadvantage, can be overcome by the use of telecommunications (see also Langdale, 1982). Since the operations which are moved out of the principal corporate complexes are mainly of a routine nature, many of the programmed and even planning contacts can be sustained by using narrow-band communications channels while continuing to travel to any contacts associated with orientation or high-level planning activities (Goddard and Morris, 1976).

Nilles *et al.* (1976) suggest that decentralization, or fragmentation, is but one phase in a four-stage evolution of the spatial organization of firms which largely handle information (Figure 10.6). The first phase is centralization which is the current dominant mode in most information-using industries. When

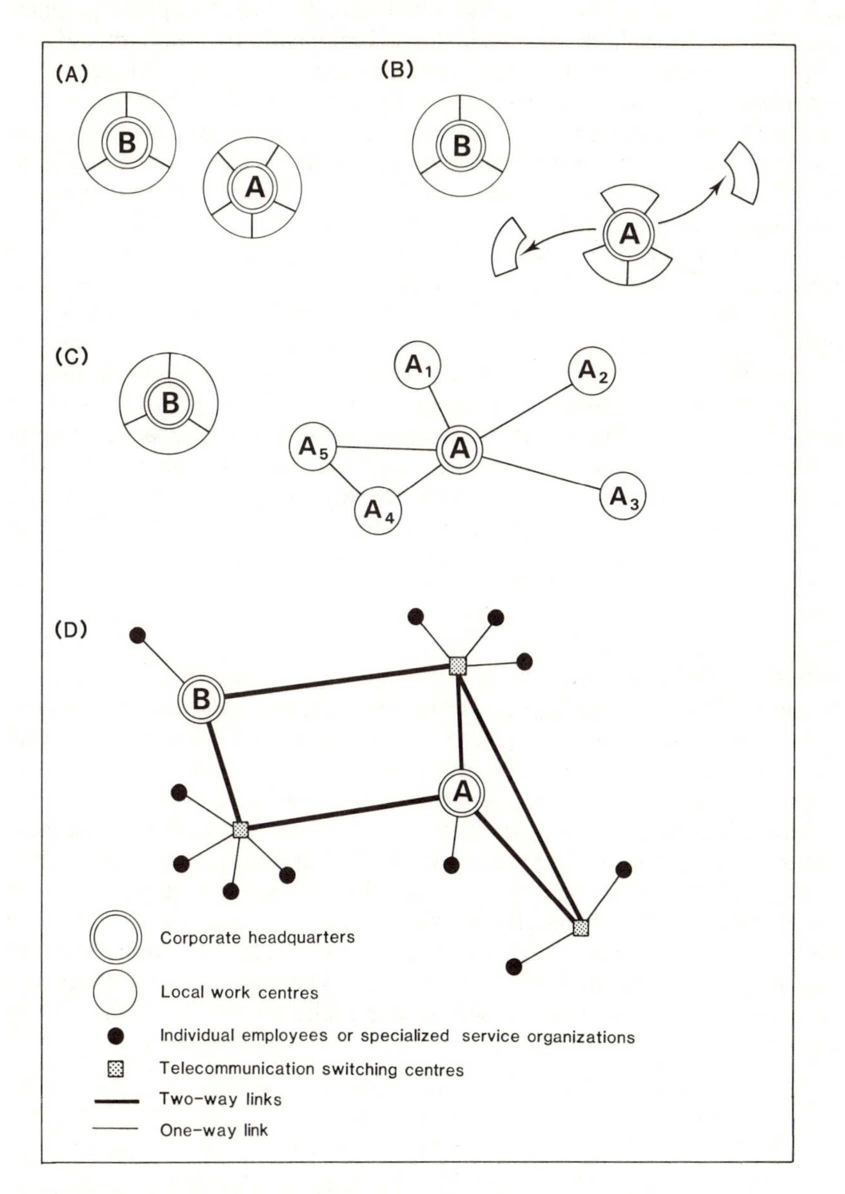

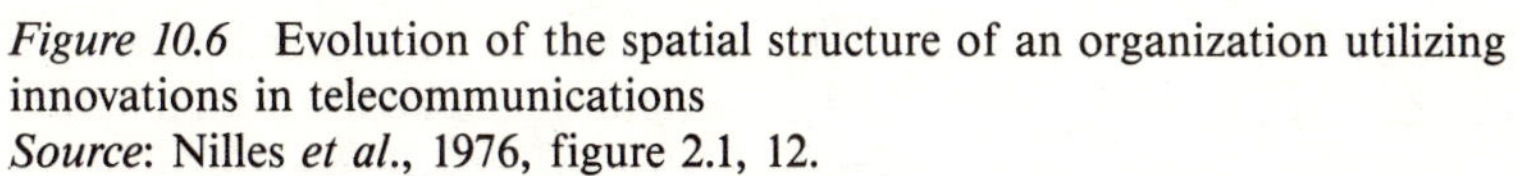

Figure 10.6 Evolution of the spatial structure of an organization utilizing innovations in telecommunications
Source: Nilles *et al.*, 1976, figure 2.1, 12.

decentralization (or fragmentation) takes place, it usually involves distinctive subunits, such as bank branches, or functional units, such as accounting, which are relocated elsewhere and push the communications boundaries of the organization outwards. Mail and telecommunications services ensure intra-organizational contacts are maintained. Decentralization does not reduce the volume of commuting, but it redistributes trips among more nodes, changes the modal split and, for some employees, reduces the time taken and the distance travelled (Daniels, 1980; Alexander, 1980). In the third stage of spatial evolution the almost random pattern of linkages and employee travel patterns is replaced by dispersion in which employees report only to that part of the organization nearest to their residential location. A central or local computer provides the information needed for their work, and supervision and co-ordination also involves greater use of telecommunications and computers. This phase depends on the use of work-stations to provide effective information transfer and employee interaction. Provided that jobs can also be redesigned so individuals can work at their nearest location, dispersion offers the prospect of substantial reductions in commuting. Diffusion is the final evolutionary stage in which individual employees work from home rather than at the nearest part of the organization. The technology and the software is available and the organization would only require a small core of senior personnel at one location or at several dispersed locations (see also Kellerman, 1984).

Is radical spatial change realistic?

While individual organizations are at various stages along this evolutionary path, very few have moved the same distance or have reacted in the same way to the possibilities presented by information technology. Herein lies one of the great imponderables: the adoption rates of new technology vary between organizations and locations, and the technology is itself changing so quickly, but there is an underlying confusion between what is technologically possible and that which is practical or feasible (Drucker, 1970). The timespans involved are asynchronous; technological change is occurring more rapidly than is the ability of its recipients, whether organizations or individual households, to adjust to the new demands on working practices and time–space relationships which it makes. The difficulty is illustrated in the context of urban land use by Nilles *et al.*, who see that a

> set of conflicting forces can be anticipated in the area of land use planning as a consequence of the increased use of telecommunications. On the one hand, greater concentrations of growth – regional activity centres or nodes – within the metropolitan areas are foreseeable and are growing in favour with urban planners. On the other hand, increased urban sprawl can be an equally likely possibility, given the increased locational flexibility afforded by telecommunications. In principal, no locational constraints are put upon an individual or organization wishing to work by means of telecommuting. (Nilles *et al.*, 1976, 149)

The most noticeable aspect of this statement is the uncertainty it portrays about the spatial outcome of information technology, although Nilles *et al.* seem to favour rejection of total dispersal or 'domestication' (see Kellerman, 1984) of services which are users of information and high technology.

Social and infrastructural inertia will inhibit the kinds of locational changes which have been briefly outlined. Remote working is a concept that will most readily be accepted by those who start their working lives, and indeed are familiar with the technology involved before they start work, in such environments. The majority of those currently employed in office and similar services expect to have to travel to work, use work as a social as well as economic source of well-being, and are suspicious of the 'value' of the new technology with respect both to job satisfaction and job security. Working at home may help to avoid the congestion and other disadvantages of coming together to work but it requires greater self-motivation and, from the employer's viewpoint, presents problems of supervision and ensuring a smooth flow of work. For the self-employed and salesmen in insurance, for example, working from home is a long-established practice which has evolved gradually; a similar process will probably need to precede the widespread appearance of home-based working by the full-time, office-based employees of medium- and large-sized organizations.

Infrastructural inertia refers both to the capacity of telecommunications channels to cope with the volume of information flows implied by remote working and to the limitations imposed by the built environment already used by service activities. The property development industry has been slow to respond to the changing requirements of office space users who wish to utilize the new technology. In existing buildings flexible access to a battery of electrical and telephone points requires the construction of false floors or ceilings to accommodate the wiring conduits. But floor-to-ceiling heights in most office buildings completed during the last twenty years make this impossible without contravening building regulations. Airconditioning is not a common feature in most British office buildings completed before 1980, but it has now become essential for the purpose of dissipating the heat generated by computers, word processors, photocopiers, printers and other electronic equipment in offices. The cost of converting the existing stock of office buildings (some 44 million m^2.) has been estimated at £350 per square metre gross floorspace, comparing unfavourably with new construction costs which average £400 m^2 (Phillips, 1983).

The spectre of empty and outmoded office buildings in city centres and suburban areas seems hardly likely to materialize for some years yet; the technology may suggest that it is possible (Phillips, 1981) but the vested interests such as development companies, financiers, landowners, corporate owners and landlords, and public authorities responsible for utilities and related infrastructure, will combine to protect their vast investments. This can be done by providing new buildings fully equipped to allow companies to use information technology, or by adjusting rents or the length of leases (the latter have become much shorter during the last ten years) in the hope of persuading organizations to continue operating along 'conventional' lines, or allow for refurbishment. Some office space users may be prepared to cover this expenditure themselves and to write off adaptation costs over a period of time.

Gottman has, therefore, proposed that urban settlements

> will not dissolve under the impact of this technology, although they may evolve and, indeed, are evolving. For transactional activities, the new technology is rather helping concentration; first, in large urban centres, already well established transactional crossroads; and, second, in a great number of smaller centres of regional scale or highly specialized character. (Gottman, 1983, 28)

The CBD may, therefore, be reinforced rather than undermined as high-order producer services in headquarters and regional offices, surrounded by publishing and broadcasting services which need the centrality in order to serve ever-expanding metropolitan areas, make the area a 'telecommunications junction' (Kellerman, 1984) for inter-urban and international, as well as intra-urban, information exchange and transactional activity (see also Edgington, 1982a).

Similar constraints will affect the degree to which consumer services such as retailing, parts of banking, building societies, or post offices, will actually undergo the structural and spatial changes consequent upon the possibilities for home-based selection of goods, holiday bookings, travel, or banking transactions. Much will depend on the adaptiveness of consumers; an example of the difficulty of getting the acceptance of new technology among consumers is the experience of Prestel (Witcher, 1982), which is a collection of independent services brought together by British Telecom and accessible via a modified television set or computer terminal. About 800 organizations supply services on Prestel and there are some 200,000 pages of information (not all available to the home user). When introduced in late 1979 Prestel was described as a 'major new medium . . . comparable with print, radio, and television' which would lead to 'major changes in social habits and styles of life, and [will] have long-lasting as well as complex economic effects' (Fedida and Malik, 1979, 1). British Telecom anticipated tens of thousands of customers, increasing to millions in the 1980s; in March 1982 there were less than 14,000 registered users, increasing at 400–500 per month (Witcher, 1982). Many of the users are businesses rather than households, and although the cost of the service (rents for Prestel televisions, for example, are £20–30 per month) is too high to establish a mass market (and the existing market is too small to bring costs down), for a nationally available facility the impact of Prestel is very disappointing. There are cheaper alternatives such as teletext for which normal household televisions can be adapted and, in contrast to Prestel, once this additional cost has been borne there are no further charges for obtaining the information available (information from Prestel involves a charge for the use of the telephone line, a charge for using Prestel per unit of time and a charge for each page viewed). Apart from educating people to use services like Prestel, it may also be necessary to provide a public subsidy during the early years in order to make such systems available to the full spectrum of potential users rather than only those able to afford the charges. According to this argument, viewdata facilities are a social service and should be provided on this basis.

It seems clear that the rationale underlying the existing location patterns of consumer services will remain for the foreseeable future. Radical change of the kind which will involve the closure of retail units and rationalization of the pattern of location into fewer places which are well positioned to act as delivery points to consumers (as demonstrated by the growth of mail-order services or direct marketing) will only evolve further if users perceive real benefits from home-based access to information of the kind available through viewdata. Many consumers will need persuading that the inability to touch, smell, compare, assess the colour and size, or to take immediate delivery of the majority of the goods which they purchase is outbalanced by the ability to sit at home to view the goods on their television screens and to place their orders via the telephone line. There may be a threshold beyond which remote shopping, currently represented by direct marketing catalogues, becomes unacceptable because, for many consumers, the act of undertaking a shopping trip is a pleasant, social use of time. Of course, some services such as entertainment, education, or security systems are more readily consumed at home than others and users are more readily adaptable to them. Entertainment services, particularly cinemas, have been losing audiences ever since television started to become widespread in the early 1950s; more recent technology such as the video recorder has simply reinforced a long-established trend. Even if more education is conveyed over the air to students working at home in the way pioneeered by the Open University, there will be a requirement for academic staff to prepare the material for transmission, to provide a wide variety of support services and to update courses and their content. Research activities will still require buildings, equipment and other resources, so that it seems unlikely that higher education centres will be displaced on a large scale even if the manpower is deployed in a different way.

Likewise local area networks are seen to be important by producer and information-oriented services because they provide rapid-communication facilities. To a large degree, however, it remains to be seen what really can and should be communicated and how to use the large volumes of information which they can make available. Information technology does have positive effects such as creating an environment for the supply of completely new services or refinements of those currently supplied (Gershuny and Miles, 1983), but it also has disruptive effects on the hierarchical relationships between departments and divisions and individual workers in executive, administrative, or clerical occupations (Crozier, 1983; Simon, 1977; Wynne, 1983). Adoption of information technology by producer and consumer services is, therefore, likely to be a very gradual process; it follows that the spatial impacts will be equally slow to become manifest. Indeed such is the pace of change in information technology that the use of keyboards as the main mode of interaction between individuals and computers could prove to be the 'wrong' way towards service industry automation. Touch-driven computer software is already a reality but speech-driven units may yet be developed (the so-called fifth generation of computers), so making this innovation much more 'user-friendly' than a typewriter keyboard. Such an innovation may not only reduce the hardware requirements for individual workers, but also make home-based working more practicable on operational, if not other, grounds.

Changes in the geography of services may indeed occur as a result of recent innovations in computer and telecommunications technology but speculation seems to predominate over fact. Gershuny and Miles, therefore, observe that it

> is hard to forecast what uses may be made of information technology by the end of this century. We can certainly make intelligent guesses, and can use these to deepen our analysis. But in many respects we are in the same position as our predecessors in 1930. (Gershuny and Miles, 1983, 233)

If this is the case, then it seems clear that the pace and outcome of the possible spatial changes should not be exaggerated; the same technology is responsible for creating forces which promote decentralization but which, at the same time, could lead to centralization of service industries. Of course, this paradox only serves to confuse further any attempt at assessing likely spatial outcomes. Higher-order producer services, mainly concerned with control rather than providing customer service, can become more centralized but in order to be able to deliver services effectively to customers many activities still require localized, dispersed establishments. There are, therefore, two schools of thought which offer polarized views on the impact of new information technology on society and its spatial organization (Kellerman, 1984). The 'maximizers' believe that all new technology will be accepted and used with consequent employment and spatial effects, while the 'minimizers' advance the view that technological feasibility does not – for cost, social and inertia reasons – equate with utilization, and change is therefore much more gradual. In addition, information technology will not, on its own, produce locational changes; rather it should be seen as but one enabling agent among many other prerequisites for gradual, evolutionary changes in the behaviour and location of service industries.

SUMMARY

One of the assumptions about the growth of service industries has been that the employment they generate will make good the losses experienced in manufacturing and, in developing economies, the primary sector. This also explains the interest, in several countries, in the use of public policies to ensure that as many regions and cities as possible will benefit from the continuing expansion of the service sector. The purpose of this chapter has been to illustrate the growing uncertainty about the merits of current assumptions because of the inroads on service sector output and productivity made by technology. The impact of technological innovation on labour demand in manufacturing is well known and it seems certain that it will retard, if not actually reverse, the labour inputs required by service industries. If technology does bring about a reduced demand for labour, then the spatial problems already apparent in the distribution of some services may well be exaggerated while adequate access to the more ubiquitous consumer services will depend on the availability of the telephone, television, or home computer as the main link with the suppliers of such services.

But there remains a great deal of uncertainty about how technology will affect the employment and occupational characteristics of services. There is already evidence showing that it causes de-skilling of certain kinds of service work, but it also creates opportunities for improving the skills of others and the appearance of completely new occupations based upon the application of new technology in a variety of service activities. It is much more difficult to assess the effects on employment levels and the forecasts which have been cited show little agreement. Much will depend on which industries adopt the new technology, at what rate and for what kinds of task. There have been relatively few case studies of the detailed consequences for organizations and their workers, and more will be necessary if more general forecasts which suggest 30 or 40 per cent reductions in employment levels are viewed as plausible. It may also be the case that new technology will allow a larger proportion of services to be provided by the self-employed, so that any job shedding would need to be considered in the context of trends elsewhere in the provision of services.

The technology available to service industries or to their customers can be divided into telecommunications, electronics and computing. Together they comprise information technology and this has greatly increased the vulnerability of service industries to the mechanization of routine tasks, to the use of computer aids for internal and external communication, for developing information systems which permit quicker responses to unexpected events, or to the application of decision-making aids which can substitute for some of the tasks performed by middle or even upper management. Telecommunications, which now offer high-speed, large-volume transfer of voice, digital, document, or photographic information over very large distances (using satellites), either within or between organizations, have the greatest potential for generating important changes in the spatial pattern of service industries.

As information technology has evolved and become – as the result of decreasing costs – within reach of more service activities the spatial organization of firms has become more dispersed. A sequence of centralization, fragmentation, dispersion, diffusion and ultimately the domestication of workplaces has been suggested as the logical outcome of using in particular telecommunications technology. However, the last part of this chapter has urged caution: that which is technologically feasible for service industries may not be economically, socially, or even politically practicable. It is stressed that technological change has been much too rapid for the capacity of the social or spatial systems to absorb or adapt to it. A good deal of inertia is incorporated within the existing distribution of service industries, attaching to the buildings they occupy and the cities within which they congregate, together with the locational traditions some of them value and the expectations which customers or clients have about access or viewing goods – or the value perceived by workers of a collective rather than isolated existence in a 'wired' office at home. These are just a few of the reservations which can be cited for the more extreme forecasts of radical spatial change in the location and operation of service industries. Of course, some changes have taken place and will continue to do so but the timescale is uncertain and perhaps will be longer than expected. Some of the changes might be counterintuitive, with information technology causing recentralization of consumer and producer services rather than decentralization of the kind which might be expected in view of the locational trends over the past twenty-five years. There are clearly far more questions than answers about the impact of information technology on the geography of services; a point which can also be made in relation to many of the other aspects of service activities outlined elsewhere in this book.

APPENDIX 1

Alternative bases for classifying service industries

THE PRIMARY/NON-PRIMARY CLASSIFICATION

Sometimes a basic division is made between primary and non-primary sectors, and the latter is then subdivided into three sets of activities: first, those involving intensive use of capital and skills such as mining, 'modern' industry, water and light and power and transport; and secondly, activities involving only limited use of capital and skills such as construction, cottage industries, trade and personal services; and thirdly, activities in which skills are of primary importance, including banking, insurance, finance, government services, community services and business services (see, for example, Sabolo, 1975). This is a useful schema, but unfortunately activities generally considered to be part of the service sector are located in all three subgroups, along with others which are not.

CLASSIFICATION OF SERVICES USING OCCUPATIONAL CHARACTERISTICS

A definition of the services sector by industry order, that is by the kind of product which comprises output, and which is then used to collate employment or output information, is likely to be too rigid and therefore misleading. This arises from the problem which has already been alluded to: the imprecise division or distinction between goods and services. It is quite possible that service activities will take place in goods-producing establishments and vice versa. Therefore, rather than define the service sector relative to what is produced, it is often more useful to think of it with reference to the tasks performed by workers in each industry order, that is the occupational characteristics of the labourforce.

Service occupations are defined in the International Standard Classification of Occupations (ISCO) and most national censuses collect occupation data along similar lines and can be cross-tabulated by industry. These occupations are also described as white collar (Mills, 1953; Crozier, 1965) because the workers perform supervisory, administrative, analytical, planning, book-keeping and other functions which can be distinguished from the blue-collar unskilled, semi-skilled and skilled manual occupations.

Given the changing character of modern industry, for example, in microelectronics or pharmaceuticals, this two-way classification may be inappropriate since work-tasks involve a mix of intellectual and manual skills which may merit the designation of a third category of 'purple-collar' occupations. Using this system of classification, the sales and marketing or product planning staff of an engineering company (secondary sector) can be identified as the service workers within manufacturing (and likewise for businesses in the primary sector) as well as part of the total population of service sector workers. Clearly, under the industry order classification the headquarters staff of the engineering company, who are likely to be principally white-collar workers, would be overlooked. Such an oversight is not insignificant; almost 31 per cent of the employees in British manufacturing industries were in administrative, technical and clerical occupations in 1982 (see Figure 3.4, p. 61).

FORMAL AND INFORMAL SERVICES

In view of the attributes of service industries in developing countries some observers have made extensive use of the distinction between informal and formal sectors (International Labour Office, 1972; Soussan, 1980; Lozano, 1983; Bromley and Gerry, 1979). Such a division applies to all types of industry or employment but it seems reasonable to use it to distinguish two groups of services. The informal services are characterized by enterprises relying on small-scale operation, family ownership, indigenous capital and labour and, above all, ease of entry into markets which may already be saturated and therefore highly competitive. Indeed Lozano (1983) suggests that entry into informal sector work is largely involuntary and will lead to growing rates of unemployment and a worsening of the quality of life for the labourforce as a whole. As the formal sector becomes less able to provide jobs and competitively priced services (as well as goods) it is feasible for individuals to give more of their time to informal work (Gershuny, 1978). Services are prominent in, and well suited to, the informal economy the growth of which represents a loss of revenue for national taxes and levies. Many of the economic activities generated in and around rapidly growing urban fringe areas (Soussan, 1980) are of this kind.

By contrast, the formal sector requires a more sophisticated approach to enterprise formation, relying on high initial capital investment and modern technology. Much of the latter, along with labour and control of the enterprise, will probably come from outside the developing country and, thus far at least, is more likely to be manufacturing than service-type industry. At present the way in which the formal and informal sectors co-exist within a single system of production, distribution and consumption is not well understood (Lozano, 1983).

MODERN AND TRADITIONAL SERVICES

There is much disagreement about both the meaning and the significance of the formal–informal dichotomy for the development of urban areas and economies in general (for a review see Soussan, 1980); matters are not helped when yet another classification employs the terms 'modern' and 'traditional'

(Doctor and Gallis, 1964). The modern sector is characterized by the production of services (and goods) for the market. This might suggest that the informal sector could be included but the supply and demand for labour is organized through the labour market rather than kinship or family ties. The modern sector comprises wage-earning employment, higher wage levels and higher levels of productivity. Again it is conceivable that some informal sector services would conform to at least a part of this definition. Economic organization in the traditional sector is based on household or family enterprise, including a variety of unorganized services, and is clearly orientated towards the informal category.

OTHER CLASSIFICATIONS

Other classifications which rely on a basic dichotomy include foot-loose and tied services, and office/non-office services. It should be apparent that if some services are considered to be consumer orientated, then it is likely that their choice of location is governed by the distribution of population as well as by the durable/semi-durable/non-durable status of the service being provided. Such services are tied, therefore, with reference to the degree of flexibility open to them; retailing is obviously the best example but certain non-market services such as health care, dental, transport, or education services are also representative. Tied services are mainly directed at local markets, while foot-loose services serve regional, national and even international markets with which they have contact through postal, telecommunications and business travel. Location is constrained far less by the distribution of population but rather more by the distribution of potential clients and competitors, mainly in major centres of employment (see Chapter 7). In theory, the location behaviour of these, largely producer, services is 'flexible', that is they can operate with equal effectiveness from several alternative locations. In practice, this is a gross oversimplification. Nevertheless, if some services can be shown to be more foot-loose than others, there may be an opportunity to take advantage of this in the interests of regional development objectives or the restructuring of urban economies.

It is possible to distinguish between office and non-office services because, using occupation statistics or information on the kind of accommodation occupied, some service industry orders are found to be largely located in office buildings occupied by white-collar workers. Insurance, banking and finance and professional services are major office-using services, while retailing and distribution and many of the miscellaneous services can be considered non-office since they are more likely to be found in specialist premises such as warehouses, individual shop units, or shopping malls. Some office services, connected with the day to day administration of a department store, for example, may be attached to such premises but they will utilize only a minor proportion of the total floorspace or occupy a small proportion of the total labourforce.

MORE ON THE PRODUCER–CONSUMER CLASSIFICATION

Although this dichotomous classification is used in this book, it is worth elaborating its limitations. As mentioned in Chapter 1, unfortunately such a compact classification is difficult to operate in practice because most services do not fit exclusively into the producer or consumer categories. Certainly, almost all the receipts generated by the wholesale trade come from other businesses (for a detailed case study of wholesaling see Vance, 1970) and certain business services, such as advertising, market research, computer bureaux, or accounting, will also have a large proportion of their income from other businesses. But such activities are a relatively small number in absolute and employment terms among a much larger number of other services, such as estate agents, insurance brokers, building societies, credit companies, leisure and entertainment services, architects and surveyors, or public services which meet both intermediate and final demand from the same establishment or with the same workforce.

The difficulty is nicely explained by Stigler as follows:

> The industries that provide services to the business community do not form a category wholly distinct from those providing services to consumers; there is indeed, a fairly continuous array of industries between the limits. At one extreme, service industries, such as the consulting construction engineers serve only business; in the middle of the array, independent lawyers receive approximately equal shares of income from business and non-business clients; and at the other extreme, teachers serve individuals in their non-business capacity. Enough important industries fall in the category of those services serving chiefly business, however, to justify separate discussion. (Stigler, 1956, 138)

Despite this very real difficulty of classifying individual services as belonging to the producer or consumer category, there is a pragmatic reason for persisting with this schema: there is evidence for spatial variation in the location patterns and behaviour of the two groups and this immediately lends itself to geographical enquiry. The simplicity of the classification is also attractive and it can be supplemented by other classifications.

Perhaps the strongest advocate of the significance of the dichotomy between producer–consumer services is Greenfield (1966, 10) who believes that the distinction 'will facilitate both the description and the analysis of an advanced economy, and, by way of contrast, should serve to illuminate some problems of developing economies as well'. His enthusiasm seems to arise from research undertaken in the early 1960s into the connection between economic growth in the USA and investment in human rather than physical resources. Such investment on the part of companies involves the purchase of services of the kind provided by management consultants or economists which allows them to harness knowledge to make their organizational practices more streamlined and cost-effective or to anticipate macroeconomic trends significant for the availability of their raw-material supplies or the appearance of new markets. Spending of this kind is no less an investment than that made in new machinery and when 'knowledge is produced in order that or in the expectation that, as a result, the productivity of resources – human, natural, or man-made – will increase in the foreseeable future, the production of knowledge can be regarded as an investment' (Machlup, 1962, quoted in Greenfield, 1966, 12).

Another important reason for employing the distinction between consumer and producer services is that the latter are often found in business enterprises which would not, on other criteria, be classed as service activities. A growing proportion of wholesaling activity is being undertaken by the manufacturers or publicity and product advertising will be prepared in-house rather than by service sector companies specializing in such work. Research and development for an oil company is usually undertaken by its own highly specialized staff rather than by external consultants. The list of such examples is a long one and it would, therefore, be misleading to make assessments of output or employment trends purely on the basis of information derived from independent businesses providing such services.

APPENDIX 2

Structure of the labourforce, by occupation status and ratio of female to male employees in each occupation group, selected countries

Country		*Occupation*[1]						*Total economically active*
		(1)	*(2)*	*(3)*	*(4)*	*(5)*	*(6)*	
LESS DEVELOPED COUNTRIES								
Ecuador	a[2]	5.2	1.0	3.8	7.8	7.0	75.2	1,940,628
(1974)	b[3]	0.75	0.14	0.56	0.37	1.95	0.10	0.20
Guatemala	a	3.6	1.1	2.7	6.2	8.3	77.4	1,545,658
(1973)	b	0.68	0.23	0.52	0.54	1.90	0.43	0.16
India	a	2.6	0.07	3.0	4.2	3.2	86.3	180,373,400
(1971)	b	0.22	0.01	0.04	0.07	0.20	0.22	0.21
Mauritius	a	5.2	0.5	5.7	6.2	9.6	72.7	260,749
(1972)	b	0.58	0.06	0.25	0.17	1.06	0.18	0.52
Philippines	a	5.4	1.1	3.1	6.4	7.2	76.8	12,416,450
(1970)	b	1.31	0.40	0.61	1.32	1.94	0.26	0.40

Country		*Occupation*[1]						*Total economically active*
		(1)	*(2)*	*(3)*	*(4)*	*(5)*	*(6)*	
DEVELOPED COUNTRIES								
Argentina	a[2]	7.5	2.5	11.4	11.9	12.6	55.1	9,011,450
(1970)	b[3]	1.21	0.07	0.55	0.31	1.52	0.14	0.34
Austria	a	8.7	0.6	15.3	8.4	11.7	54.7	3,097,987
(1971)	b	0.59	0.24	0.97	1.26	2.4	0.37	0.63
Belgium	a	11.4	4.7	13.2	10.5	7.0	51.6	3,524,559
(1970)	b	0.79	0.12	0.88	1.02	1.67	0.19	0.42
Bulgaria	a	17.2	2.2	8.0	3.1	8.0	61.4	4,447,784
(1975)	b	1.36	0.21	1.13	1.78	2.72	0.66	0.88
Canada	a	13.2	2.5	16.7	9.6	11.1	45.9	8,626,930
(1971)	b	0.75	0.18	1.90	0.43	1.09	0.23	0.52
Finland	a	11.9	0.3	9.6	8.2	10.5	59.1	2,118,257
(1970)	b	1.10	0.32	2.60	1.41	4.21	0.35	0.73
France	a	16.1	3.4	14.6	7.5	8.3	49.0	20,939,800
(1975)	b	0.80	0.20	2.0	0.99	2.00	0.25	0.58
Hong Kong	a	5.3	2.1	9.4	1.2	14.6	57.2	1,952,000
(1976)	b	0.71	0.10	0.73	0.31	0.51	0.59	0.54
Japan	a	7.8	4.3	16.8	13.2	8.4	49.6	53,140,700
(1975)	b	0.65	0.06	0.99	0.65	1.17	0.46	0.59
New Zealand	a	14.1	3.2	16.1	9.8	6.7	49.1	1,272,333
(1976)	b	0.71	0.08	1.94	0.61	1.37	0.18	0.47
Spain	a	5.5	0.7	9.2	8.3	9.3	65.7	11,908,062
(1970)	b	0.49	0.04	0.43	0.38	1.0	0.14	0.24
Sweden	a	22.1	2.4	11.6	8.7	12.6	42.0	3,850,451
(1975)	b	0.99	0.28	3.51	0.96	0.33	0.22	0.73
West Germany	a	10.6	2.4	18.9	9.6	10.3	48.3	24,606,700
(1970)	b	0.52	0.16	1.20	1.11	1.20	0.31	0.57

Notes: 1 (1) Professional, technical and related workers; (2) administrative and managerial workers; (3) clerical and related workers; (4) sales workers; (5) services workers; and (6) primary and secondary sector workers. Members of the armed forces are not included, but are represented in the total economically active in some countries).
2 Proportion of total economically active by occupation.
3 Ratio of female to male workers in each occupation group.

Source: United Nations, *Demographic Yearbook, 1979*, New York, UN, 1979, table 42, 1044–92.

APPENDIX 3

Shift–share analysis

Shift–share analysis is a technique which provides a measure of the total change in a region's performance over a given time period relative to that of the nation: the region's actual growth less its 'expected' growth if it has grown at the same rate as the nation (Thirlwall, 1967; Stilwell, 1969). This is usually referred to as the 'shift', and the change which it represents is then divided into two components: that part of it which results from the concentration of particular industries in the area or region, together with the extent to which these industries are growing or declining nationally at rates different from the overall national growth rate (the structural component); and that part which results from the degree to which employment change in a particular industry in the region is greater or less than would have occurred had the national growth rate taken place in the region (the differential component). The differential effect is to some extent a surrogate for the relative locational advantages of some regions over others which causes industries to be attracted there or to grow faster *in situ*. The technique is most frequently applied to employment data, although it can also be used with other types of economic data.

While the technique may help the task of isolating the factors which contribute to the relative growth of service industries, it is important to assess the underlying weaknesses of shift–share analysis. Richardson (1978, 202) describes it as 'the most overvalued tool of analysis in regional economics'. One of the most serious difficulties is the sensitivity of the analysis to industrial disaggregation; for example, Richardson suggests that the size of the structural component will tend to increase in line with greater disaggregation. Shift–share is often used as the basis for projections but some studies have shown that the differential component is unstable over time. The technique also represents a way of looking at data which does not necessarily give a useful insight to the ability of a region to attract new service industries or to retain those which may be performing well. Richardson cites a study of Merseyside by Buck (1970), which showed that positive differential shifts were not connected with locational advantages, but were caused by the subsidies available to firms to assist inter-regional relocation or the reorganization of branches. The opposite view is that 'in the context of the British regions' shift–share analysis 'is a highly robust technique' (Fothergill and Gudgin, 1982, 50). It is argued that the level of disaggregation or the choice of base year, for example, has no significant effect on the subsequent results.

Bibliography

(1975) 'A view of industrial employment in 1981', *Department of Employment Gazette*, 83, 400–5.

(1983) 'How many self-employed?', *Department of Employment Gazette*, 91, 55–6.

(1984) 'Flexible high-flyers', *The Economist*, 29 September, 79.

Abe, K. (1984) 'Head and branch offices of big private enterprises in major cities of Japan', *Geographical Review of Japan*, 57, 43–67.

Abernathy, W.J. and Hershey, J.C. (1971) 'A spatial allocation model for regional health service planning', *Operations Research*, 19, 629–42.

Abler, R.F. (1975) 'Effects of space-adjusting technologies on the human geography of the future', in Abler, R.F., Janelle, D., Philbrick, A. and Sommer, J. (eds) *Human Geography in a Shrinking World*, North Scituate, Mass., Duxbury Press.

Abler, R.F., Adams, J.S. and Gould, P. (1971) *Spatial Organization*. Englewood Cliffs, N.J., Prentice-Hall.

Adrian, C. (1983) 'Analysing service equity: fire services in Sydney', *Environment and Planning A*, 15, 1083–1100.

Akehurst, G.P. (1983) 'Concentration in retail distribution: measurement and significance', *Service Industries Journal*, 3, 161–79.

Alexander, I. (1978) 'Office relocation in Sydney, Australia', *Town Planning Review*, 49, 402–16.

Alexander, I. (1979) *Office Location and Public Policy*, London, Longman.

Alexander, I. (1980) 'Office dispersal in metropolitan areas. II, case study results and conclusions', *Geoforum*, 11, 249–75.

Alexander, I. and Dawson, J.A. (1981) 'Employment in retailing: a case study of employment in suburban shopping centres', *Geoforum*, 10, 407–25.

Alonso, W. (1960) 'A theory of the urban land market', *Papers and Proceedings of the Regional Science Association*, 6, 149–57.

Alonso, W. (1964) *Location and Land Use*, Cambridge, Mass., Harvard University Press.

Alperovich, G. (1982) 'Density gradients and the identification of the central business district', *Urban Studies*, 19, 313–20.

Ambrose, P. and Colenutt, B. (1975) *The Property Machine*, Harmondsworth, Penguin.

Anderson, B.L. (1983) 'Financial institutions and the capital market on Merseyside in the eighteenth and nineteenth centuries', in Anderson, B.L. and Stoney, P.J. (eds) *Commerce, Industry and Transport: Studies in Economic Change on Merseyside*, Liverpool, Liverpool University Press, 26–59.

Anderson, G. (1983) 'The service occupations of nineteenth century Liverpool', in Anderson, B.L. and Stoney, P.J. (eds) *Commerce, Industry and Transport: Studies in Economic Change on Merseyside*, Liverpool, Liverpool University Press, 77–94.

Applebaum, W. (1932) *The Secondary Commercial Centres of Cincinnati*, Cincinnati,

Ohio, University of Cincinnati.
Applebaum, W. (1965) 'Can store location research be a science?', *Economic Geography*, 41, 234–7.
Archer, B.H. (1977) *Tourism Multipliers: The State of the Art*, Bangor, University of Wales Press.
Armstrong, R.B. (1972) *The Office Industry: Patterns of Growth and Location*, Cambridge, Mass., MIT Press.
Armstrong, R.B. (1979) 'National trends in office construction, employment and headquarter location in US metropolitan areas', in Daniels, P. W. (ed.) *Spatial Patterns of Office Growth and Location*, London, Bell, 61–94.
Arrow, K., Chenery, H., Minhas, B. and Solow, R. (1961) 'Capital labour substitution and economic efficiency', *Review of Economics and Statistics*, XLIII, 225–50.
Aschmann, M.P. (1976) 'Three miles from LA and New York: a study of intra-metropolitan office location near Chicago-O'Hare International Airport', Ph.D. thesis, University of Syracuse.
Ashcroft, B. and Swales, J.K. (1982a) 'The importance of the first round in the multiplier process: the impact of civil service dispersal', *Environment and Planning A*, 14, 429–44.
Ashcroft, B. and Swales, J.K. (1982b) 'Estimating the effects of government office dispersal', *Regional Science and Urban Economics*, 12, 81–98.
Ashcroft, R.J. (1981) 'Intra-metropolitan branch banking: Ottawa, 1930–1979', *Ontario Geography*, 17, 1–18.
Azimov, I. (1978) *The Naked Sun*, London, Granada.
Bach, L. (1980) 'Locational models for systems of private and public facilities based on concepts of accessibility and access opportunity', *Environment and Planning A*, 12, 301–20.
Bacon, R. and Eltis, W. (1976) *Britain's Economic Problem: Too Few Producers*, London, Macmillan.
Bagby, W.S. (1957) 'Automation in insurance', *Journal of Insurance*, 24, 158–67.
Bagdikian, B.H. (1971) *The Information Machines: Their Impact on Man and Media*, London, Harper & Row.
Bairoch, P. (1973) *Urban Unemployment in Developing Countries*, Geneva, International Labour Office.
Bank of Montreal (1956) *The Service Industries*, Hull, Clautier.
Bannon, M.J. (1972) 'The changing centre of gravity of office establishments within central Dublin, 1940 and 1970', *Irish Geography*, 6, 480–4.
Bannon, M.J. (ed.) (1973) *Office Location and Regional Development*, Dublin, An Foras Forbartha.
Bannon, M.J. (1978) 'Service functions, occupational change and regional policy for the 1980s', *Journal of the Institute of Public Administration in Ireland*, 26, 180–96.
Bannon, M.J. and Eustace, J.G. (1978) *The Role of the Tertiary Sector in Regional Policy: A Comparative Study – the Republic of Ireland*, Brussels, report prepared for the EEC Commission.
Bannon, M.J., Eustace, J.G. and Power, M. (1977) *Service-Type Employment and Regional Development*, Dublin, Stationery Office.
Barber, L. (1982) 'When pen pushing had to stop', *Sunday Times*, 24 October.
Barras, R. (1979a) 'The returns from office development and investment', Research Series No. 35, London, Centre for Environmental Studies.
Barras, R. (1979b) 'The development cycle in the City of London', Research Series No. 36, London, Centre for Environmental Studies.

Barras, R. (1981) 'The causes of the London office boom', paper presented at the First CES London Conference, February, mimeo.

Barras, R. (1983) 'A simple theoretical model of the office development cycle', *Environment and Planning A*, 15, 1381–94.

Barras, R. (1984) 'The office development cycle in London', *Land Development Studies*, 1, 35–50.

Barras, R. and Swann, J. (1983) The adoption and impact of information technology in the UK insurance industry, Report No. TCCR-83–014, London, The Technical Change Centre.

Barron, T. and Curnow, R. (1979) *The Future with Microelectronics*, London, Frances Pinter.

Bateman, M. (1976) 'Office location policy in the Paris region', paper presented at IBG Urban Geography Study Group Meeting, Keele, September, mimeo.

Bateman, M. (1985) *Office Development: A Geographical Analysis*, London, Croom Helm.

Batty, M. and Saether, E. (1972) 'A note on the design of shopping models', *Journal of the Royal Town Planning Institute*, 58, 303–6.

Baumol, W.J. (1965) *Business Behaviour, Value and Growth*, New York, Harcourt Brace.

Bearse, P.J. (1978) 'On the intra-regional diffusion of business services activity', *Regional Studies*, 12, 563–78.

Beaujeu-Garnier, J. (1974) 'Toward a new equilibrium in France?', *Annals, Association of American Geographers*, 64, 113–25.

Beaujeu-Garnier, J. and Bouveret-Gauer, M. (1979) 'Retail planning in France', in Davies, R.L. (ed.) *Retail Planning in the European Community*, Farnborough, Saxon House, 99–112.

Bell, D. (1974) *The Coming of Post-Industrial Society*, London, Heinemann.

Bell, D. (1979) 'Communications technology – for better or for worse', *Harvard Business Review*, 57, 20–45.

Bennet, P.R. (1980) *The Impact of Toronto International Airport on the Location of Offices*, Joint Program in Transportation, Research Report No. 72, Toronto, University of Toronto/York University.

Bennett, W.A. and Tucker, T.A. (1979) *Structural Determinants of the Size of the Service Sector: An International Comparison*, Working Paper No. 4, Canberra, Bureau of Industry Economics.

Bennison, D.J. and Davies, R.L. (1980) 'The impact of town centre shopping schemes in Britain: their impact on traditional retail environments', *Progress in Planning*, 14, 1–104.

Bergson, A (1964) *The Economics of Soviet Planning*, New Haven, Conn., Yale University Press.

Berry, B. and Cohen, Y. (1973) 'Decentralization of commerce and industry: the restructuring of metropolitan America', in Masott, L. and Hadden, J. (eds) *The Urbanization of the Suburbs*, Beverly Hills, Calif., Sage, 431–56.

Berry, B.J.L. (1967) *Geography of Market Centres and Retail Distribution*, Englewood Cliffs, N.J., Prentice-Hall.

Berry, B.J.L. (1973) *The Human Conseqences of Urbanization*, New York, St Martin's Press.

Berry, B.J.L. (1965) 'The retail component of the urban model', *Journal of the American Institute of Planners*, 31, 150–5.

Berry, B.J.L. (1975) 'The decline of the ageing metropolis: cultural bases and social processes', in Sternlieb, G. and Hughes, J. (eds) *Post Industrial America:*

Metropolitan Decline and Inter-Regional Job Shifts, State University of New Jersey, Rutgers, Centre for Urban Policy Research, 175–86.

Berry, B.J.L. and Garrison, W.L. (1958) 'Recent developments in central place theory', *Papers and Proceedings, Regional Science Association*, 4, 107–20.

Beyers, W.B. (1983) 'Services and industrial systems', paper presented at the Annual Meeting of the Association of American Geographers, Denver, Colo, 24–27 April.

Beyes, W.B., Alvine, M.J. and Johnson, E.K. (1985) *The Service Economy: Export of Services in the Central Puget Sound Region*, Seattle, Central Puget Sound Development District.

Bhalla, A.S. (1970) 'The role of services in employment expansion', *International Labour Review*, 101, 519–39.

Bies, S.C. (1977) 'The future of CBDs as financial centres in metropolitan areas: a demand analysis', *Journal of Regional Science*, 17, 431–40.

Bird, E. (1980) *Information Technology and the Office*, London, EOC.

Bird, J. (1977) *Centrality and Cities*, London, Routledge & Kegan Paul.

Birg, H. (1978) *Ein Ansatz zur Erklarung des regionalen Arbeitseinsatzes in den Dienstleistungssektoren aus der Siedlungsstruktur einer Region*, Berlin, Deutsches Institut fur Wirtschaftsforschen.

Bjorn-Anderson, E.M., Holst, O. and Mumford, E. (1982) *Information Society: For Richer, for Poorer*, Amsterdam, North Holland.

Black, J.T. (1978) *The Changing Economic Role of Central Cities*, Washington, DC, Urban Land Institute.

Black's Guides (1983) *Black's Guide to the Office Space Market: Suburban Manhattan*, Vol. 1, Red Bank, NJ, Black's Guides.

Blackaby, F. (ed.) (1978) *De-industrialization*, London, Heinemann.

Blackburn Health District (1979) *A Strategic Plan for Health Care in the Blackburn Health District*, Preston, Lancashire Area Health Authority.

Bluestone, B. and Harrison, B. T. (1982) *The Deindustrialization of America*, New York, The Free Press.

Bohland, J.R. and Frech, P. (1982) 'Spatial aspects of primary health care for the elderly', in Warnes, A.M. (ed.) *Geographical Perspectives on the Elderly*, New York, Wiley, 339–54.

Borchert, J.R. (1978) 'Major control points in American economic geography', *Annals of the Association of American Geographers*, 68, 214–32.

Boswell, J. (1969) *JS100 – The Story of Sainsbury's*, London, J. Sainsbury.

Bosworth, D. L. (ed.) (1983) *The Employment Consequences of Technological Change*, London, Macmillan.

Boulding, K. (1971) 'The boundaries of social policy', *Social Work*, 12, 7.

Boulding, K.E. (1956) 'General systems theory – the skeleton of science', *Management Science*, 2, 197–208.

Bowlby, S.R. (1979) 'Accessibility, mobility and shopping provision', in Goodall, B. and Kirby, A. (eds) *Resources and Planning*, Oxford, Pergamon, 293–324.

Bradley, J.E., Kirby, A.M. and Taylor, P.J. (1978) 'Distance decay and dental decay: a study of dental health among primary school children in Newcastle upon Tyne', *Regional Studies*, 12, 529–40.

Braun, E. and Macdonald, S. (1978) *Revolution in Miniature: The History and Impact of Semi-conductor Electronics*, London, Cambridge University Press.

Bridges, M. (1976) *The York Asda: A Study of Changing Shopping Patterns around a Superstore*, University of Manchester, Centre for Urban and Regional Planning.

British Tourist Authority (1975) *The Economic Significance of Tourism*, London, BTA.

Britton, J.N.H. (1975) 'Environmental adaptation of industrial plants: service linkages, locational environment and organization', in Hamilton, F.E.I. (ed.) *Spatial Perspectives on Industrial Organization and Decision Making*, London, Wiley, 363–90.

Brocard, M. (1972) 'Recherche scientifique et régions français', *Progrès Scientifique*, 152, 4–33.

Bromley, R. (1978) 'The urban informal sector: why is it worth discussing?', *World Development*, 6, 1033–9.

Bromley, R. and Gerry, C. (eds) (1979) *Casual Work and Poverty in Third World Cities*, New York, Wiley.

Brooks, R. (1983) 'Switch-on time at last for the electronic office', *Sunday Times*, 23 January.

Brown, L.A. (1968) *Diffusion Processes and Location: A Conceptual Framework and Bibliography*, Regional Science Research Institute Bibliography series, No. 4.

Browne, L.E. (1983) 'High technology and business services', *New England Economic Review*, July–August, 5–17.

Browning, H.L. and Singelmann, J. (1975) *The Emergence of a Service Society: Demographic and Sociological Aspects of the Sectoral Transformation of the Labour Force in the USA*, Springfield, Va, National Technical Information Service.

Buck, T.W. (1970) 'Shift and share analysis: a guide to regional policy?', *Regional Studies*, 4, 445–50.

Buckley, P.J. and Casson, M.C. (1976) *The Future of Multinational Enterprise*, London, Macmillan.

Bureau of Industry Economics (1980) *Features of the Australian Service Sector*, Canberra, Australian Government Publishing Service.

Burns, L.S. (1977) 'The location of the headquarters of industrial companies', *Urban Studies*, 14, 211–15.

Burns, L.S. and Healy, R.G. (1978) 'The metropolitan hierarchy of occupations: an economic interpretation of central place theory', *Regional Science and Urban Economics*, 8, 381–93.

Burns, S. (1976) *The Household Economy*, Boston, Mass., Beacon Press.

Burns, W. (1959) *British Shopping Centres: New Trends in Layout and Distribution*, London, Leonard Hill.

Burrows, E.M. and Town, S. (1971) *Office Services in the East Midlands*, Nottingham, East Midlands Economic Planning Council.

Burtenshaw, D., Bateman, M. and Ashworth, G.J. (1981) *The City in West Europe*, Chichester, Wiley.

Bussiere, Y. (1974) *Le Secteur tertiaire et les mécanismes de création d'emplois*, Toulouse, Institut d'Étude de l'Emploi de Toulouse.

Buswell, R.J. and Lewis, E.W. (1970) 'The geographical distribution of industrial research activity in the United Kingdom', *Regional Studies*, 4, 297–306.

Butera, F. and Bartezzaghi, E. (1983) 'Creating the right organizational environment', in Otway, H.J. and Peltu, M. (eds) *New Office Technology: Human and Organizational Aspects*, London, Frances Pinter, 102–19.

Cameron, R.E. (1963) 'The banker as entrepreneur', *Explorations in Entrepreneurial History*, 1, 50–5.

Carlstein, T., Parkes, D. and Thrift, N. (eds) (1978) *Human Activity and Time Geography*, London, Edward Arnold.

Cartwright, A. (1967) *Patients and their Doctors*, London, Routledge & Kegan Paul.

Catalano, A. and Barras, R. (1980) 'Office development in central Manchester', Research

Series 37, London, Centre for Environmental Studies.

Central Statistical Office (1979) *Standard Industrial Classification, Revised 1980*, London, HMSO.

Central Statistical Office (1983) *Social Trends, 14*, London, HMSO.

Champion, A.G. (1983) 'Population trends in the 1970s', in Goddard, J.B. and Champion, A.G. (eds) *The Urban and Regional Transformation of Britain*, London, Methuen, 187–214.

Chandler, G. (1968) *Four Centuries of Banking: Vol. 2, The Northern Constituent Banks*, London, Batsford.

Channon, D.F. (1978) *The Service Industries: Strategy, Structure and Financial Performance*, London, Macmillan.

Christaller, W. (1933) *Die Zentralen Orte in Suddeutschland*, Jena, Fischer; trans. Baskin, C.W. (1966) *Central Places in Southern Germany*, Englewood Cliffs, NJ, Prentice-Hall.

City of Liverpool (1982) *Economy 1982*, Planning Information Digest, Liverpool, City Planning Department.

City of Liverpool (1983) *City Centre Offices: Trends and Opportunities*, Liverpool, City Planning Department.

Civil Service Department (1973) *The Dispersal of Government Work from London*, Cmnd 5322, London, HMSO.

Clapp, J.M. (1980) The intrametropolitan location of office activities, *Journal of Regional Science*, 2, 387–99.

Clapp, J.M. (1983) 'A model of public policy toward office relocation', *Environment and Planning A*, 15, 1299–1310.

Clark, C. (1940) *The Conditions of Economic Progress*, London, Macmillan.

Clark, D. (1985) *Post-Industrial America: A Geographical Perspective*, London, Methuen.

Clarke, A.C. (1977) 'Communications in the second century of the telephone', in Poole, I.S. (ed.) *The Telephone's First Century – and Beyond*, New York, Thomas Crowell, 83–111.

Clarke, W.M. (1965) *The City in the World Economy*, London, Institute of Economic Affairs.

Clarke, B. and Bolwell, L. (1968) 'Attractiveness as part of retail potential models', *Journal of the Royal Town Planning Institute*, 54, 477–8.

Cleland, E.A., Stimson, R.J. and Goldsworthy, A.J. (1977) *Suburban Health Care Behaviour in Adelaide*, Monograph Series 2, Adelaide, Flinders University, Centre for Applied Social and Survey Research.

Coates, J.F. (1977) 'Technological change and future growth: issues and opportunities', *Technological Change and Forecasting*, 11, 49–74.

Code, W.R., Morris, P. and Wilder, K. (1981) 'The decentralization of office space in metropolitan Toronto', London, University of Western Ontario, Department of Geography, mimeo.

Cohen, R.B. (1977) 'Multinational corporations, international finance, and the sunbelt', in Perry, D.C. and Watkins, A.J. (eds) *The Rise of the Sunbelt Cities*, Beverly Hills, Calif., Sage, 191–210.

Cohen, R.B. (1979) 'The changing transactions economy and its spatial consequences', *Ekistics*, 274, 7–15.

Committee on Invisible Exports (1976) *Annual Report, 1975–76,* London, HMSO.

Committee on Invisible Exports (1981) *Annual Report, 1980–81*, London, HMSO.

Cookson, C. (1983a) 'Home shopping via TV gets under way', *The Times*, 2 March.

Cookson, C. (1983b) 'The telephone to put a smile on your deal', *The Times*, 25 June.

Coombs, R.W. and Green, K. (1981) 'Microelectronics and the future of service employment', *Service Industries Review*, 1, 4–21.

Cooper, L. (1963) 'Location – allocation problems', *Operations Research*, 11, 331–43.

Corey, K.E. (1982) 'Transactional forces and the metropolis', *Ekistics*, 297, 416–23.

Cottrell, P.L. (1975) *British Overseas Investment in the Nineteenth Century*, London, Macmillan.

Cowan, P., Fine, D., Ireland, J., Jordan, C., Mercer, D. and Sears, A. (1969) *The Office: A Facet of Urban Growth*, London, Heinemann.

Cowell, D.W. (1983) 'International marketing of services', *Service Industries Journal*, 3, 308–28.

Cowing, T.G. and Holtmann, A.G. (1976) *The Economics of Local Public Service Consolidation*, Farnborough, Teakfield.

Creamer, D. (1976) *Overseas Research and Development by United States Multinationals*, New York, Conference Board.

Crozier, M. (1965) *The World of the Office Worker*, Chicago, University of Chicago Press.

Crozier, M. (1983) 'Implications for the organization', in Otway, H.J. and Peltu, M. (eds) *New Office Technology: Human and Organizational Aspects*, London, Frances Pinter, 86–101.

Crum, R. and Gudgin, G. (1977) *Non-Production Activities in UK Manufacturing Industry*, Brussels, European Economic Commission.

Cuzzort, R.P. (1955) *Suburbanization of Service Industries within Standard Metropolitan Areas*, Oxford, Ohio, Scripps Foundation for Research in Population Problems.

Cyert, R.M. and March, J.G. (1963) *A Behavioural Theory of the Firm*, Englewood Cliffs, N.J., Prentice-Hall.

Daly, M.T. (1982) *Sydney Boom, Sydney Bust*, Sydney, Allen & Unwin.

Damesick, P.J. (1979) 'Office location and planning in the Manchester conurbation', *Town Planning Review*, 50, 436–60.

Damesick, P.J., Howick, C. and Key, T. (1982) 'Economic regeneration of the inner city: manufacturing industry and office development in inner London', *Progress in Planning*, 18, 133–267.

Daniels, P.W. (1975) *Office Location*, London, Bell.

Daniels, P.W. (1977) 'Office location in the British conurbations: trends and strategies', *Urban Studies*, 14, 261–74.

Daniels, P.W. (1980) *Office Location and the Journey to Work: A Comparative Study of Five Urban Areas*, Farnborough, Gower.

Daniels, P.W. (1982) 'An exploratory study of office location behaviour in Greater Seattle', *Urban Geography*, 3, 58–78.

Daniels, P.W. (1983a) 'Business services in British provincial cities: location and control', *Environment and Planning A*, 15, 1101–20.

Daniels, P.W. (1983b) 'Service industries: supporting role or centre stage', *Area*, 15, 301–9.

Daniels, P.W. (1983c) 'Modern technology in provincial offices: some empirical evidence', *Service Industries Journal*, 3, 21–41.

Daniels, P.W. (1984) 'CBD versus suburbs: metropolitan office location in North America', *Liverpool Geographer*, 8, 5–19.

Daniels, P.W. (1985a) 'Some changing horizons for the geography of service industries', in Pacione, M. (ed.) *Progress in Industrial Geography*, London, Croom Helm, 111–41.

Daniels, P.W. (1985b) 'Office location in Australian metropolitan areas: concentration or dispersal?', *Australian Geographical Studies* (in press).

Daniels, P.W. (1985c) 'Producer services in the post-industrial space economy', in Martin, R. and Rowthorn, B. (eds) *Deindustrialization and the British Space Economy*, London, Macmillan.

Danzin, A. (1983) 'The nature of new office technology', in Otway, H.J. and Peltu, M. (eds) *New Office Technology: Human and Technological Aspects*, London, Pinter, 19–36.

Davey, J. (1972) *The Office Industry in Wellington: A Study of Contact Patterns, Location and Employment*, Wellington, Ministry of Works.

Davies, R.L. (1972a) 'The location of service activity', in Chisholm, M. and Rodgers, B. (eds) *Studies in Human Geography*, London, Heinemann, 125–71.

Davies, R.L. (1972b) 'Structural models of retail distribution: analogies with settlement and land-use theories', *Transactions of the Institute of British Geographers*, 57, 59–82.

Davies, R.L. (1976) *Marketing Geography: With Special Reference to Retailing*, London, Methuen.

Davies, R.L. (ed.) (1979) *Retail Planning in the European Community*, Farnborough, Saxon House.

Davies, R.L. (1984) *Retail and Commercial Planning*, London, Croom Helm.

Davies, R.L. and Bennison, D.J. (1978) 'Retailing in the city centre: the characters of shopping streets', *Tijdschrift voor Economische en Sociale Geografie*, 69, 270–5.

Davies, R.L. and Kirby, D.A. (1980) 'Retail organization', in Dawson, J.A. (ed.) *Retail Geography*, London, Croom Helm, 156–92.

Davis, C.D. and Hutton, T.A. (1981) 'Some planning implications of the expansion of the urban service sector', *Plan Canada*, 21, 15–23.

Dawson, J.A. (1974) 'The suburbanization of retail activity', in Johnson, J.H. (ed.) *Suburban Growth: Geographical Processes at the Edge of the Western City*, London, Wiley, 155–73.

Dawson, J.A. (ed.) (1979) *The Marketing Environment*, London, Croom Helm.

Dawson, J.A. (ed.) (1980) *Retail Geography*, London, Croom Helm.

Dawson, J.A. (1983a) *Shopping Centre Development*, London, Longman.

Dawson, J.A. (1983b) 'Independent retailing in Great Britain: dinosaur or chameleon?', *Retail and Distribution Management*, 11, 29–32.

Dawson, J.A. and Kirby, D.A. (1979) *Small Scale Retailing in the UK*, Farnborough, Saxon House.

Deakin, B.M. and George, K.D. (1965) *Productivity Trends in the Service Industries, 1948–63*, London, Cambridge University Press.

Delehanty, G. (1968) *Non-Production Workers in US Manufacturing*, Amsterdam, North Holland.

Department of Employment (1969) *British Labour Statistics, 1886–1968*, London, HMSO.

Department of Employment (1982) 'AT&C employment in manufacturing industries, Great Britain', *Department of Employment Gazette*, 90, 1.10 (S.14).

Department of Employment (1983) 'Standard Industrial Classification – revised 1980', *Department of Employment Gazette*, 91, 118–22.

Department of Health and Social Security (1980) *Health and Personal Social Services: Statistics for England, 1978*, London, HMSO.

Department of Trade and Industry (1983) *British Business*, 10.

De Smidt, M. (1983) *Offices in the Netherlands: Locational Tendencies and Planning*,

Utrecht, University of Utrecht, Department of Geography.

De Smidt, M. (1984) 'Office location and the urban functional mosaic: a comparative study of five cities in the Netherlands', *Tijdschrift voor Economische en Sociale Geografie*, 75, 110–22.

De Smidt, M. (1985) 'Relocation of Government services in the Netherlands', *Tijdschrift voor Economische en Sociale Geografie*, 76, 232–6.

De Vise, P. (1973) 'Misused and misplaced hospitals and doctors: a locational analysis of the urban health care crisis', CCG Resource Paper No. 22, Washington, DC, Association of American Geographers.

De Vito, M.J. (1980) 'Retailing plays a key role in downtown renaissance', *Journal of Housing*, 37, 197–200.

Dhillon, H., Doerman, A. and Walcoff, P. (1978) 'Tele-medicine and rural primary health care: an analysis of the impact of telecommunications technology', *Socio-Economic Planning Sciences*, 12, 37–41.

Doctor, K.C. and Gallis, H. (1964) 'Modern sector employment in Asian countries', *International Labour Review*, XC, 544–68.

Douglas, S. (1981) 'Business service provision in Newcastle upon Tyne', M.Phil. thesis, Newcastle upon Tyne Polytechnic, Department of Geography.

Driscoll, J.W. (1979) *Office Automation: The Organizational Re-Design of Office Work*, Cambridge, Mass., MIT Centre for Information Systems Research.

Drucker, P.F. (1970) *Technology, Management and Society*, London, Heinemann.

Drury, P. (1983) 'Some spatial aspects of health service developments: the British experience', *Progress in Human Geography*, 7, 60–77.

Dunning, J.H. and Norman, G. (1979) *Factors Influencing the Location of Offices of Multinational Enterprise*, Research Paper No. 8, London, Location of Offices Bureau.

Dunning, J.H. and Norman, G. (1983) 'The theory of the multinational enterprise: an application to multinational office location', *Environment and Planning A*, 15, 675–92.

Dymmel, M.D. (1979) 'Technology and telecommunications: its effect on labour and skills', *Monthly Labour Review*, 102, 181–90.

Earickson, R. (1970) *The Spatial Behaviour of Hospital Patients: A Behavioural Approach to Spatial Interaction in Metropolitan Chicago*, Research Paper No. 124, University of Chicago, Department of Geography.

Edgington, D.W. (1982a) 'Organizational and technological change and the future of the central business district: an Australian example', *Urban Studies*, 19, 281–92.

Edgington, D.W. (1982b) 'Changing patterns of central business district office activity in Melbourne', *Australian Geographer*, 15, 231–42.

Edwards, L.E. (1983) 'Towards a process model of office-location decision making', *Environment and Planning A*, 15, 1327–42.

Egan, D. (1983) 'The location of service outlets: an economic perspective', *Service Industries Journal*, 3, 180–90.

Emi, K. (1978) *Essays on the Service Industry and Social Security in Japan*, Tokyo, Kinokuniya.

Erickson, R.A. (1983) 'The evolution of the suburban space economy', *Urban Geography*, 4, 95–121.

Evans, A.W. (1967) 'Myths about employment in central London', *Journal of Transport Economics and Policy*, 1, 214–25.

Evans, A.W. (1973) 'The location of headquarters of industrial companies', *Urban Studies*, 10, 387–95.

Everson, J.A. and Fitzgerald, B.P. (1969) *Settlement Patterns*. London, Longman.
Ezeikel, H. (1976) *Services*, Delhi, Macmillan Co. of India.
Facey, M. and Smith, G. (1968) *Offices in a Regional Centre: A Study of Office Location in Leeds*, Research Report No. 2, London, Location of Offices Bureau.
Fedida, S. and Malik, R. (1979) *The Viewdata Revolution*, London, Associated Business Press.
Fernie, J. (1977) 'Office linkages and location: an evaluation of patterns in three cities', *Town Planning Review*, 48, 78–89.
Fernie, J. (1979) 'Office activity in Edinburgh', *Ekistics*, 46, 25–33.
Fernie, J. and Carrick, R.J. (1983) 'Quasi-retail activity in Britain: planning issues and policies', *Service Industries Journal*, 3, 93–104.
Firth, K. (1976) *The Distribution Services Industry: Operator and User Attitudes*, Cranfield, National Materials Handling Centre.
Fisher, A.G.B. (1935) *The Clash of Progress and Security*, London, Macmillan.
Forester, T. (ed.) (1982) *The Microelectronics Revolution*, Oxford, Blackwell.
Forester, T. (ed.) (1985) *The Information Technology Revolution*, Oxford, Blackwell.
Fothergill, S. and Gudgin, G. (1982) *Unequal Growth: Urban and Regional Employment Change in the UK*, London, Heinemann.
Freeman, C. (1974) *The Economics of Industrial Location*, Harmondsworth, Penguin.
Freestone, R. (1977) 'Provision of child care facilities in Sydney', *Australian Geographer*, 13, 318–25.
Frey, L. (1975) *L'occupazione terziara: con particolare riguardo all'Italia*, Milano, F. Angeli.
Friedrich, P. (1984) *Erfahrungsbericht über Behördenverlagerung*, Munich, Bayerisches Statsministerium für Landesentwickelung und Umweltfragen.
Frobel, F., Heinrichs, J. and Kreize, O. (1980) *The New International Division of Labour: Structural Unemployment in Industrialized Countries and Industrialization in Developing Countries*, London, Cambridge University Press.
Fuchs, C. (1983) 'Developers and users of office space: an examination of the location decision processes in the Washington DC area, 1981–83', M.A. thesis, University of Maryland, Department of Geography.
Fuchs, V.R. (1965) 'The growing importance of the service industries', *Journal of Business of the University of Chicago*, 38, 360–62.
Fuchs, V.R. (1968) *The Service Economy*, New York, Bureau of Economic Research.
Fuchs, V.R. (ed.) (1969) *Production and Productivity in the Service Industries*, New York, National Bureau of Economic Research.
Fuchs, V.R. (1977) *The Service Industries and US Economic Growth since World War II*, Working Paper No. 211, Stanford, Calif., National Bureau of Eonomic Research.
Gad, G.H.K. (1979) 'Face-to-face linkages and office decentralization potentials: a study of Toronto', in Daniels, P.W. (ed.) *Spatial Patterns of Office Growth and Location*, London, Wiley, 277–324.
Galbraith, J.K. (1967) *The New Industrial State*, London, Hamish Hamilton.
Galenson, W. (1963) 'Economic development and the sectoral expansion of employment', *International Labour Review*, 87, 505–19.
Galtman, R.E. and Weiss, T.J. (1969) 'The service industries in the nineteenth century', in Fuchs, V.R. (ed.) *Production and Productivity in the Service Industries*, New York, National Bureau of Economic Research, 287–382.
Garner, B.J. (1966) 'The internal structure of retail nucleations', Research Series No. 12, Northwestern University, Department of Geography.
Garner, B.J. (1967) 'Models of urban geography and settlement location', in Chorley,

R.J. and Haggett, P. (eds) *Models in Geography*, London, Methuen, 303–60.
Gartner, A., Nixon, R.A. and Reissman, F. (1973) *Public Service Employment: An Analysis of its History, Problems and Prospects*, New York, Praeger.
Gemmell, N. (1982) 'Economic development and structural change: the role of the service sector', *Journal of Development Studies*, 19, 37–66.
George, R.E., Diphand, C.R. and Storey, R.G. (1980) 'The location of offices', *La Revue Canadienne des Sciences Régionales*, 3, 71–91.
Gerschenkron, A. (1962) *Economic Backwardness in Historical Perspective*, Cambridge, Mass., Harvard University Press.
Gershuny, J. (1978) *After Industrial Society: The Emerging Self-Service Economy*, London, Macmillan.
Gershuny, J. and Miles, I. (1983) *The New Service Economy: The Transformation of Employment in Industrial Societies*, London, Frances Pinter.
Gersuny, W. and Rosengren, W.R. (1973) *The Service Society*, Cambridge, Mass., MIT Press.
Ginsberg, E. (1968) *Manpower Strategy for the Metropolis*, New York, Columbia University Press.
Goddard, J.B. (1967) 'Changing office location patterns within central London', *Urban Studies*, 4, 276–84.
Goddard, J.B. (1971) 'Office communication and office location: a review of current research', *Regional Studies*, 5, 263–80.
Goddard, J.B. (1973) 'Office linkages and location: a study of communications and spatial patterns in central London', *Progress in Planning*, 1, 109–232.
Goddard, J.B. (1975) *Office Location in Urban and Regional Development*, London, Oxford University Press.
Goddard, J.B. (1980) 'Technology forecasting in a spatial context', *Futures*, 12, 90–105.
Goddard, J.B. and Morris, D. (1976) 'The communications factor in office decentralization', *Progress in Planning*, 6, 1–80.
Goddard, J.B. and Smith, I.J. (1978) 'Changes in corporate control in the British urban system, 1972–77', *Environment and Planning A*, 10, 1073–84.
Goddard, J.B. and Spence, N.A. (1976) 'A national perspective on employment change in urban labour markets and questions about the future of the provincial conurbations', paper presented at the Centre for Environmental Studies Conference on Employment in the Inner City, University of York, June.
Gold, J.R. (1980) *An Introduction to Behavioural Geography*, Oxford, Oxford University Press.
Goodwin, W. (1965) 'The management center in the United States', *Geographical Review*, 55, 1–16.
Gottman, J. (1983) *The Coming of the Transactional City*, College Park, Md., University of Maryland Institute of Urban Studies.
Gould, P.R. and Leinbach, T.R. (1966) 'An approach to the geographical assignment of hospital services', *Tijdschrift voor Economische en Sociale Geografie*, 57, 203–6.
Gould, W.T.S. and Hodgkiss, A.G. (eds) (1982) *The Resources of Merseyside*, Liverpool, Liverpool University Press.
Government of India (1969) *A Survey of Expenditure, Composition and Reaction Pattern of Foreign Tourism in India*, Delhi, Ministry of Tourism and Civil Aviation, Department of Tourism.
Government of Sweden (1982) *Swedish Regional Policy*, Stockholm, Ministry of Industry.
Gower Economic Publications (1977) *Retail Trade Developments in Great Britain*,

London, Gower.

Greater Manchester Council (1975) *County Structure Plan, Report of Survey: Employment and the Economy*, Manchester, County Planning Department.

Green, K., Coombs, R.W. and Molroyd, K. (1980) *The Effects of Microelectronic Technologies on Employment Prospects*, Westmead, Gower.

Greenfield, H.I. (1966) *Manpower and the Growth of Producer Services*, New York, Columbia University Press.

Greenhut, M.L. (1956) *Plant Location in Theory and Practice: The Economics of Space*, Chapel Hill, NC, University of North Carolina Press.

Grit, S. and Korteweg, P.J. (1976) 'Perspectives on office relocation in the Netherlands', *Tijdichrift voor Economische en Sociale Geografie*, 67, 2–14.

Guiliano, V.E. (1982) 'The mechanization of office work', *Scientific American*, 247, 149–64.

Gujerati, D. and Dors, L. (1972) 'Production and non-production workers in US manufacturing industries', *Industrial and Labour Relations Review*, 26, 660–69.

Guy, C.M. (1977) 'A method of examining and evaluating the impact of major retail developments upon existing shops and their users', *Environment and Planning A*, 9, 491–504.

Hagerstrand, T. (1967) *Innovation Diffusion as a Spatial Process*, Chicago, University of Chicago Press.

Hagerstrand, T. (1970) 'What about people in regional science?', *Papers and Proceedings of the Regional Science Association*, 24, 7–21.

Hall, J.M. (1970) 'Industry grows where the grass is greener', *Area*, 1, 40–6.

Hall, P. and Markusen, A. (1985) *Silicon Landscapes*, Winchester, Mass., Allen & Unwin.

Hall, P.G., Thomas, R., Gracey, H. and Drewett, R. (1973) *The Containment of Urban England*, 2 vols, London, Allen & Unwin.

Hall, R.K. (1972) 'The movement of offices from central London', *Regional Studies*, 6, 385–92.

Hamilton, F.E.I. (1979) 'Spatial structure in East European cities', in French, R.H. and Hamilton F.E.I. (eds) *The Socialist City*, Chichester, Wiley, 195–261.

Hanna, M. (1978) *Tourism Multipliers in Britain*, London, English Tourist Board.

Harkness, R.C. (1977) *Technology Assessment of Telecommunications/Transportation Interactions*, Menlo Park, Calif., Stanford Research Institute.

Harper, R.A. (1982) 'Metropolitan areas as transactional centres', in Christian, C.M. and Harper, R.A. (eds) *Modern Metropolitan Systems*, Columbus, Ohio, Charles Merrill, 87–109.

Harris, B. (1964) 'Models of locational equilibrium for retail trade', *Journal of Regional Science*, 5, 31–5.

Hart, J.T. (1971) 'The inverse care law', *Lancet*, 1, 405–12.

Hartwell, R.M. (1973) 'The service revolution: the growth of services in the modern economy 1750–1914', in Cipolla, C.M. (ed.) *The Fontana Economic History of Europe*, Vol. 3, London, Fontana, 358–96.

Harvey, D. (1973) *Social Justice and the City*, London, Edward Arnold.

Haynes, R.M. and Bentham, C.G. (1979) *Community Hospitals and Rural Accessibility*, Farnborough, Saxon House.

Head, J.G. (1974) *Public Goods and Public Welfare*, Durham, NC, Duke University Press.

Heenan, D.A. (1977) 'Global cities of tomorrow', *Harvard Business Review*, 55, 79–92.

Henize, J. (1981) 'Evaluating the employment impact of information technology', *Technological Forecasting and Social Change*, 20, 41–61.

Herbert, D.T. (1976) 'Urban education: problems and policies', in Herbert, D.T. and Johnston, R.J. (eds) *Social Areas in Cities, Vol. 2, Spatial Perspectives on Problems and Policies*, Chichester, Wiley, ch. 3.

Herbert, D.T. and Thomas, C.J. (1982) *Urban Geography: A First Approach*, Chichester, Wiley.

Heyel, C. (1969) *Computers, Office Machines and the New Information Technology*, London, Macmillan.

Hiestand, D.L. (1977) 'Recent trends in the not-for-profit sector', in *Research Papers: Sponsored by the Commission on Private Philanthropy and Public Needs*, Washington, DC, US Department of the Treasury, 333–7.

Hill, D.M. (1965) 'A growth allocation model for the Boston region', *Journal of the American Institute of Planners*, 31, 111–20.

Hillman, M. and Whalley, A. (1977) 'Fair play for all: study of access for sport and informal recreation', Political and Economic Planning Broadsheet No. 57, London, Political and Economic Planning.

Hills, P.J. (ed.) (1982) *Trends in Information Transfer*, London, Frances Pinter.

Hirschorn, L. (1974) *Toward a Political Economy of the Service Society*, Berkeley, Calif., Institute of Urban and Regional Development.

Hoare, A.G. (1973) 'International airports as growth poles: a case study of Heathrow airport', *Transactions of the Institute of British Geographers*, 63, 75–97.

Hodgart, R.L. (1978) 'Optimizing access to public services: a review of problems, models and methods of locating central facilities', *Progress in Human Geography*, 2, 17–48.

Hogg, J.M. (1968) 'The siting of fire stations', *Operations Research Quarterly*, 19, 275–87.

Holly, B.P. and Cadigan, J.L. (1981) 'Real estate development and the office location decision', Kent, Ohio, Kent State University, Department of Geography, mimeo.

Hood, N. and Young, S. (1979) *The Economics of the Multinational Enterprise*, Oxford, Pergamon.

Hoover, E.M. (1948) *The Location of Economic Activity*, New York, McGraw-Hill.

Hopkins, E.J., Pye, A.M., Soloman, M. and Soloman, S. (1968) 'The relation of patients' age, sex and distance from surgery to the demand on the family doctor', *Journal of the Royal College of General Practitioners*, 16, 368–78.

Hopkins, M. (1983) 'Employment trends in developing countries, 1960–80 and beyond', *International Labour Review*, 122, 461–78.

Hornik, J. and Feldman, L. (1982) 'Retailing implications of the do-it-yourself movement', *Retail and Distribution Management*, 10, 44–9.

Hotelling, H. (1929) 'Stability in competition', *Economic Journal*, 39, 41–57.

Howells, J.R.L. (1984) 'The location of research and development: some observations and evidence from Britain', *Regional Studies*, 18, 13–29.

Hoyt, H. (1933) *One Hundred Years of Land Values in Chicago*, Chicago, University of Chicago Press.

Hubbard, R.K.B. and Nutter, D.S. (1982) 'Service sector employment on Merseyside', *Geoforum*, 13, 209–35.

Huff, D.L. (1963) 'A probability analysis of shopping centre trade areas', *Land Economics*, 53, 81–9.

Hughes, J. and Thorne, E. (1975) *Service Industries in Metropolitan Areas*, University of Glasgow, Department of Social and Economic Research.

Husain, M.S. (1980) 'Office relocation in Hamburg: the City-Noord project', *Geography*, 65, 131–4.

Hymer, S. (1972) 'The multinational corporation and the law of uneven development', in Bhagwati, J.N. (ed.) *Economics and World Order*, London, Macmillan, 113–40.

International Labour Office (1972) *Employment, Incomes and Equality*, Geneva, International Labour Office.

International Labour Office (1977) *Labour Force Estimates, 1950–70. Vol. V, World Summary*, Geneva, International Labour Office.

Isard, W. (1956) *Location and Space-Economy: A General Theory Relating to Industrial Location, Market Areas, Land Use, Trade and Urban Structure*, New York, Wiley.

Isard, W. (1960) *Methods of Regional Analysis*, Cambridge, Mass., MIT Press.

Isserman, A.M. (1977) 'The location quotient approach to measuring regional economic impacts', *Journal of the American Institute of Planners*, 43, 33–41

Jeanneret, P.H., Hussy, J., Bailly, A., Maillat, D. and Rey, M. (1984) *Le Tertiare Moteur dans la petite et moyenne ville en Suisse: le bas d'Aigle et de Delemont*, Lausanne, Communauté d'Etudes pour l'Aménagement du Territoire.

Johnson, P. (1978) 'Policies towards small firms: time for caution', *Lloyds Bank Review*, 129, 1–11.

Johnston, R.J. (1964) 'The measurement of a hierarchy of central places', *Australian Geographer*, 9, 315–17.

Jones, B.D. (1977) 'Distribution considerations in models of government service provision', *Urban Affairs Quarterly*, 12, 291–312.

Jones, D. (1984) 'US experiments with home banking', *Banker*, 134, 61, 63, 65–7.

Jones, S.R. (1981) *Accessibility Measures: A Literature Review*, Report 967, Crowthorne, Transport and Road Research Laboratory.

Joseph, A. and Phillips, D. (1984) *Accessibility and Utilization: Geographical Perspectives on Health Care Delivery*, London, Harper & Row.

Jussawalla, M. (1978) *T3: Transportation/Telecommunications/Trade-offs*, Honolulu, East–West Communications Institute.

Kaldor, N. (1966) *Causes of the Slow Rate of Economic Growth in the UK*, London, Cambridge University Press.

Kay, W. (1985) 'How the City laid its foundations', *The Times*, 4 January.

Keeble, D.E. (1968) 'Industrial decentralization and the metropolis: the NW London case', *Transactions of the Institute of British Geographers*, 44, 1–54.

Keeble, D.E., Owens, P.L. and Thompson, C. (1982) *Centrality, Peripherality and EEC Regional Development*, London, HMSO.

Kellerman, A. (1984) 'Telecommunications and the geography of metropolitan areas', *Progress in Human Geography*, 8, 222–46.

Kellerman, A. (1985) 'The evolution of service economies: a geographical perspective', *Professional Geographer*, 37, 133–43.

Kelley, E. (1956) *Locating Planned Regional Shopping Centres*, Connecticut, Eno Foundation for Traffic Control.

Kendall, P.M.M. (1979) *The Impact of Chip Technology on Employment and the Labour Market*, London, Metra Consulting.

Kent County Council (1975) *Kent County Structure Plan – Aspect Report on Employment,* Maidstone, County Planning Department.

King, L.J. (1962) 'A quantitative expression of the pattern of urban settlements in selected areas of the US', *Tijdschrift voor Economische en Sociale Geografie*, 53, 1–7.

Kirby, A. (1979) *Education, Health and Housing: An Empirical Investigation of Resource Accessibility*, Saxon House, Farnborough.

Kirby, D.A. (1975) 'The small shop in Britain', *Town and Country Planning*, 43, 496–500.

Kivell, P.T. and Shaw, G. (1980) 'The study of retail location', in Dawson, J.A. (ed.) *The Study of Retail Geography*, London, Croom Helm, 95–155.

Knox, F. (1969) 'Service employment in the West Midlands', *Town Planning Review*,

40, 69–79.

Knox, P. (1978) 'The intraurban ecology of primary medical care: patterns of accessibility and their policy implications', *Environment and Planning A*, 10, 415–35.

Knox, P. (1982a) *Urban Social Geography: An Introduction*, London, Longman.

Knox, P. (1982b) 'Residential structure, facility location and patterns of accessibility', in Cox, K.R. and Johnston, R.J. (eds) *Conflict, Politics and the Urban Scene*, London, Longman, 62–87.

Knox, P. and Pacione, M. (1980) 'Locational behaviour, place preferences and the inverse care law in the distribution of primary health care', *Geoforum*, 11, 43–55.

Kornblau, C. (ed.) (1968) *Guide to Store Location Research*, New York, Addison-Wesley.

Koutsopolos, K.C. and Schmidt, C.G. (1976) 'Mobility constraints of the carless', *Traffic Quarterly*, 30, 67–83.

Kumar, K. (1978) *Prophecy and Progress: The Sociology of Industrial and Post-industrial Society*, Harmondsworth, Penguin.

Kuznets, S. (1938) *Commodity Flow and Capital Formation*, New York, National Bureau of Economic Research.

Kuznets, S. (1957) 'Quantitative aspects of the economic growth of nations: iii, industrial distribution of income and labour force by States, 1919–21 to 1955', *Economic Development and Cultural Change*, 7, 150–60.

Kuznets, S. (1966) *Modern Economic Growth: Rate, Structure and Spread*, New Haven, Conn., Yale University Press.

Kuznets, S. (1971) *Economic Growth of Nations: Total Output and Production Structure*, Cambridge, Mass., Belknap Press.

Lakshmanan, T.R. and Hansen, W.G. (1965) 'A retail market potential model', *Journal of the American Institute of Planners*, 31, 134–43.

Langdale, J.V. (1982) 'Telecommunications in Sydney: towards an information economy', in Cardew, R.V. (ed.) *Why Cities Change: Urban Development and Economic Change in Sydney*, Sydney, Allen & Unwin, 77–94.

Lankford, P.M. (1971) 'The changing location of physicians', *Antipode*, 3, 68–72.

Lathrop, G.T. and Hamburg, J.R. (1965) 'An opportunity–accessiblity model for allocating regional growth', *Journal of the American Institute of Planners*, 31, 95–103.

Lawton, R. (1974) 'England must find room for more', *Geographical Magazine*, 46, 179–84.

Lawton, R. and Pooley, C.G. (1976) *The Social Geography of Merseyside in the Nineteenth Century*, University of Liverpool, Department of Geography, Final Report to SSRC.

Lea, A.C. (1973) *Location–Allocation Systems: An Annotated Bibliography*, Discussion Paper No. 13, Toronto, University of Toronto, Department of Geography.

Lee, C.H. (1979) *British Regional Employment Statistics, 1841–1971*, London, Cambridge University Press.

Lee, C.H. (1984) 'The service sector, regional specialization, and economic growth in the Victorian economy', *Journal of Historical Geography*, 10, 139–56.

Lee, M., Jones, P. and Peach, C. (1973) *Caerphilly Hypermarket Study*, Research Report No. 1, London, Donaldson.

Leigh, R. (1970) 'The use of location quotients in urban economic base studies', *Land Economics*, 48, 202–6.

Leigh, R. and North, D. (1978) 'The spatial consequences of takeovers in some British industries and their implications for regional development', in Hamilton, F.E.I. (ed.) *Contemporary Industrialization: Spatial Analysis and Regional Development*,

London, Longman, 158–81.

Lengelle, M. (1966a) *The Growing Importance of the Service Sector in Member Countries*, Geneva, OECD.

Lengelle, M. (1966b) *La Révolution Tertiare*, Paris, Genin.

Leontief, W. (1966) *Input–Output Economics*, New York, Oxford University Press.

Levitt, T. (1976) 'The industrialization of service', *Harvard Business Review*, September/October, 63–74.

Lewis, J.P. and Traill, A.L. (1968) 'The assessment of shopping potential and the demand for shops', *Town Planning Review*, 38, 317–26.

Lewis, R. (1973) *The New Service Society*, London, Longman.

Lineberg, R. (1977) *Equality and Urban Policy*, Beverly Hills, Calif., Sage.

Livesey, F. and Hall, R.J. (1981) *Retailing: Development and Prospects to 1985*, London, Staniland Hall.

Lloyd, P.E. and Dicken, P. (1977) *Location in Space: A Theoretical Approach to Economic Geography*, London, Harper & Row.

Lluch, C. (1975) *Los Servicios españoles del futuro*, Madrid, Guadlaña de Pubblicaciónes.

Location of Offices Bureau (1975) *Office Relocation: Facts and Figures*, London, LOB.

Location of Offices Bureau (1976) *Annual Report, 1974–75*, London, LOB.

Location of Offices Bureau (1977) *Annual Report, 1975–76*, London, LOB.

Location of Offices Bureau (1978) *Annual Report, 1976–77*, London, LOB.

Logan, M.I. (1966) 'Location behaviour of manufacturing firms in urban areas', *Annals of the Association of American Geographers*, 56, 451–66.

London Borough of Hammersmith (1979) *Junior Office Jobs in Hammersmith*, Research Report No. 34, London, London Borough of Hammersmith.

Long, S.K., Witte, A.D., Tauchen, H. and Archer, W. (1984) 'The location of office firms', paper presented at Annual Meeting of the Regional Science Association, Denver, Colo., November.

Lord, T.D. (1984) 'Shifts in the wholesale trade status of US metropolitan areas', *Professional Geographer*, 36, 51–63.

Lösch, A. (1954) *The Economics of Location*, New Haven, Conn., Yale University Press.

Lowe, J. (1983) 'Cash and carry: what prospects for growth?', *Retail Distribution and Management*, 11, 21–5.

Lowry, I.S. (1964) *A Model of Metropolis*, Santa Monica, Calif., Rand Corporation.

Lozano, B. (1983) 'Informal sector workers: walking out the system's front door?', *International Journal of Urban and Regional Research*, 7, 340–62.

McAllister, D. (1976) 'Equity and efficiency in public facility location', *Geographical Analysis*, 8, 47–64.

McDonald, J.F. (1975) 'Some causes of the decline of central business district retail sales in Detroit', *Urban Studies*, 12, 229–33.

McKee, W. (1981) 'Office development in the London Borough of Hammersmith and Fulham', paper presented at the First CES London Conference, February.

McKeever, J.R. (1957) *Shopping Centres Revisited*, Technical Bulletin No. 30, Washington, DC, Urban Land Institute.

McKinnon, A.C. (1981) *The Historical Development of Food Manufacturers' Distribution Systems*, Occasional Paper No. 7, University of Leicester, Department of Geography.

McKinnon, A.C. (1983) 'The development of warehousing in England', *Geoforum*, 14, 389–99.

Malecki, E.J. (1979) 'Locational trends in R & D by large US corporations, 1965–77', *Economic Geography*, 55, 309–23.

Malecki, E.J. (1980) 'Corporate organization of R and D and the location of technological activities', *Regional Studies*, 14, 219–34.

Malecki, E.J. (1982) 'Federal R and D spending in the United States of America: some impacts on metropolitan economies', *Regional Studies*, 16, 19–35.

Malone, P. (1981) *Office Development in Dublin, 1960–80*, Dublin, Trinity College, Department of Geography.

Mandeville, T. (1983) 'Spatial effects of information technology', *Futures*, 15, 65–72.

Manners, G. and Morris, D. (1981) 'Does London need an office policy?', paper presented at Centre for Environmental Studies London Conference, London, February.

Mansfield, R. (1984) 'Changes in information technology, organizational design and managerial control', in Piery, N. (ed.) *The Management Implications of New Information Technology*, London, Croom Helm.

Marquand, J. (1978) *The Role of the Tertiary Sector in Regional Policy: Comparative Report*, London, Centre for Environmental Studies, University of Louvain-le-Neuf.

Marquand, J. (1979) *The Service Sector and Regional Policy in the United Kingdom*, Research Series No. 29, London, Centre for Environmental Studies.

Marquand, J. (1983) 'The changing distribution of service employment', in Goddard, J.B. and Champion, A.G. (eds) *The Urban and Regional Transformation of Britain*, London, Methuen, 99–134.

Marriott, O. (1967) *The Property Boom*, London, Hamish Hamilton.

Marshall, J.N. (1979) 'Corporate organization and regional office employment', *Environment and Planning A*, 11, 553–63.

Marshall, J.N. (1980) *Spatial Variations in Manufacturing Industry Demand for Business Services: Some Implications for Government Economic Policies*, Discussion Paper No. 35, Newcastle upon Tyne, Centre for Urban and Regional Development Studies.

Marshall, J.N. (1981) *Business Service Activities in Provincial Conurbations: Implications for Regional Economic Development*, Discussion Paper No. 37, Newcastle upon Tyne, Centre for Urban and Regional Development Studies.

Marshall, J.N. (1982) 'Linkages between manufacturing industry and business services', *Environment and Planning A*, 14, 1523–40.

Marshall, J.N. and Bachtler, J.F. (1984) 'Spatial perspectives on technological changes in the banking sector of the United Kingdom', *Environment and Planning A*, 16, 437–50.

Marshall, J.U. (1969) *The Location of Service Towns*, Toronto, University of Toronto, Department of Geography, Research Publication No. 3.

Martin, J. (1978) *The Wired Society: A Challenge for Tomorrow*, Englewood Cliffs, NJ, Prentice-Hall.

Massam, B. (1975) *Location and Space in Social Administration*, New York, Wiley.

Mathieson, A. and Wall, G. (1982) *Tourism: Economic, Physical and Social Impacts*, London, Longman.

Mills, C.W. (1953) *White Collar*, New York, Oxford University Press.

Mitchell, J.E. (1980) 'Small firms: a critique', *Three Banks Review*, 126, 50–61.

Moore, B.C., Rhodes, J. and Tyler, P. (1981) *The Growth of Employment in the Inner and Outer Cities of the Six Largest Conurbations in the United Kingdom, 1951–76*, Discussion Paper No. 7, University of Cambridge, Department of Land Economy.

Moroz, A.R. (1983) 'Trading new services', *Policy Options*, 4, 46–8.

Morrill, R.L., Earickson, R.J. and Rees, P. (1970) 'Factors affecting distances travelled to hospitals', *Economic Geography*, 46, 161–71.

Morris, J. (1976) 'Access to community health facilities in Melbourne', in *Papers of the First Australian New Zealand Regional Science Association Conference*, Brisbane, University of Queensland, 143–7.

Moseley, M.J. (1979) *Accessibility: The Rural Challenge*, London, Methuen.

Muller, T. (1981) 'Regional malls and central city retail sales: an overview', in Sternlieb, G. and Hughes, J.W. (eds) *Shopping Centers: USA*, Rutgers, NJ, Center for Urban Policy Research, 177–200.

Nagai, S. and Myaji, O. (1967) 'Economic management functions and urban restructuring', in Okita, S. (ed.) *Economics of Regional Development*, Tokyo, Chikuma-Shobo, 91–126.

National Economic Development Office (1982) *Technology: The Issues for the Distributive Trades* London, National Economic Development Office.

Nelson, R.L. (1958) *The Selection of Retail Locations*, New York, Dodge.

Nilles, J.M., Carlson, R., Grey, P. and Heineman, G. (1976) *Telecommunications–Transportation Trade-Offs: Options for Tomorrow*, New York, Wiley.

Norcliffe, G.B. (1983) 'Using location quotients to estimate the economic base and trade flows', *Regional Studies*, 17, 161–8.

North West Economic Planning Council (1975) *Strategic Plan for the North West*, London, HMSO.

Northern Region Strategy Team (1977) *Strategic Plan for the Northern Region*, London, HMSO.

Oberai, A.S. (1978) *Changes in the Structure of Employment with Economic Development*, 2nd edn, Geneva, International Labour Office.

O'Hara, D.J. (1977) 'Location of firms within a square central business district', *Journal of Political Economy*, 85, 1189–1207.

Ofer, G. (1973) *The Service Sector in Soviet Economic Growth: A Comparative Study*, Cambridge, Mass., Harvard University Press.

Office of Population Censuses and Surveys (1981) *Census 1981, Preliminary Report*, London, HMSO.

Organization for Economic Co-operation and Development (1973) *International Tourism and Tourism Policy in OECD Countries*, Paris, OECD.

Organization for Economic Co-operation and Development (1978a) *Regional Policies and the Services Sector*, Paris, OECD.

Organization for Economic Co-operation and Development (1978b) *Tourism Policy and International Tourism in OECD Member Countries*, Paris, OECD.

Organization for Economic Co-operation and Development (1982) *Innovation Policy: Trends and Perspectives*, Paris, OECD.

Organization for Economic Co-operation and Development (1983) *The Internationalization of Banking*, Paris, OECD.

Otway, H.J. and Peltu, M. (eds) (1983) *New Office Technology: Human and Organizational Aspects*, London, Frances Pinter.

Pacione, M. (1982) 'Space preferences, locational decisions, and the dispersal of civil servants from London', *Environment and Planning A*, 14, 323–33.

Pare, S. (1981) 'La localisation des fonctions tertiares informatique en France et leur rôle dans l'organisation de l'espace', *Information Géographique*, 45, 142–3.

Parker, A.J. (1975) 'Hypermarkets: the changing pattern of retailing', *Geography*, 60, 120–4.

Parliamentary Committee on Scottish Affairs (1980) *Dispersal of Civil Service Jobs to Scotland*, London, HMSO.

Pascal, A.H. and McCall, J.J. (1980) 'Agglomeration economies, search costs and industrial location', *Journal of Urban Economics*, 8, 383–8.

Pateman, C. (1981) 'The concept of equity', in Troy, P.N. (ed.) *A Just Society*, Sydney, Allen & Unwin, 21–36.

Perry, D.C. and Watkins, A.J. (eds) (1977) *The Rise of the Sunbelt Cities*, Beverly Hills, Calif., Sage.

Pesola, J. (1978) *Palveluelingkeinozen investoinnit vicosina, 1953–1975*, Helsinki, Suomen Pankhi.

Peston, M. (1972) *Public Goods and the Public Sector*, London, Macmillan.

Peston, M. (1978) 'Public tasks: necessities, claims, problems of financing', in Mukalski, W. (ed.) *The Future of Industrial Societies: Problems – Prospects – Solutions*, Alphen aun den Rijn, Sitskoff and Nordhoff, 137–53.

Philippe, J. (1984) 'Les services aux enterprises et la politique de développement régional en France', paper presented at ASDRLF Colloquy, Lugano, September.

Phillips, B. (1981) 'Does automation threaten city-based offices?', *The Times*, 26 October.

Phillips, B. (1983) 'Britain's tower blocks not designed for the office of the future', *The Times*, 28 April.

Phillips, D.R. (1979a) 'Public attitudes to general practitioner services: a reflection of an inverse care law in intra-urban primary medical care', *Environment and Planning A*, 11, 315–24.

Phillips, D.R. (1979b) 'Spatial variations in attendance at general practitioner services', *Social Science and Medicine*, 13D, 169–81.

Phillips, J.D. (1964) *Some Industrial and Community Conditions for Small Retailer Survival*, Urbana, Ill., University of Illinois Press.

Phillips, P. (1982) *Regional Disparities*, Toronto, Lorimer.

Piercy, N. (1984) 'The impact of new technology on services marketing', *Services Industries Journal*, 4, 193–204.

Polese, M. (1981) 'Inter-regional service flows, economic integration and regional policy: some considerations based on Canadian survey data', *Revue d'Economie Régionale et Urbaine*, 4, 489–503.

Polese, M. (1982) 'Regional demand for business services and inter-regional service flows in a small Canadian region', *Papers of the Regional Science Association*, 50, 151–63.

Polese, M. (1983) 'Montreal's role as an international business centre: cultural images versus economic realities', paper presented at Workshop on Non-Capital Cities, Hosts to International Organizations, New York, City University of New York, October.

Polese, M. and Stafford, R. (1982), 'Une estimation des exportations de services des régions urbaine: l'application d'un modèle simple au Canada', *Canadian Journal of Regional Science*, 5, 313–31.

Poole, I. de S. (1977) *The Social Impact of the Telephone*, Cambridge, Mass., Massuchusetts Institute of Technology.

Potter, R.B. (1979) 'The morphological characteristics of urban retailing areas: a review and suggested methodology', Papers in Geography No. 2, Bedford College (University of London), Department of Geography.

Potter, R.B. (1980) 'Spatial and structural variations in the quality characteristics of intra-urban retailing centres', *Transactions of the Institute of British Geographers, n. s.*, 5, 207–28.

Pred, A.R. (1967) *Behaviour and Location: Foundations for a Geographic and Dynamic Location Theory*, Pt I, Lund, Gleerup.

Pred, A.R. (1974) *Major Job-Providing Organizations and Systems of Cities*, Washington, DC, Association of American Geographers, Commission in College Geography, Resource Paper No. 27.

Pred, A.R. (1976) 'The interurban transmission of growth in advanced economies: empirical findings versus regional planning assumptions', *Regional Studies*, 10, 151–71.

Pred, A.R. (1977) *City Systems in Advanced Economies*, London, Hutchinson.

Pred, A.R. and Palm, R. (1978) 'The status of American women: a time-geographic view', in Lanegran, D.A. and Palm, R. (eds) *An Invitation to Geography*, New York, McGraw-Hill, 99–109.

Pritchard, M. (1982) *The Spatial Implications of Technological Innovations in the Office Sector*, Working Paper No. 4, Liverpool Papers in Human Geography, Department of Geography, University of Liverpool.

Pye, R. (1977) 'Office location and the cost of maintaining contact', *Environment and Planning A*, 9, 149–68.

Pye, R. (1979) 'Office location: the role of communications and technology', in Daniels, P.W. (ed.) *Spatial Patterns of Office Growth and Location*, London, Wiley, 239–75.

Pye, R. and Williams, E. (1977) 'Teleconferencing: is video valuable or is audio adequate', *Telecommunications Policy*, 1, 230–41.

Quante, W. (1976) *The Exodus of Corporate Headquarters from New York City*, New York, Praeger.

Rees, J. (1979) 'Manufacturing headquarters in the post-industrial context'. *Economic Geography*, 54, 337–54.

Reilly, W.J. (1931) *The Law of Retail Gravitation*, New York, Knickerbocker Press.

Richardson, H.W. (1978) *Regional and Urban Economics*, Harmondsworth, Penguin.

Richardson, J.F. (1982) 'The evolving dynamics of American urban development', in Gappert, G. and Knight, R.V. (eds) *Cities in the 21st Century*, Beverly Hills, Calif., Sage, 37–46.

Rider, K.L. (1979) 'The economics of the distribution of municipal fire protection services', *Review of Economics and Statistics*, 61, 249–58.

Robertson, K.A. (1983) 'Downtown retail activity in large American cities 1954–77', *Geographical Review*, 73, 314–23.

Robinson, H. (1976) *A Geography of Tourism*, London, Macdonald & Evans.

Robson, W.A. (1976) *Welfare State and Welfare Society*, London, Allen & Unwin.

Rogers, D. (1983) 'The changing pattern of American retailing', *Retail and Distribution Management*, 11, 8–13.

Roggero, M.A. (1976) *Urbanización, industrialización y crecimiento del sector servicios en América Latina*, Buenos Aires, Educiónes Nueva Visión.

Rosenberg, M. (1983) 'Accessibility of health care: a North American perspective', *Progress in Human Geography*, 7, 78–87.

Rudolph, R.L. (1976) *Banking and Industrialization in Austria-Hungary: The Role of Banks in the Industrialization of the Czech Crownlands, 1873–1914*, London, Cambridge University Press.

Sabolo, Y. (1975) *The Service Industries*, Geneva, International Labour Office.

Samuelson, P.A. (1954) 'The pure theory of public expenditure', *Review of Economics and Statistics*, 36, 387–9.

Schaff, A. and Friedrichs, G. (eds) (1982) *Microelectronics and Society: For Better, for Worse*, Oxford, Pergamon.

Schannon, G.W. and Dever, G.E.A. (1974) *Health Care Delivery: Spatial Perspectives*, New York, McGraw-Hill.

Schannon, G.W., Spurlock, C.W. and Skinner, J.L. (1975) 'A method for evaluating the geographic accessibility of services', *Professional Geographer*, 27, 30–6.
Schiller, R.K. (1971) 'Location trends of specialist services', *Regional Studies*, 5, 1–10.
Schiller, R.K. (1972) 'The measurement of attractiveness of shopping centres to middle class luxury consumers', *Regional Studies*, 6, 291–7.
Schiller, R.K. (1977) 'What the census says about shops', *Chartered Surveyor*, 109, 190–2.
Schneider, M. (1959) 'Gravity models and trip distribution theory', *Papers of the Regional Science Association*, 5, 51–6.
Scott, A.J. (1971) *Combinatorial Programming, Spatial Analysis and Planning*, London, Methuen.
Scott, M. (1980) 'The co-op superstore versus small shop', *Town and Country Planning*, 49, 119–21.
Scott, P. (1970) *Geography and Retailing*, London, Hutchinson.
Seeley, J.E. (ed.) (1981) 'New directions in public services', *Economic Geography*, 57, (special issue); including 'Introduction: new directions in public services' by editor, 1–9.
Semple, R.K. (1973) 'Recent trends in the concentration of corporate headquarters', *Economic Geography*, 49, 309–18.
Semple, R.K. (1977) 'The spatial concentration of domestic and foreign multinational corporate headquarters in Canada', *Cahiers de Géographie de Québec*, 21, 33–51.
Semple, R.K. and Green, M.B. (1983) 'Interurban corporate headquarters relocation in Canada', *Cahiers de Géographie du Québec*, 27, 389–406.
Semple, R.K. and Phipps, A.G. (1982) 'The spatial evolution of corporate headquarters within an urban system', *Urban Geography*, 3, 258–79.
Shannon, G.W. and Dever, G.E. (1974) *Health Care Delivery: Spatial Perspectives*, New York, McGraw-Hill.
Shapero, A., Howell, R.P. and Tombaugh, J.R. (1969) *An Exploratory Study of the Structure and Dynamics of the R and D Industry*, Menlo Park, Calif., Stanford Research Institute.
Shelp, R.K. (1983) *Beyond Industrialization: Ascendancy of the Global Service Economy*, New York, Praegar.
Sim, D. (1982) *Change in the City Centre*, Aldershot, Gower.
Simmons, J. (1964) *The Changing Pattern of Retail Locations*, Research Paper No. 92, University of Chicago, Department of Geography.
Simon, H.A. (1959) 'Theories of decision making in economics', *American Economic Review*, 49, 253–83.
Simon, H.A. (1977) *The New Science of Management Decision-Making*, Englewood Cliffs, NJ, Prentice-Hall.
Sleigh, J., Boatwright, B., Irwin, P. and Stanyon, R. (1979) *The Manpower Implications of Micro-Electronic Technology*, London, HMSO.
Smith, A. (1776) *The Wealth of Nations*, London, Dent; Everyman edn.
Smith, A. (1972) 'The future of downtown retailing', *Urban Land*, 31, 3–10.
Smith, D.M. (1966) 'A theoretical framework for geographical studies of industrial location', *Economic Geography*, 42, 95–113.
Smith, D.M. (1977) *Human Geography: A Welfare Approach*, London, Edward Arnold.
Smith, D.M. (1982) *Industrial Location*, 2nd edn, Chichester, Wiley.
Smith, I.J. (1979) 'The effects of external takeover and manufacturing employment change in the Northern Region, 1963–73', *Regional Studies*, 13, 421–37.
Smith, R. and Selwood, D. (1983) 'Office location and the distance decay relationship', *Urban Geography*, 4, 302–16.

Smith, S. and Rodgers, P. (1984) 'The Post Office is ready to drop its quill pen image and join the 21st century', *Guardian*, 28 January.

Smith, S.L.J. (1983) *Recreation Geography*, London, Longman.

Soussan, J. (1980) *The Dual Economy Debate and Patterns of Economic Organization in the Urban Fringe of Delhi*, Working Paper No. 288, University of Leeds, Department of Geography.

South East Economic Planning Council (1967) *A Strategy for the South-East*, London, HMSO.

Spence, N.A. and Frost, M.E. (1983) 'Urban employment change', in Goddard, J.B. and Champion, A.G. (eds) *The Urban and Regional Transformation of Britain*, London, Methuen, 71–98.

Spink, F.H. (1981) 'Downtown malls: prospects, design, constraints', in Sternlieb, G. and Hughes, J.W. (eds) *Shopping Centers: USA*, Rutgers, NJ, State University of New Jersey, Center for Urban Policy Research, 201–18.

Stanback, T.M. (1979) *Understanding the Service Economy: Employment, Productivity, Location*, Baltimore, Md, Johns Hopkins University Press.

Stanback, T.M., Bearse, P.J., Noyelle, T.J. and Karasek, R.A. (1981) *Services: the New Economy*, Totowa, NJ, Allanheld, Osmun.

Stanback, T.M. and Noyelle, T.J. (1980) *Economic Transformation in Selected American Cities: A Study of Processes with Implications for Development Policy*, first interim report to the Economic Research Division of the Economic Development Administration, US Department of Commerce, New York, Columbia University Press.

Stanback, T.M. and Noyelle, T.J. (1982) *Cities in Transition: Changing Job Structures in Atlanta, Denver, Buffalo, Phoenix, Columbus (Ohio), Nashville and Charlotte*, Totowa, NJ, Allanheld, Osmun.

Statistical Office of the European Communities (1970) *General Industrial Classification of Economic Activities within the European Communities (NACE)*, Luxembourg, SOEC.

Steele, L.W. (1975) *Innovation in Big Business*, New York, Elsevier.

Stephens, J.D. and Holly, B.P. (1981) 'City system behaviour and corporate influence: the headquarters location of US industrial firms, 1955–75', *Urban Studies*, 18, 285–300.

Sternlieb, G. and Hughes, J.M. (eds) (1981) *Shopping Centers: USA*, Rutgers, NJ, State University of New Jersey, Center for Urban Policy Research.

Sternlieb, G. and Hughes, J.M. (1983) 'The uncertain future of the central city', *Urban Affairs Quarterly*, 18, 455–72.

Stigler, G.J. (1956) *Trends in Employment in the Service Industries*, Baltimore, Md, Johns Hopkins University Press.

Stilwell, F.J.B. (1969) 'Regional growth and adaptation', *Urban Studies*, 6, 162–78.

Stimson, R.J. (1980) 'Spatial aspects of epidemiological phenomena and of the provision and utilization of health care services in Australia: a review of methodological problems and empirical analyses', *Environment and Planning A*, 12, 881–907.

Stonier, T. (1983) *The Wealth of Information,* London.

Strassman, P.A. (1980) 'The office of the future: information management for the new age', *Technology Review*, 80, 54–65.

Sumner, G. (1971) 'Trends in the location of primary medical care in Britain', *Antipode*, 3, 46–53.

Sundqvist, J.L. (1975) *Dispersing Population: What America Can Learn from Europe*, Washington, DC, Brookings Institution.

Swedish Government Decentralization Commission (1978) *Move Out the Decisions Closer to the People*, Stockholm, Decentralization Commission.

Syron, R.F. (1984) 'The New England experiment in interstate banking', *New England Economic Review*, March–April, 5–17.

Tauchen, H. and Witte, A.D. (1983) 'An equilibrium model of office location and contact patterns', *Environment and Planning A*, 15, 1311–26.

Tauchen, H. and Witte, A.D. (1984) 'Socially optimal and equilibrium distributions of office activity: models with exogenous and endogenous contacts', *Journal of Urban Economics*, 15, 66–86.

Taylor, M.J. and Thrift, N. (1982) 'Models of corporate development and the multinational corporations', in Taylor, M.J. and Thrift, N. (eds) *The Geography of Multinationals*, London, Croom Helm, 14–32.

Taylor, M.J. and Thrift, N. (1983) 'Business organization, segmentation and location', *Regional Studies*, 17, 445–65.

Taylor, P.J. (1977) *Quantitative Methods in Geography*, Boston, Mass., Houghton Mifflin.

Teitz, M. (1968) 'Towards a theory of urban public facility location', *Papers of the Regional Science Association*, 13, 35–51.

Thirlwall, A.P. (1967) 'A measure of the proper distribution of industry', *Oxford Economic Papers*, 1, 46–58.

Thirlwall, A.P. (1982) 'De-industrialization in the United Kingdom', *Lloyds Bank Review*, 144 (April), 22–37.

Thomas, C.J. (1974) 'The effects of social class and car ownership on intra-urban shopping behaviour in Greater Swansea', *Cambria*, 2, 98–126.

Thomas, R.W. (1975) 'Some functional characteristics of British city centre areas', *Regional Studies*, 9, 369–78.

Thorngren, B. (1970) 'How do contact systems affect regional development?', *Environment and Planning A*, 2, 409–27.

Thorngren, B. (1973) 'Communications studies for Government office dispersal in Sweden', in Bannon, M.J. (ed.) *Office Location and Regional Development*, Dublin, An Foras Forbartha, 47–58.

Thorpe, D. (1977) *Shopping Trip Patterns and the Spread of Superstores and Hypermarkets in Great Britain*, Research Paper No. 10, Manchester, Manchester Business School, Retail Outlets Research Unit.

Thorpe, D. and Kirby, D.A. (1971) *The Density of Cash and Carry Wholesaling: A Study in Comparative Market Potential*, Research Report No. 2, Manchester, Manchester Business School, Retail Outlets Research Unit.

Thwaites, A.T. (1978) *The Future Development of R & D Activity in the Northern Region: A Comment*, Discussion Paper No. 12, University of Newcastle upon Tyne, Centre for Urban and Regional Development Studies.

Toby, J. (1973) 'Regional development and government office relocation in the Netherlands', in Bannon, M.J. (ed.) *Office Location and Regional Development*, Dublin, An Foras Forbartha, 37–46.

Toffler, A. (1981) *The Third Wave*, New York, Bantam.

Tornqvist, G. (1970) *Contact Systems and Regional Development*, Lund Studies in Geography No. 35, series B, Lund, Gleerup.

Tornqvist, G. (1973) 'Contact requirements and travel facilities: contact models of Sweden and regional development alternatives in the future', in Pred, A. and Tornqvist, G. *Systems of Cities and Information Flows*, Lund Studies in Geography No. 38, series B, Lund, Gleerup, 81–121.

Torrens, P.R. (1978) *The American Health Care System: Issues and Problems*, St. Louis, Mis., C.V. Mosby.
Touraine, A. (1971) *The Post-Industrial Society* New York, Random House.
Uhlig, R.P., Forber, D.J. and Bair, J.H. (1982) *The Office of the Future*, Amsterdam, North Holland.
United Nations (1948) *International Standard Industrial Classification of All Economic Activities*, Statistical Papers No. 4, series M, New York, UN; 1st rev., 1958; 2nd rev., 1968.
University Of Manchester (1966) *Regional Shopping Centres in North West England*, Manchester, University of Manchester, Department of Town and Country Planning.
Urban Land Institute (1977) *Shopping Center Development Handbook*, Washington DC, Urban Land Institute.
Vance, J.E. Jr. (1970) *The Merchant's World: The Geography of Wholesaling*, Englewood Cliffs, NJ, Prentice-Hall.
Van Dinteren, J.H.J. (1984) *De Kantorensector in Middelgrote Steden in Zuid en Oost Nederland*, Nijmegen, Geografisch Instituut KU.
Vernon, R. (1960) *Metropolis 1985*, Cambridge, Mass., Harvard University Press.
Virgo, P. (1979) *Cashing in on the Chips*, London, Conservative Political Centre.
Walker, S.R. (1979) 'Educational services in Sydney: some spatial variations', *Australian Geographical Studies*, 17, 175–92.
Walters, D. (1975) 'Physical distribution features for the UK food industry', *Retail and Distribution Management*, 3, 42–7.
Wanhill, S.R.C. (1983) 'Measuring the economic impact of tourism', *Service Industries Journal*, 3, 9–20.
Warnes, A.M. (1975) 'Commuting towards city centres: a study of population and employment density gradients in Liverpool and Manchester', *Transactions of the Institute of British Geographers*, 64, 77–96.
Warnes, A.M. and Daniels, P.W. (1979) 'Spatial aspects of an intrametropolitan central place hierarchy', *Progress in Human Geography*, 3, 384–406.
Watts, D. (1977) 'The impact of warehouse growth', *Planner*, 63, 105–7.
Weatheritt, L. and John, O.N. (1979) 'Office development and employment in Greater London, 1967–76', Research Memorandum RM556, London, Greater London Council.
Webber, M.M. (1963) 'Order in diversity: community without propinquity', in Wingo, L., Jr (ed.) *Cities and Space: The Future Use of Urban Land*, Baltimore, Md, Johns Hopkins University Press, 22–54.
Weber, A. (1909) *Theory of the Location of Industries*, Chicago, University of Chicago Press.
Weir, M. (1977) 'Are computer systems and humanized work compatible?', in Ottaway, R.N. (ed.) *Humanizing the Workplace*, London, Croom Helm, 44–64.
Wels, A. (1984) 'How the US banking equation is changing', *Banker*, 134, 27–9.
White, J.A. and Case, K.E. (1974) 'On covering problems and the central facilities location problem', *Geographical Analysis*, 6, 281–93.
Whitehouse, B.P. (1964) *Partners in Property*, London, Birn Shaw.
Whitelegg, J. (1982) *Inequalities in Health Care: Problems of Access and Provision*, Retford, Straw Barnes.
Wilson, A.G. (1976) 'Retailers' profits and consumers' welfare in a spatial interaction shopping model', in Masser, I. (ed.) *Theory and Practice in Regional Science*, London, Pion, 42–59.
Witcher, B.J. (1982) 'Videotext in the UK: problems of public service viewdata and

implications for publishers', in Hills, P.J. (ed.) *Trends in Information Transfer*, London, Frances Pinter, 65–86.

Wood, P.A. (1983) 'The regional significance of manufacturing–service sector links: some thoughts on the revival of London's docklands', paper presented at Anglo-Canadian Symposium on Industrial Geography, Calgary, mimeo.

Woolnough, R. (1983) 'Super telex comes in fast', *The Times*, 21 October.

Wright, M.W. (1967) 'Provincial office development', *Urban Studies*, 4, 213–57.

Wynne, B. (1983) 'The changing role of managers', in Otway, H.J. and Peltu, M. (eds) (1983) *New Office Technology: Human and Organizational Aspects*, London, Frances Pinter, 138–51.

Yannopoulos, G. (1973) 'Local income effects of office relocation', *Regional Studies*, 7, 33–46.

Zimmerman, M.M. (1955) *The Supermarket: A Revolution in Retailing*, New York, McGraw-Hill.

Name index

Subject index